Biology of Extracellular Matrix

Volume 10

Extracellular matrix (ECM) biology, which includes the functional complexities of ECM molecules, is an important area of cell biology. Individual ECM protein components are unique in terms of their structure, composition and function, and each class of ECM macromolecule is designed to interact with other macromolecules to produce the unique physical and signaling properties that support tissue structure and function. ECM ties everything together into a dynamic biomaterial that provides strength and elasticity, interacts with cell-surface receptors, and controls the availability of growth factors. Topics in this series include cellular differentiation, tissue development and tissue remodeling. Each volume provides an in-depth overview of a particular topic, and offers a reliable source of information for post-graduates and researchers alike.

All Chapters are systematically reviewed by the series editor and respective volume editor(s).

"Biology of Extracellular Matrix" is published in collaboration with the American Society for Matrix Biology and the International Society for Matrix Biology.

More information about this series at http://www.springer.com/series/8422

Michel Goldberg • Pamela Den Besten
Editors

Extracellular Matrix Biomineralization of Dental Tissue Structures

Editors
Michel Goldberg
Faculty of Fundamental and Biomedical Sciences
Paris Cite University INSERM
UMR-S 1124
Paris, France

Pamela Den Besten
Department of Orofacial Sciences, UCSF School of Dentistry
University of California
San Francisco, USA

ISSN 0887-3224 ISSN 2191-1959 (electronic)
Biology of Extracellular Matrix
ISBN 978-3-030-76285-8 ISBN 978-3-030-76283-4 (eBook)
https://doi.org/10.1007/978-3-030-76283-4

This Springer imprint is published by the registered company Springer Nature Switzerland AG.
The registered company address is: Gewerbestrasse 11, 6330 Cham, Switzerland

Preface

Once upon a time, in Italy, I was in a traditional restaurant. I asked for a plate of spaghetti. The waiter brings a plate of long and thin, cylindrical spaghettis, forming something that was very similar to a collagen network. This reminds me of the dense collagen network of the intertubular dentin and intermingled scaffold in bone. Then, the waiter brings me tomato sauce and he mixed the two components. The tomato sauce glued to the surface of the collagen pasta. Finally, the waiter brings the Parmigiano. He spread the cheese on the pasta as crystal-like seeds, adhering to the surface of the collagen coated with the tomato sauce. Then, being a dental surgeon, I recognize the three major components of the intertubular bulk of the dentin-like structure and also of the bone-like structure: the collagen network, coated by the tomato sauce, and bearing at the surface of the noddle the mineral crystals, similar to the initial nuclei of apatite formation. I was not alone in this restaurant, and my fellow traveler mixed the spaghetti into a pile, and this reminded me of the wavy structures of mineral crystals that form tooth enamel.

The roles of extracellular matrix components in tooth and/or tooth-supporting tissue (alveolar bone) are crucial to mineral formation, and this is the main topic of the present book. These components have a regulatory role, some of these complex proteins being either promoter of mineralization, whereas others act as inhibitors or/even display both functions. In addition, some ECM molecules behave as growth factors or transcription factors. These molecules in bone and calcified cartilage play structural and regulatory roles.

ECM constitutes between 5 and 30 percent of the mineralized tissues, and the mineral forms between 70 and 90% of the tissues. Collagen expression is an essential matrix component of all calcified tissues, except it does not participate in enamel structure. Collagens form a microenvironment that facilitate the initiation, needle-like formation, and growth of hydroxyapatite crystallites within calcified tissues. It is obvious that ECM components have a crucial role in bone and teeth mineralization. ECM limits or regulates the mineralization process.

Metalloproteinases play role in ECM turnover, as well as TIMPs and other matrix molecules. Molecular cleavage allows complex structures such as dentin sialophosphoproteins (DSPP) to become three major sub-components: dentin phosphoproteins (DPP), dentin sialoprotein (DSP), and dentin glycoprotein (DGP). The spliced molecules after fragmentation become functional and play crucial factors for the potential role of ECM.

A series of non-collagenous proteins phosphorylated (phosphophoryn, DMP-1, osteopontin, dentin MEPE, and sialophosphoprotein) and/or non-phosphorylated molecules (osteocalcin, SLRPs) are the consecutive elements of the mineralized tissues. They are reviewed in the chapters written specifically by qualified authors. I thank them warmly for their superb contribution.

In enamel, which is uniquely different from all other mineralized tissues, extracellular matrix proteins (amelogenin, ameloblastin (AMBN), amelotin, enamelin, heparin-binding domains, CD63-interaction domains, are matrix proteins that influence crystal growth. Proteolytic processing of amelogenin by MMP20 and Klk4 allows the growth of crystals during the secretory stage and also later, during enamel maturation.

Collagens, SIBLINGs, non-phosphorylated matrix proteins, SLRPs, enamel proteins, growth and transcription factors, MMPs proteolytic enzymes, are the specific topic of the different chapters of the present book that is devoted to extracellular matrix molecules and their role in biomineralization.

We thank the authors who agreed to share their skill to write their different chapters in due time. And first of all, I wish to express all my gratitude to my co-editor Pamela DenBesten who is an efficient co-worker, also giving me advice and correcting my language mistakes. We have a long-standing collaboration, publishing together articles, chapters, and now the present book, covering the complex field of ECM and mineralization.

Paris, France
March 2021

Michel Goldberg

Contents

Part I
Extracellular Matrix Molecules of Mineralized Structures

Chapter 1
Structure of Collagen-Derived Mineralized Tissues (Dentin, Cementum, and Bone) and Non-collagenous Extra Cellular Matrix of Enamel

Yukiko Nakano, Pamela DenBesten, and Michel Goldberg

Abstract Most of the human body's biomineralization is due to collagenous extracellular matrix (ECM) of such as dentin, cementum, bone, and cartilage (lacking in mature dental tissues). The ECM of these tissues is formed by mesenchymal cells, odontoblasts, cementoblasts, osteoblasts, and at some specific period by chondrocytes. When these tissues are fully formed, ECM and cellular components remain in the matrix to provide bioreactivity via occasional modeling and remodeling.

Unlike the mesenchymally derived mineralized tissues, the ECM of dental enamel is non-collagenous and synthesized by epithelial cells (ameloblasts). Mineralization of the enamel matrix is unique in that it ends in removing the majority of the organic matrix. This process makes the enamel matrix highly mineralized. Fully formed enamel is an acellular structure that covers the erupted tooth crown.

1.1 Mineralization in Collagen-Derived Tissues

1.1.1 Matrix Vesicle (MV) Initiated Mineralization

MVs are extracellular vesicles with ~ 100 nm in diameter and are located at sites of initial mineralization of collagen-derived tissue (cartilage, bone/osteoid, cementum, and predentin). They are derived from osteoblasts, chondrocytes, odontoblasts, and cementoblasts as a type of extracellular microparticles. Condensed calcium and

Y. Nakano · P. DenBesten
Department of Orofacial Sciences, School of Dentistry, University of California San Francisco, San Francisco, CA, USA
e-mail: Yukiko.Nakano@ucsf.edu; Pamela.DenBesten@ucsf.edu

M. Goldberg (✉)
Faculté des Siences Fondamentales et Médicales, Université Paris Cité, Paris, France

INSERM UMR-S 1124, Paris, France

M. Goldberg, P. Den Besten (eds.), *Extracellular Matrix Biomineralization of Dental Tissue Structures*, Biology of Extracellular Matrix 10,
https://doi.org/10.1007/978-3-030-76283-4_1

phosphate in the MVs initiate the formation of hydroxyapatite (HAp) crystals in ECM (phase 1). As the crystals grow and extend, they perforate the membrane of MVs. The forming HAp crystals are then extruded from MVs, and merge with other mineral needles, expanding to form mineralized areas (phase 2). Phase 1 is the initiation of the mineralization within MVs, regulated by the molecules, including pyrophosphatase and calcium-binding molecules (annexin I, phosphatidylserine). During phase 2, HAp crystals are released through the MV membrane. Crystals are exposed to extracellular fluids containing Ca^{2+} and $PO4^{3-}$, and support the continuous crystal proliferation, on performed crystals serving as nuclei (template) for the formation of new crystals (Anderson 1995, 2003).

Though the mechanisms of formation and release into the ECM are not entirely understood, the composition of MVs is similar to the plasma membrane of cells. Currently, MVs are thought to be formed either within corresponding cells containing amorphous calcium phosphate or by budding/blebbing of the plasma membrane of the cells. The ultrastructural observation supports both possibilities indicating subtypes of MVs (Boonrungsiman et al. 2012; Chaudhary et al. 2016; Garces-Ortiz et al. 2013; Thouverey et al. 2009). The difference in the formation mechanisms likely reflects the functional stage of the cells or the role of MVs.

Further recent studies demonstrate the similarities of proteins and lipids components in MVs and exosomes, suggesting the contribution of the endosomal pathway in the formation of MVs (Wuthier and Lipscomb 2011). The similarities of molecular profile between exosomes and MVs also indicate possible roles of MVs in extracellular signaling besides their roles in mineralization.

MV-associated ECM proteins are the following: Type VI and X collagens, proteoglycans and aggrecan, fibrillin −1 and −2. Also included are a series of annexins (A5, A6, A2, A1, A11, and A4), and enzymes (tissue-nonspecific alkaline phosphatase (TNAP), nucleotide pyrophosphate phosphodiesterase, PHOSPHO-1, acid phosphatase, lactate dehydrogenase, carbonic anhydrase, phospholipases A, C, sphingomyelinase) (Wuthier and Lipscomb 2011). MVs also include a series of proteins: osteopontin (OPN), bone sialoprotein (BSP), matrix extracellular phosphoglycoprotein (MEPE), dentin matrix protein (DMP), and dentin sialoprotein (DSP), which are components of a group of non-collagenous extracellular mineralization-regulating proteins termed SIBLING (small integrin-binding ligand N-linked glycoprotein) proteins. These proteins share a conserved arginine-glycine-aspartic acid (RGD) motif, which mediates their cell attachment and signaling functions (Anderson 1995; Anderson 2003; Wuthier and Lipscomb 2011).

1.1.2 Collagen and Extracellular Matrix (ECM) Proteins-Derived Mineralization

Matrix proteins that direct the mineralization of collagenous tissue originate as pro-collagen chains synthesized in the cell bodies (see Table 1.1). The non-helical

Table 1.1 The successive steps of collagen synthesis

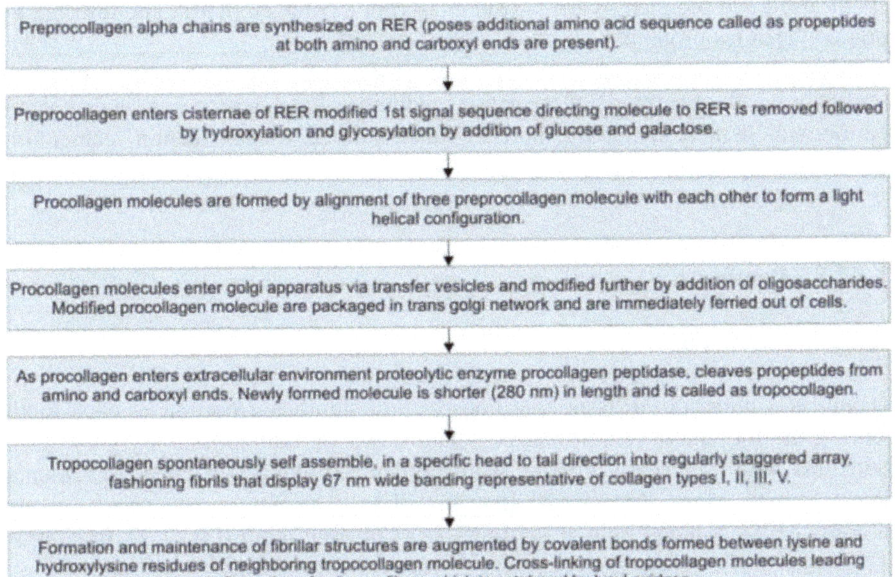

Reprinted from Sandhu et al. (2012)

C- and N-terminal extensions are cleaved by proteases, and helical collagens ([α1(I)]2 and α2(I) type I collagen- 95% and 5% type I trimer collagen [α1(I)]3) are secreted into the extracellular matrix.

To form dentin, collagen fibrillation induces the tropocollagen fibrils to be transported toward the predentin mineralization front together with a group of non-collagenous proteins, secreted in the distal part of predentin. The secretion of another group of non-collagenous proteins occurs within the lumen of canaliculi, inside the dentin tubule, close to the odontoblast processes. These ECM components contribute to crystal nucleation inside collagen fibers, and in the inter-collagen spaces, mineralization occurring between collagen fibers. Fibers increase in number, diameter, and density and contribute to forming a homogenous layer underlying the mineralization front (also called metadentin) (Goldberg and Septier 1996). These crucial events are associated with tissue mineralization.

Bone is another example of collagen-dependent mineralization. Bone is also synthesized and secreted by cells (osteoblasts) in a collagen matrix and directed by non-collagenous proteins.

We review here the different structures of the collagen-implicated mineralized tissues or layers, their mode of formation, and mineralization processes in dental and periodontal tissues, including three collagen-derived tissues (dentin, cementum, and bone).

1.1.2.1 Dentin

Anatomy of Dentin

Dentin is a mineralized tissue that constructs a significant portion of each tooth. During the formation, dentin matrix is laid down by odontoblasts by growing in thickness with time (daily growth: 4 μm; organized along von Ebner incremental lines, formed every 5 days, 20 μm, Fig. 1.1). The dentin layer encloses the non-mineralized dental pulp at the inner most of each tooth (Fig. 1.2). In contrast

Fig. 1.1 In dentin, tubules are bended, especially near the amelo-dentinal junction. In enamel, the Retzius lines are visible

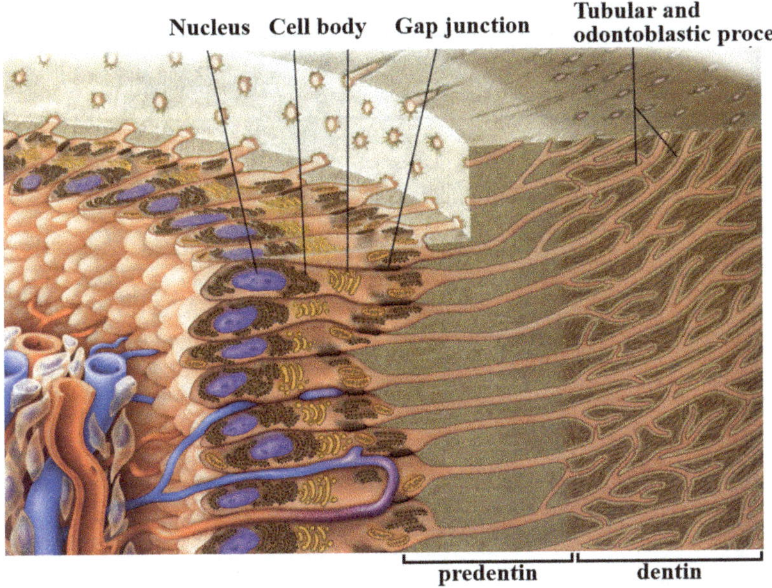

Fig. 1.2 Schematic aspect of the odontoblastic layer. Odontoblast cell bodies are lining the predentin and odontoblastic processes penetrate into dentin tubules. Modified from Nanci (2007)

Table 1.2 Types of dentin

Classification Criteria	
Order of Production	Primary Dentin
	Secondary Dentin
	Tertiary Dentin / Reparative Dentin
Macro Distribution	Mantle Dentin
	Circumpulpal Dentin
Micro Distribution	Intertubular Dentin
	Peritubular Dentin

with dentin, the dental pulp is vascularized and innervated. Capillaries infiltrate the odontoblast layer. Nerves come in contact with the basal part of odontoblast cell bodies.

Anatomically, dentin is further categorized according to the order of formation, macroscopic distribution, and microscopic localization patterns (Tables 1.2 and 1.3, Fig. 1.3).

Table 1.3 Types of dentin and properties

	intertubular dentin		peritubular dentin
	mantle dentin	circumpulpal dentin	
mechanism of mineralization	matrix vesicle (+)	matrix vesicle (-) additional mineralization	matrix vesicle (-) highly mineralized
collagen	thick (up to 1 μm)	thin	NA
non-collagenous protein	very low	rich (DPP>DSP>DMP-1 etc.)	DSP polysaccharide

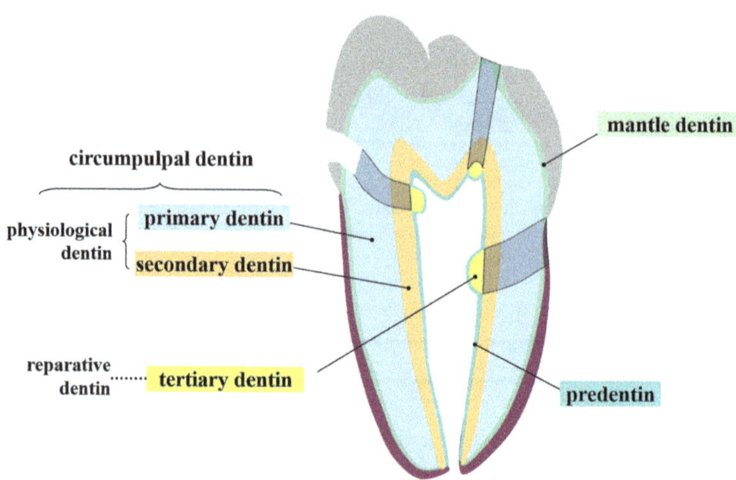

Fig. 1.3 Types of dentin (primary, secondary, and tertiary) and distribution

Matrix Vesicles Initiated Mineralization in Mantle Dentin

The outer layer of coronal dentin, the so-called mantle dentin, is formed through MV-mediated mineralization. MVs are formed by early differentiating odontoblasts and deliver either crystals or a high concentration of ions to the mineralization front either as calcification nodules or extrafibrillar calcospherites mineralization. As HA crystals expand to form mineralized areas, these areas continue to enlarge and diffuse laterally along with the whole layer of the mantle dentin. In the roots, a similar sequence of MV mineralization occurs in the tubular hyaline Hopewell-Smith layer and Tomes' granular layer (about 8–15 μm thick, for a total outer dentin layers thickness of about 30 μm) (Goldberg et al. 2011), as well as in the cementum (namely the inner cementum intermediate layer).

Collagen and ECM Proteins-Derived Mineralization in Circumpulpal Dentin

Mineralization of circumpulpal dentin, sub-adjacent to the outer dentin layers, is directed by collagen fibers and non-collagenous matrix components. This process begins with the formation of the primary dentin (mantle dentin) near the dentin–enamel junction (DEJ). Later, when the tooth becomes functional, dentinogenesis continues with odontoblasts directing the formation of a secondary dentin layer.

Dentin Formation

Cells, Structure and Extracellular Matrix Components of Dentin

Odontoblasts are terminally differentiated post-mitotic cells. During the tooth organ development, cells located at the periphery of the dental papilla forming a border with inner enamel epithelium undergo terminal differentiation through the pre-odontoblast stage (polarizing odontoblasts) to mature odontoblasts (polarized odontoblasts). This differentiation is characterized by a sequence of cytostructural and functional changes (Ruch et al. 1995; Ruch 1998). The post-mitotic polarized odontoblasts locate at the outermost layer of the pulp and are implicated in dentin ECM synthesis, secretion, and turnover, initiating biomineralization. In young teeth, after the root formation is completed, the cell bodies of odontoblasts partially overlap each other and form a pseudostratified layer with 4–5 rows. The number of odontoblasts decreases with time, and the number of pseudo layers decreases finally to one single row of cells in the adult (aged) teeth (Murray et al. 2002).

Hoehl's cells are undifferentiated odontoblast progenitors, residing adjacent to the mature odontoblasts. They are a reservoir of reparative cells in case of wound or aging (Goldberg and Smith 2004).

The dental pulp encompasses a series of cells, namely fibroblasts, mast cells, histiocytes/macrophages, dendritic cells, immunoglobulin-synthesizing cells, nerves (axons), and capillaries. B-lymphocytes (CD7+, CD8+) are rarely found, but immunologically related cells (endothelial cells, epithelial and mesenchymal cells) are present. Pulp cells express high motility, the capacity of antigen presentation to T lymphocytes, which are more abundant at the pulp's periphery.

Odontoblast cell bodies are linked by intercellular junctions (tight and gap junctions and desmosome-like junctions). Connexin 43, reelin, and osteoadherin are identified in differentiation-associated genes in odontoblast cells, playing a role in the recruitment of odontoblasts when dentin repair is initiated. The primary cilium of odontoblasts has been described near the Golgi apparatus, close to the centrioles. A cross-talk occurs between odontoblasts and nerve axons.

Two distinct parts are recognized in polarized odontoblasts: the cell body and processes protruding into the dental tubule up to the outer atubular dentin layer (mantle dentin). The cell bodies are implicated in extracellular matrix synthesis and secretion. In contrast, the processes are involved in secretion in the predentin (fibrillation of type I collagen, together with proteoglycans) and secretion inside the tubules of non-collagenous matrix proteins at the mineralization front (namely phosphorylated ECM together with re-internalization of degraded ECM components (promoters or inhibitors of dentin mineralization). Several laminins have been identified in odontoblasts (laminin α1, α2, β1, and γ1 subunits) (Yuasa et al. 2004).

The cell body includes the rough endoplasmic reticulum (RER), one (or more) saccules and vesicles of the Golgi apparatus, secretory and endosome vesicles, transported to lysosomes. The helicoidal extensions or matrix components, cleavaged by MMPs (MMP-2, MMP-9, and other proteases) undergo re-internalization through coated vesicles (clathrin-coated) that transport these vesicles back to the distal part of the cell bodies and incorporated in lysosomes. The vesicle degradation resulting from endocytosis processes occurs throughout the entire odontoblast processes running through the predentin and the dentinal tubules.

Odontoblast processes contain microtubules (α-tubulin), intermediary filaments (vimentin, nestin), and actin microfilaments (forming the sub-plasmalemmal undercoat). The main trunk and lateral branches are connected, and differ in composition. Secretory vesicles are convoyed and released first in predentin and second in dental tubule beneath the thin layer of metadentin, the first mineralizing layer located near the mineralization front. Clathrin-coated vesicles are implicated in the transport back to the cell body of re-internalized vesicles containing degraded extracellular matrix components.

Fibrillation of the just-released collagen occurs in the predentin. The native collagen fibers form a thick three-dimensional collagen fiber after coiling and transportation toward the distal pre-dentin near the dentin mineralization front (Figs. 1.4, 1.5, 1.6 and 1.7).

Calcospherites structures, measuring 1.5–10 μm in diameter, are formed at the mineralization front, whereas interglobular areas are less mineralized. In vitamin-D rickets hypo-phosphatase, large interglobular structures display deficient mineralization. They contain some non-collagenous proteins, and microprobe investigations

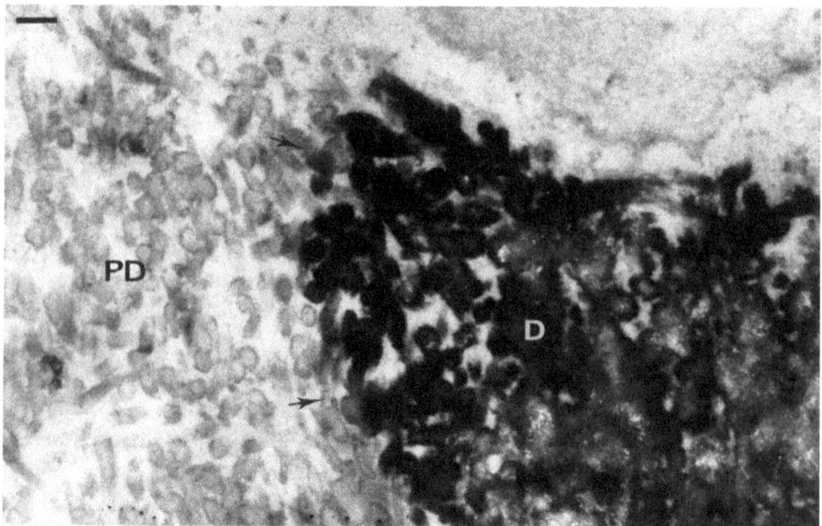

Fig. 1.4 Predentin (PD)/dentin (D) junction- unstaines section

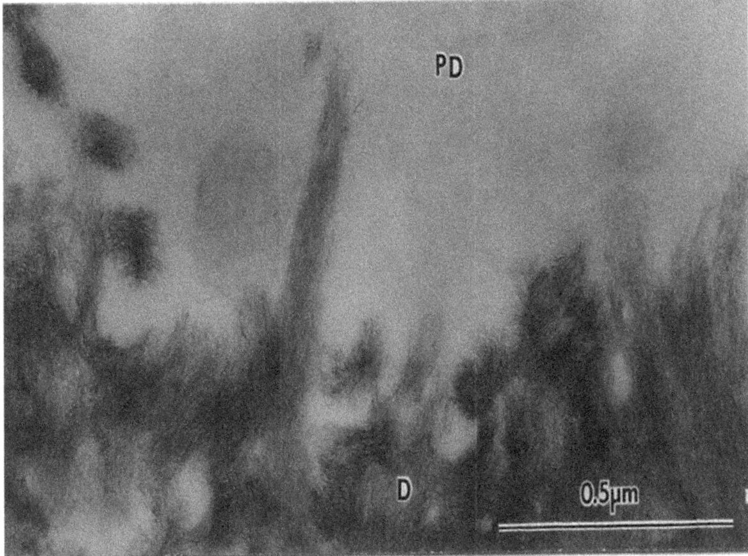

Fig. 1.5 Mineralization front. Unstained section. HAp crystals are attached along the collagen fibrils that underwent mineralization in intertubular dentin, located between the predentin (PD) and dentin (D)

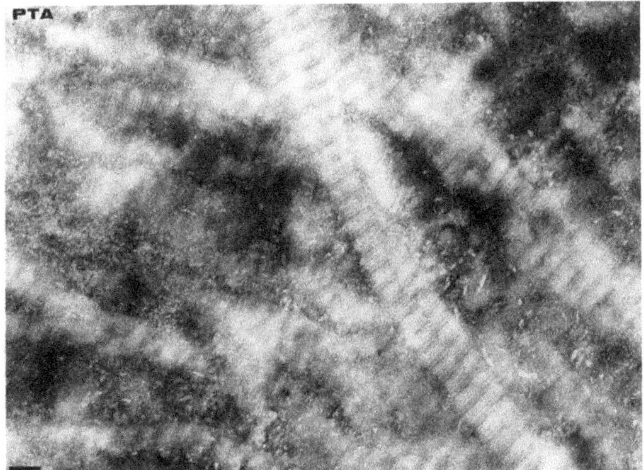

Fig. 1.6 Transition between the predentin (PD—which is electron lucent—and the mineralization front (D) stained with phosphotungstic acid and chromic acid (a specific staining for DPP). In the thin metadentin layer (1–2.5 mm thick), collagen fibers are stained, but not the intercollagenic empty spaces. In dentin the collagen fibers and intercollagenic spaces are densely stained

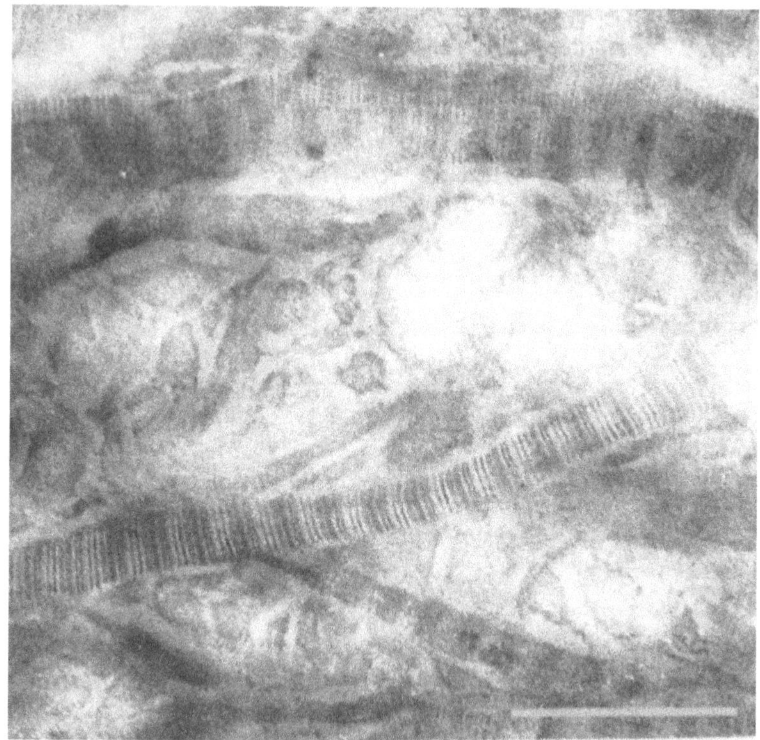

Fig. 1.7 Type I collagen in intertubular dentin

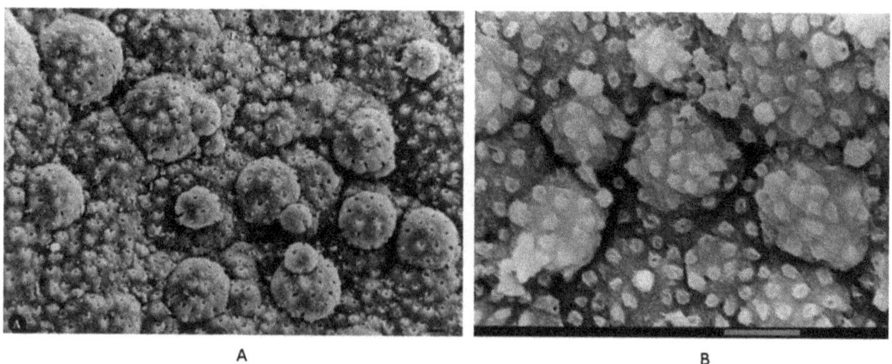

Fig. 1.8 Calcospherites at the dentin mineralization front (Goldberg and Escaig 1981)

indicate low mineralization compared to the well-mineralized content of globular calcospherites (Figs. 1.8 and 1.9).

Fig. 1.9 Calcosherite at the pre-dentin/dentin junction (Goldberg et al. 1987)

Metalloproteinases (MMP-2, MMP-9, stromelysin (MMP-3), MT-MMP, TIMP-1 to 3) enamelysins (MMP-20), tissue-nonspecific alkaline phosphatase (TNAP), and serum albumin-derived proteins (αHS2-glycoprotein, albumin) are suggested to function in ECM degradation and re-internalization (Chaussain-Miller et al. 2006).

Dentinal tubules (Fig. 1.10) measuring about 1.21 ± 0.08 μm (2–4 μm) in diameter are surrounded by a hyper-mineralized peritubular dentin (Fig. 1.11), and between tubules, a softer collagen-rich intertubular dentin constitutes the bulk of dentin. Mineral analysis indicates carbonate-substituted hydroxyapatite. Peritubular dentin crystallites are characterized by small 250 Å equal-sized microcrystals (Lester and Boyde 1968), or a = 36.00 nm (mean length), b = 25 nm (width), and c = 9.76 nm (thickness) (Schroeder and Frank 1985).

Circumpulpal (and Secondary) Dentin Formation

Circumpulpal dentin includes intertubular and peritubular dentin and intratubular mineralization (sclerotic dentin), filling totally or partially the lumens of tubules. The tubular density varied from 18,000 to 21,000/mm^2, from the inner dentin to the outer layer, according to Schilke et al. (2000) (Fig. 1.1).

In the root, the peripheral of circumpulpal dentin includes Tomes' granular layer, which is an 8–15mm thick hypomineralized layer formed by calcospherites and interglobular spaces. At the boundary with cementum, the innermost layer of cementum is deprived of dentinal tubules. At this border, a thin layer of matrix is distinguished from both cementum and dentin (named intermediate cementum, or the 7–15 mm thick hyaline layer of Hopewell-Smith). Spherical bodies, which may

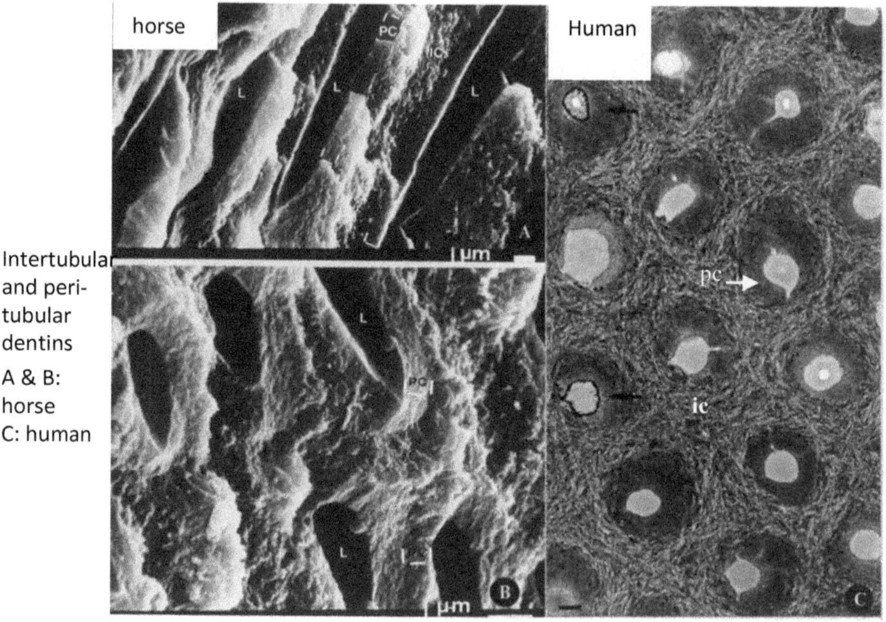

Fig. 1.10 Longitudinal and transverse sections of horse and human dentin/intertubular dentin is collagen-rich, peritubular dentin appears around the lumen of canaliculi

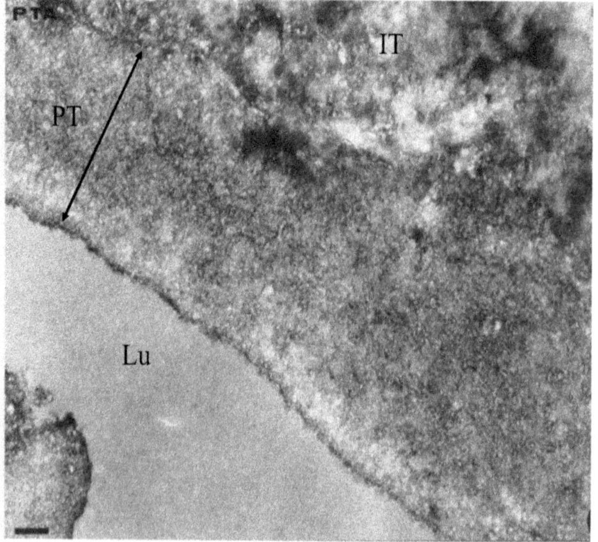

Fig. 1.11 Phospholycoprotein stained by PTA in the peritubular dentin IPTI. LU lumen, IT intertubular dentin

be matrix vesicles, together with a few collagen fibrils are present in this layer. It contains proteoglycans and enamel matrix proteins and is possibly to be a product of Hertwig's epithelial root sheath (see also the cementum section).

During tooth development, the dentin secreted prior to completing the root formation is defined as the primary dentin and comprises the main bulk of the circumpulpal dentin. This forms at a rate of 4 μm per day. Secondary dentin is the dentin bordering the pulp. It starts to be formed after root completion and continues throughout the life of the tooth, albeit at a slow rate. Secondary dentin may contain fewer tubules than primary dentin, and usually, there is a bend in the tubules at the primary and secondary dentin interface. The third type of dentin is tertiary dentin, which is laid down at specific loci of the pulp–dentin interface in response to environmental stimuli. Mild effects stimulate an increased rate of matrix secretion by existing odontoblasts, resulting in reactionary dentin formation. Stronger stimuli will probably lead to the death of the odontoblasts, and if the conditions in the dentin–pulp complex are favorable, specific pulp cells will differentiate into a generation of odontoblast-like cells secreting a reparative dentin matrix that will further differentiate. Both reactionary dentin and reparative dentin are considered tertiary dentin. Primary and secondary dentin is physiologic, whereas tertiary dentin is formed in response to pathologic alteration after a carious lesion or a trauma. The primary and secondary dentin are tubular. In comparison, the tertiary dentin may be tubular or atubular and even may have a bonny appearance (osteodentin).

Dentin mineralization is initiated with the deposition of amorphous calcium phosphate (ACP). As mineralization progresses, inorganic crystalline consist of whitlockite, with the presence of hydroxyapatite at later stages, are deposited. In the dentinal tubular lumen, aggregates of calcium phosphate crystallites (Mg-substituted beta-tricalcium phosphate, beta-TCP), brushite (calcium hydrogen phosphate dehydrate) and amorphous and crystalline calcium phosphate are deposited. Dentin HAp mineral appears as plate-like crystallites, 2–5 nm in thickness and 60 nm in length, and display a needle-like appearance. Inside the lumen of the tubules, rhombohedric whitlockite crystallite presents with the occurrence of dissolution and re-precipitation has been found in aging dentin (sclerotic dentin) (Fig. 1.3).

In case of trauma or carious lesion, intraluminal mineralization occurs, which is due to calcium/phosphate re-precipitations. Tiny flat, needle-like, hydroxyapatite crystals (34 Å thick, 120–135 Å wide, and 250–600 Å long) are formed in the intertubular dentin. They are mostly located along and between type I collagen fibers (90%), including type I trimer (11%). As minor components, type III and type V collagen are found. Mineral crystals include hydroxyapatite, calcium phosphate, and in tertiary dentin, occluding the lumen, whitlockite, and amorphous octacalcium phosphates. In the peritubular dentin, rhombohedric crystals have 9–10 m in height and $a = b$ 25 nm appearing isodiametric on sections (Lester and Boyde 1968; Schroeder and Frank 1985).

LIM mineralization protein (LMP-1) may play a role in odontoblast differentiation and dentin matrix mineralization of human teeth with normal and pathologic conditions. LMP-1, which is known to regulate osteoblast differentiation and bone formation, is also expressed by predentin, odontoblasts, and endothelial cells of the blood vessels of healthy teeth. In addition, LMP-1 expression is found in unmineralized reparative dentin, odontoblast-like cells, and pulp fibroblasts in teeth with caries and pulpitis. Furthermore, LMP-1 expression is found on the surface of the pulp stone or in the residual predentin of teeth with pulp calcification.

1.2 Cementum

Cementum, together with the periodontal ligament (PDL) and the alveolar bone, contribute to the attachment of the tooth to the socket.

1.2.1 Cementum Anatomy

Cementum is divided into acellular and cellular cementum. Acellular cementum is located from the cervical margin to the half or one-third, and cellular cementum is observed from the apical half or two-thirds to the apical portion (Fig. 1.12). From the

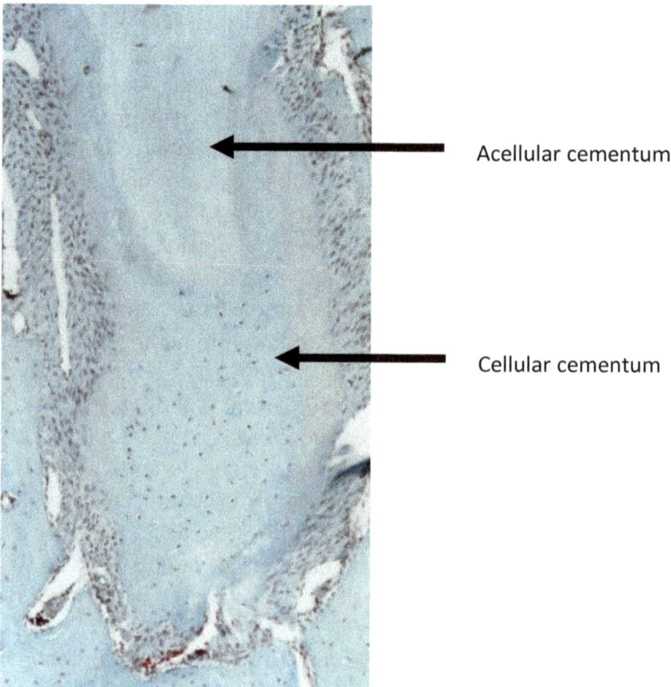

Fig. 1.12 Acellular (upper part of the root) and cellular cementum (lower part of the root) (Goldberg et al. 2008)

Table 1.4 The different types of human root cementum: anatomic characterization

Terms	Abbreviation	Organic components	Location	Function
Acellular, afibrillar cementum	AAC	Homogeneous matrix, no cells, no collagen fibrils	At dentino–enamel junction, on enamel	Unknown
Acellular, extrinsic fiber cementum	AEFC	Collagen fibrils as Sharpey's fibers, no cells.	Cervical to middle root	Tooth anchorage
Cellular, intrinsic fiber cementum	CIFC	Intrinsic collagen fibrils and fibers, cementocytes	Apical and interradicular root surfaces, resorption lacunae, fractures	Adaptation, repair
Acellular, intrinsic fiber cementum	AIFC	Intrinsic collagen fibrils and fibers, no cells	Apical and interradicular root surfaces	Adaptation
Cellular, mixed, stratified cementum (AEFC + CIFC/AIFC)	CMSC	Intrinsic collagen fibrils and fibers, collagen fibrils as Sharpey's fibers, cementocytes	Apical and interradicular root surfaces	Adaptation, root anchorage

cervical to the apical portion of the tooth root surface, the acellular cementum is located in the inner part of the cementum, near the DEJ (Fig. 1.12).

Four cementum types have been characterized in human (Table 1.4):

- **Acellular afibrillar cementum** (**AAC**) is located in the cervical junction between enamel and cementum, containing proteins derived from tissue fluid or serum. Its thickness ranges between 1 and 15 μm. No collagen fibers have been identified, but fine granular or reticular meshwork, delineating more or less parallel to the tooth surface.
- **Acellular extrinsic fiber cementum** (**AEFC**) (mainly found in the cervical and middle root portions, with faster growth rates on distal (4.3 mm/year) than on mesial (1.4 mm/year) root). The extrinsic collagen fibrils are packed parallel to each other and arranged in groups of varied sizes. The diameter of these fibrils was 2–3 mm. In a 3 wk-old rat, osteocalcin was seen in cells lining the root surface. The staining was confined to the occlusal third of the root. Immunolabeling for osteocalcin was equally positive for the acellular cementum

and for the cellular cementum. Its total thickness ranges between about 30 and 230 μm. AEFC fibrils penetrate into the cementum, as Sharpey fibers. The diameter of these LAD fibers was 3.6 μm before penetrating into the AECF. Crystals appeared as plate-like or fine needles with a maximum length of about 60 nm. The lamellar appearance represents periods of activity of matrix formation, altering with mineralization. These incremental lines correspond to the past AEFC surfaces.

- **Cellular intrinsic fiber cementum** (**CIFC**) containing cementocytes embedded in a collagenous matrix. (They are deposited by cementoblasts in the space between the Hertwig's root sheath and the dentinal surface),
- **Cellular cementum** is found in the apical two-thirds. This cementum may be **mixed and stratified** lamellae or incremental lines (**CMSC**). It contains both intrinsic (40% in acellular cement and 15–40% in cellular cement) and extrinsic fibers (3–6 μm). The cells secrete intrinsic fibers around the principal fibers to result in typical intrinsic–extrinsic fiber structures.

In the early stage, when the cellular compartmentalization is not established, both the intrinsic and extrinsic fibers are irregularly arranged. Lacunae and canaliculi of cementocytes, lamellae being separated by incremental lines and a superficial layer of cementoid.

Cementum lacunae vary in diameter between 6- and -12 μm, and fine canaliculi radiate from these lacunae, and from 25 to 35 μm in length, extending up to about 15 μm. The cell bodies assume a disc- or stellate-like shape. The periphery of cementocyte lacunae and the canaliculi show strong metachromatic toluidine blue, PAS–negativity, but Sudan black-positivity (Lorber 1951), indicating PGs and lipid content on the surface of cementocyte lacunae and canaliculi. This suggests, like osteocytes, cementocytes maintain vitality and deposit the ECM.

1.2.2 Cementum Formation (Cementogenesis)

Cementum formation starts before the tooth erupt, while root formation undergoes on the outer surface of the newly formed root dentin. Cementum is formed throughout the lifetime, responding to the activity of the periodontal ligament.

Cementoblasts form both cellular and acellular cementum. Unlike other mineralized tissue forming cells, the detailed profile of cementoblasts is not well understood. It is partially because the origin of the cementoblasts is still controversial. Classically it is suggested to be the ectomesenchymal cells of the dental follicle (mesenchymal hypothesis) (Cate et al. 1971; Diekwisch 2001; Kémoun et al. 2007; Saito et al. 2005; Yamamoto et al. 2014). However, trans-differentiation of Hertwig's epithelial root sheath (HERS) into cementoblasts (referred to as an alternative epithelial hypothesis) is also suggested (Bosshardt and Nanci 2004; Bosshardt and Schroeder 1993; Chen et al. 2014; Huang et al. 2009; Li et al. 2019; Sonoyama et al. 2007; Zeichner-David et al. 2003). The numerous evidence previously shown for both hypotheses indicate a possibility that cementoblasts, in fact, derived from both tissues even with different molecular profiles, depending on the types of cementum, stages of tooth formation, tooth types, or even species. Especially morphological observations on cellular cementum showing both HERS cells and dental follicle derived cells are incorporated into the forming cementum, suggests cementoblasts derived from both tissues may even exist simultaneously.

During root formation, intermediate cementum is first laid down on top of newly formed dentin. This innermost layer of cementum is a thin layer about 2 μm thick and presumed to be rich in PGs, as it is stained with ruthenium red and chromic/phosphotungstic acid on the histological sections. This layer is located between the root dentin and cementum to establish cementodentinal junction (CDJ) by connecting the two tissues. Formation of intermediate cementum occurs while HERS covers the outer surface of root dentin. Thus the origin of cells responsible is suggested to be HERS cells (Bosshardt and Schroeder 1996).

During the early stages of cementogenesis, HERS is disintegrated, and remaining epithelial cells form a loose meshwork around the root (epithelial cell rests of Malassez/ERM). Consequently, cells extracellular matrix of the dental follicle come to the exposed root surface to initiate radicular cementum deposition. The radicular cementum is avascular. It grows in thickness to compensate for abrasion. Enamel-related proteins are involved in the formation of cementum (Hammarström 1997).

Once the tooth is in occlusion, a more rapidly formed and less mineralized variety of cementum (cellular cementum) is deposited on the unmineralized dentin surface and acellular cementum near the advancing root edge by the cells differentiated from the adjacent cells of the dental follicle. Sometimes (especially when the cementum formation speed is fast), a thin (5 μm thick) pre-cementum/cemented layer is observed. When the periodontal ligament (PDL) becomes organized, the cellular cementum continues to be deposited around the ligament fiber bundles. These PDL fiber bundles become incorporated into the cementum and partially mineralized. This way, mixed fiber cementum is formed.

At the collar of the teeth (the boundary between the crown and root), cementum also forms a boundary with enamel as the cementum-enamel junction (CEJ). There are three types of CEJ: enamel bordering cementum, cementum overlaps enamel, enamel overlaps cementum, and intermediate zone of dentin exposed. In humans, the percentage of the pattern varies according to the type of the tooth, and generally, the first two types are commonly observed.

Cementum mineralization is initiated by MVs (Takano et al. 2000) (Fig. 1.13). Later, the matrix contains a number of MVs, proteins, and lipids, including cholesterol, free fatty acids, sphingomyelin, glycolipids, lysophospholipids, and phosphatidylserine.

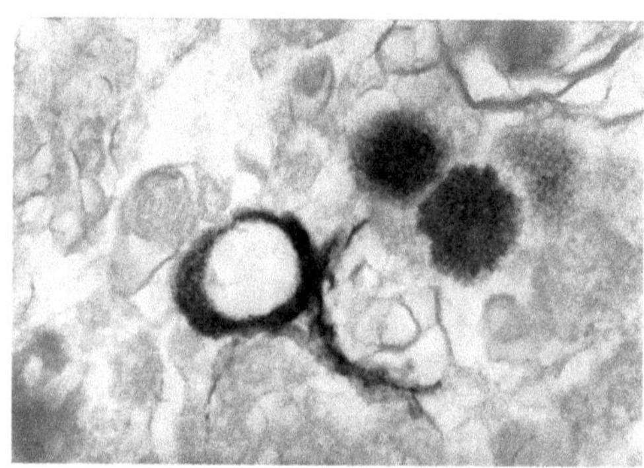

Fig. 1.13 Matrix vesicles initiate HAp crystals in the early cementum formation. Goldberg personal figure

1.3 Bone

1.3.1 Bone Anatomy

Bones can be long or short, flat or sesamoid, irregular, or associated with sutures. Different types of bone have been identified:

- Cortical bone forms the outer surface of the bone and is dense and compacted. It contains the Haversian system (osteon).
- Trabecular/cancellous bone forms a spongy structure and is typically seen inside of the long bone. It includes red marrow, where the osteogenic precursor cells are derived from, while yellow marrow is included in compact bone. Red marrow has periosteum and endosteal bone (innervated and containing blood vessels).
- Appendicular bones (including the limbs, shoulder, and pelvic bones) contrast with the axial skeletal bones, vertebral and thoracic costs.

1.3.2 Bone Cells

Bone encompass a series of cells that are present simultaneously near or at the surface and inside of the bone (Fig. 1.14). Osteoblasts are in charge of bone matrix formation and on a thin layer of the yet-to-be-mineralized osteoid. Osteoblasts are bone marrow-derived cells, and their differentiation goes through osteoprogenitors, pre-osteoblasts, bone lining cells, and osteoblasts. During differentiation, osteoblasts are closely associated with osteoclast differentiation and maturation. As located on top of osteoid, osteoblasts regulate bone ECM formation and mineralization.

Osteocytes are terminally differentiated osteoblasts and incorporated within the bone matrix between the lamellae of the matrix (osteons or Havers bone structures). The biological properties of these cells are listed in Table 1.5. They intercommunicate by thin dendritic cytoplasmic extensions in the bone canaliculi forming direct interconnections and a network inside the bone to respond to the external stimuli. Osteocytes are not only a space-holder, but they remove the perilacunar matrix

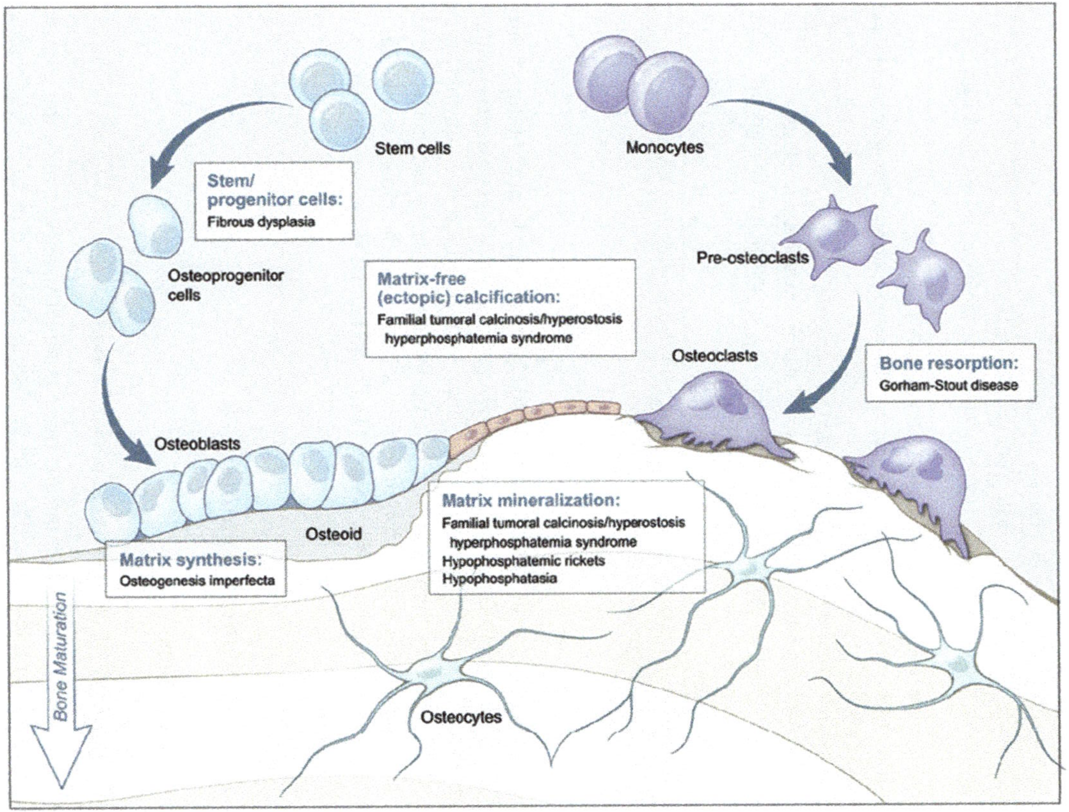

Fig. 1.14 Cells implicated in bone formation and remodeling. Bone marrow stem cells differentiate into osteoblast progenitors, that become osteoblasts and bone lining cells located over osteoid. These cells entombed into osteoplasts become osteocytes. Monocytes differentiate into pre-osteoclasts, becoming osteoclasts, implicated in remodeling bone (Howship's lacunae, filled afterward by regenerative bone). Gene-coding for matrix proteins, mutation, or excessive degradation (or defective bone synthesis) leads to pathologies of bone (Foster et al. 2014)

Table 1.5 Normal functions of osteocytes, reprinted from (Bonewald 2017)

• Control mineralization through Phex, Dmp1, and MEPE
• Regulate phosphate homeostasis through FGF23
• Play a role in calcium homeostasis in response to PTH/PTHrP
• Can recruit osteoclasts through the expression of RANKL with or without cell death
• Can regulate osteoblast activity through Sclerostin
• Are mechanosensory cells through β-catenin signaling
• Have autocrine/paracrine effects through prostaglandin production.
• Under calcium restriction, osteocytes remove calcium from the bone through the Vitamin D Receptor
• Osteocytes regulate myelopoiesis/hematopoiesis through G-CSF63
• G-CSF targets osteocytes that mediate mobilization of Hematopoietic Stem/Progenitor Cells and is prevented by surgical sympathectomy
• Osteocytes regulate primary lymphoid organs and fat metabolism
• Osteocytes can dedifferentiate to become a source of matrix-producing osteoblasts
• Can increase muscle myogenesis and muscle function and can inhibit muscle mass with aging
• Can have effects on the heart and liver through FGF23
• Play a role in fracture healing through IGF-1
• Regulate bone formation through Bmpr1a signaling, Notch activation, and ERαsignaling
• Suppress breast cancer growth and bone metastasis

called osteocytic osteolysis (Wysolmerski 2012). Osteoclasts are multinucleated cells and contain a highly developed lysosome system to resorb bone matrix. Monocytes and osteoclasts participate in local bone turnover, which goes through bone destruction followed by the filling of Howship lacunae made by osteoclasts. Bone cells are known to cross-talk with immune cells indicating the close association of bone metabolism and the immune system (Tsukasaki and Takayanagi 2019).

1.3.3 Types of Bone Formation

1.3.3.1 Primary Bone

There are two groups of primary bone mineralization:

- **Intramembranous bone formation**: Bone is directly derived from the connective tissue without a template. This mechanism is involved in bone lengthening.
- **Endochondral bone formation**: Bone is derived from the cartilage template. The epiphyseal cartilage contains combined ground substance and loose, small (10–20 nm in diameter) fibril bundles of collagen and serves as the locus for primary bone formation. The epithelial cartilage consists of zones according to the chondrocytes differentiation. It includes a resting zone, a proliferative zone that is followed by the hypertrophic zone, a calcified zone, and a zone of bone

deposition involving bone lengthening. Thickening is related to the transformation of bone marrow cells into osteoblasts. As an example of endochondral ossification, mandibular condyle requires phenotypic changes leading from a column of chondrocyte to endochondral bone formation.

1.3.4 Bone Structural Systems

1.3.4.1 Osteon

Osteon (Haversian system) is a functional unit in the compact bone containing lamellar layers (lamellar bone). Lamellar bone is composed of lamellar units, each unit consisting of five sub-layers. The orientation of the array differs in each sub-layer with respect to both collagen fibrils axes and crystal layers. The lamellar structure is multifunctional, forming a complex rotated plywood-like structure (Weiner et al. 1999). This organization allows for mechanical improvements.

1.3.4.2 Secondary Bone

Secondary bone is formed by osteoblasts. Remodeling is the basis of fibrillar or lamellar concentric lamellae bearing nanocrystals platelets of HAp. In the center of "osteon," blood vessels are localized. Collagen fibrils have a mean diameter of 78 nm. Intrafibrillar crystals are small structural properties of cortical bone most commonly employed as a surrogate for its mechanical competence include the thickness of the cortex, cortical cross-sectional area, and area moment of inertia. But microstructural properties such as cortical porosity, crystallinity, or the presence of microcracks also contribute to bone's mechanical competence. Microcracks, in particular, not only weaken the cortical bone tissue but also provide an effective mechanism for energy dissipation.

Dense external envelope (cortical bone displaying lamellae) and spongy interior (cancellous bone) contain bone marrow in its center.

1.3.4.3 Craniofacial Bones

Formation of the mandible and maxillary bones go through a unique combination of endochondral and intramembranous ossification involved in two temporary bone structures (the alveolar and basal bone). During the mandible bone early formation, the intermediate developmental cartilage (Meckel's cartilage) guides to form the basal part of the bone through intramembranous ossification. Five cartilages are formed at later stages and added to the cortical bone, including the chins (mental cartilage), alveolar bone, coronoid, angular, and temporomandibular cartilage. They contribute to the mandible formation and follow the rules and steps of endochondral ossification.

Alveolar bone formation is linked to teeth eruption. At later stages of life, the height of alveolar bone decrease, namely if they are closely associated with periodontal disease. The basal bone remains more or less intact, despite the pathologic lysis of alveolar bone. The same process occurs in the maxillary bone, the palate and basal bone being more stable. For the maxillary, the sinus remains stable, despite the periodontal alteration. Hence, alveolar and basal bones should be added to bone nomenclature.

1.3.5 Bone Composition (Also See Chap. 2)

1.3.5.1 ECM Structure

The fibrils of bone have 80–100 nm in diameter, forming fibrolamellar bone (also known as plexiform bone). MVs are seen in the osteoid during embryonic bone formation as well as appositional mineralization at the of the physiologic bone-forming site and bone at the early stages of fracture healing (Hoshi and Ozawa 2000; Sela et al. 1992).

1.3.5.2 Bone Mineral Composition

Bone is a heterogeneous composite material consisting of a mineral hydroxyapatite phase ($Ca_{10}(PO4)_6(OH)$ forming about 70% mineral, also involving an organic phase (~90% type I collagen, ~5% non-collagenous proteins (NCP), ~2% lipids (22% proteins), and water (8% by weight).

Inorganic phases consist of poorly crystalline B carbonated apatite. Bone nanocrystals are characterized by a hydrated surface layer. In particular, in trabecular bone, apatite displays reduced crystallite sizes, Ca/P molar ratio, and carbonate content and exhibits a greater extent of thermal conversion into β-tricalcium phosphate compared with cortical bone apatite. In cortical bone, bone mineral crystal size on average is larger or greater than in cancellous bone. With more carbonate substitution, the Ca/P ratio increases. Also, the mineral becomes more crystalline, and the crystal size increases. There are fewer labile phosphate groups and more incorporated carbonate ions in older bone mineral crystals. Hydroxyapatite crystals (or dahllite crystals) are plate-shaped structures 50 nm in length and 25 nm in width (Boskey and Coleman 2010).

1.4 Mineralization in Non-collagen-Derived Tissues

1.4.1 Enamel

1.4.1.1 Anatomic Organization

The Formation of Enamel from Secretion to Maturation

The formation of dental enamel involves three successive phases: (1) pre-secretory, (2) secretory, and (3) enamel maturation. Immature enamel contains about 15–20% proteins by weight, whereas, after maturation, enamel contains only 0.1% protein. Mature dental enamel is primarily (>95% by weight) composed of apatitic crystals and has a unique hierarchical structure. Critical evaluation of the present state of knowledge regarding the potential role of the assembly of enamel matrix proteins is (1) the regulation of crystal growth and (2) the structural organization of the resulting enamel structure.

Ameloblasts form the enamel organic matrix to provide the scaffold of mineralization and then remove degraded matrix protein to progress the mineralization. During the advanced bell stage of tooth germ formation, the laminin enriched basement membrane present between pre-secretory ameloblasts and odontoblasts disappears, and ameloblasts begin synthesizing and secreting matrix proteins on top of the predentin formed by odontoblasts. The basement membrane re-appears at the maturation stage. The dynamic process of enamel biomineralization occurs in the extracellular space between the ameloblasts and the mineralized dentin. The tissue grows continuously until the ameloblasts have laid down the whole thickness of the enamel matrix.

During the process of amelogenesis, ameloblasts pass through a series of differentiation stages characterized by changes in cell morphology and function. In general, cell differentiation during enamel development results in pre-secretory, secretory, transitional, and maturation phase ameloblasts. Cell morphology and chemical composition of the enamel extracellular matrix are the basis for a definition of the different stages of amelogenesis. Secretory stage ameloblasts are tall columnar epithelial cells characterized by structures called Tomes' processes, and their main function is a massive production and secretion of enamel proteins. The newly formed enamel shows a very irregular wavy boundary at the dentin–enamel junction (DEJ), where dentin and enamel interlock processes. The full thickness of enamel is established at the secretory stage following the secretion of the organic matrix and its partial, but not complete, mineralization. The first mineral phase to form is an amorphous calcium phosphate that later transforms to crystalline apatite. The size, shape, and spatial organization of these amorphous mineral particles and older crystals are the same, suggesting that the mineral morphology and organization in enamel is determined prior to its mineralization. The transition stage occurs between the secretory and maturation stages when the cells undergo structural changes, becoming shorter and losing their apical cell extension, the Tomes' process. At

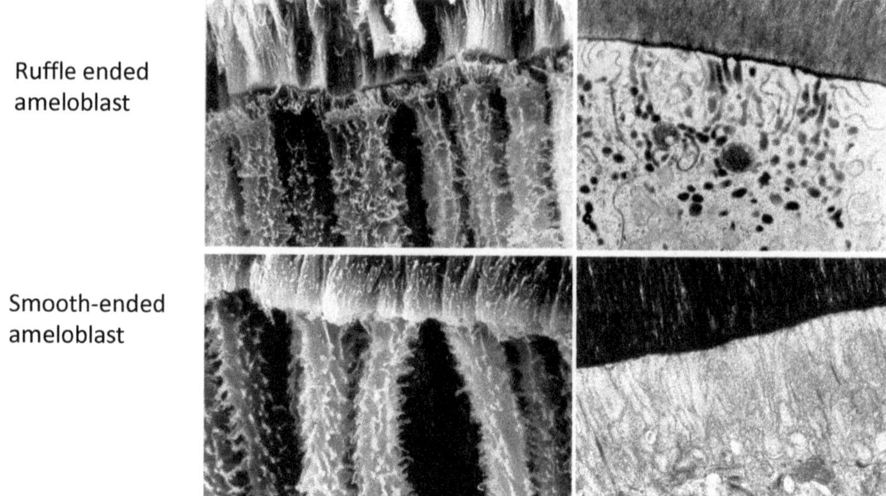

Fig. 1.15 Ruffle border and smooth ending maturation ameloblasts. Goldberg personal SEM and TEM document (Goldberg, personal images)

this time, their secretory activity is also drastically reduced. The transition stage can be visualized as the boundary between the translucent and opaque enamel in the rat's incisor. Major changes in cell morphology and function continue to occur during the maturation stage when the cells reorganize to accommodate massive protein degradation and rapid crystal growth. Their secretory activity is down-regulated but not terminated. Maturation stage ameloblasts are characterized by their periodic morphological changes from "ruffle ended" to "smooth ended" (Fig. 1.15). These periodic changes correspond to the calcium influx to the matrix and the removal of matrix proteins degraded by extracellular proteinases, such as MMP20 and KLK4 (See Chap. 11 for more detail).

1.4.1.2 Mature Enamel

Dental enamel has an internal void volume of 0.2–5%. The initial inner enamel is about 40 μm thick and comprises an inner enamel layer that consists of a layer (4 μm thick) along the DEJ, which contains a few rows of oval-shaped rod profiles.

The outer enamel layer is about 20–30 μm thick, formed by the outer enamel proper and a 2–4 μm thick final enamel layer, where all the crystallites are parallel, forming a non-prismatic or prismless (aprismatic) enamel. Prism-free enamel is 40% (mandibular) and 47% (maxillary) with a mean thickness of about 30 μm. Surface parallel laminations were frequently observed.

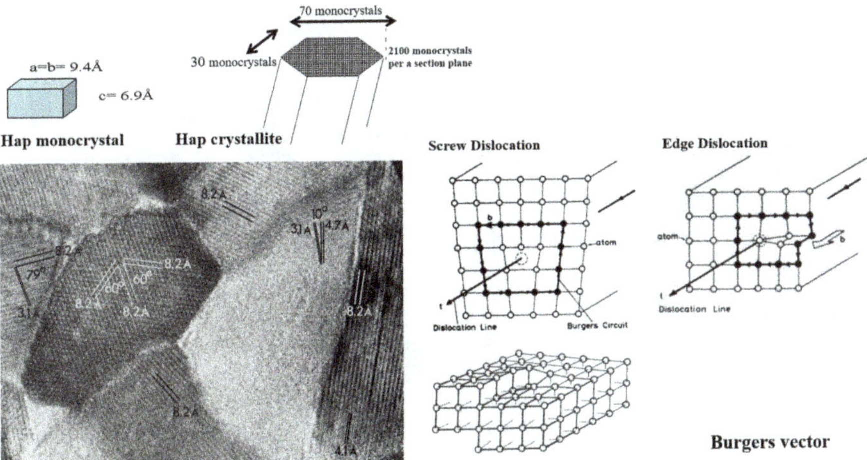

Fig. 1.16 Enamel structure: monocrystals are associated into crystallites. Defects in the Burger's vectors lead to dilocations vis & edge (Arends and Jongebloed 1979)

1.4.1.3 Enamel Mineralization

The youngest crystals have the shape of long plates. They appear as parallel ribbons with 200–300 Å in width and a thickness of about 10 Å, and later there is an increase in the size of the crystals.

Crosscut sections of crystallites implicate the association of hydroxyapatite monoclinic crystallites into hexagonal crystals. Theses subunits display the following parameters $a = b = 9.432$, and $c = 6.881$ Å (Fig. 1.16). The thickness and width of human enamel crystallites are 263 Å $\pm$ 21.9 Å and 683 Å $\pm$ 134 Å, respectively, for mature enamel crystallites. The average crystal size has a diameter of about 40 and 5 nm.

Whitlockite, brushite, and other calcium phosphate crystals are found in carious lesions. In sound enamel, crystal thickness is about 275.9 Å, and crystal width 477.6. Å. Cross-sectional dimensions of carious lesion indicate a 328.4 Å in thickness and 653.4 Å in width.

Lattice defects have been recognized during the initial crystal dissolution. Measurements of the hexagonal crystallite indicate a mean value of 50 nm (30 nm + 70 nm = 100 nm/2 = 50 nm). As the size of an enamel crystallite is about 30 nm $\times$ 70 nm, a section at right angles of a crystallite includes about 2100 monocrystals (Fig. 1.16). In length, the average HAp crystal is 254 nm, with a thickness diameter of 12 nm to ~20 nm. In longitudinal sections, the crystals form thin ribbons, they are uninterrupted between the dentin–enamel junction and enamel surface, but large fragments may result from sectioning.

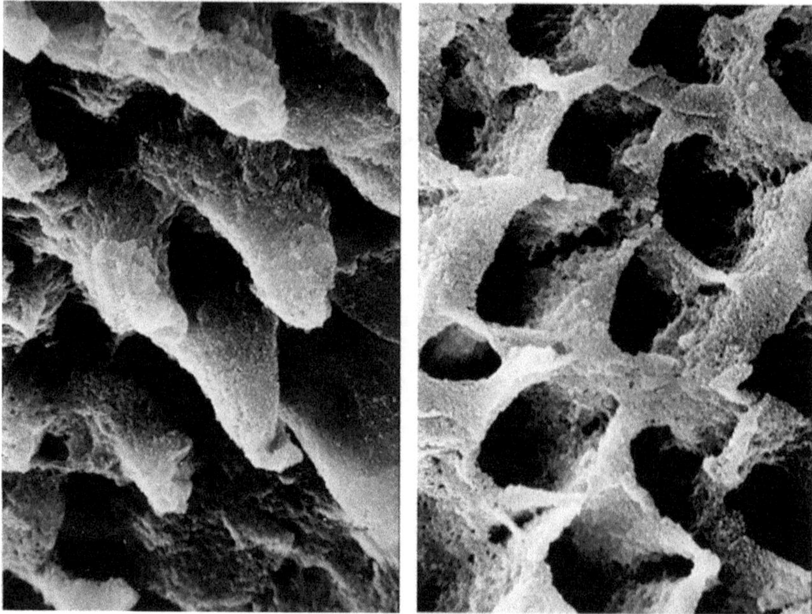

Fig. 1.17 Resin replica. Preferential acid destricution of rods (link) and interred enamel (right). Type 1 and 2 (Silverstone et al. 1975)

The presence of a dislocation line is parallel to the *c*-axis of enamel apatite. The *c*-axis is almost perpendicular to the enamel surface. For the deciduous teeth, the crystalline orientation differed from that of the underlying enamel by 26.9° ± 7.6°.

A complex series of hydroxyapatite monocrystal units are connected inside larger crystallites grouped into rods and interrods (Fig. 1.17). These structures contribute to the early carious lesion (white spot lesion), and/or to the demineralization/ remineralization process. They are contributing to cross-striations (Striae of Retzius) (Fig. 1.18) and Hunter-Schreger bands (Fig. 1.19). Micro porosities, intercrystallite resin infiltration, and the outer surface zone of enamel involve the complex enamel organization and its hierarchical structure of the enamel matrix. Hunter-Schreger (HS) alternating bands are organized in diazone and parazone, forming light and dark bands. The HS bands are visible namely in the inner prismatic enamel and vary from 1 single prism in some species to 30 prisms. The diazone and parazone differ by their orientation, at right angles between them, but not by their composition. It is a specific mode of prism decussation. Rods and interrods consisted of aggregated tubular subunits, 250 Å in diameter. Periodic modulation of the secretion-maturation process form incremental lines in enamel. The birth process influences the formation of the neonatal line in primary tooth enamel. Daily cross-striations occur in a regular period with an interval of 4–5 μm, and about every 9 days, they contribute to the formation of Striae of Retzius, terminating at the enamel surface as perikymata

Fig. 1.18 Retzius lines or bands are joining enamel surface. They constitute perikymatia mostly seen in young enamel, whereas they look abraded in old, flat, enamel (Goldberg, personal collection)

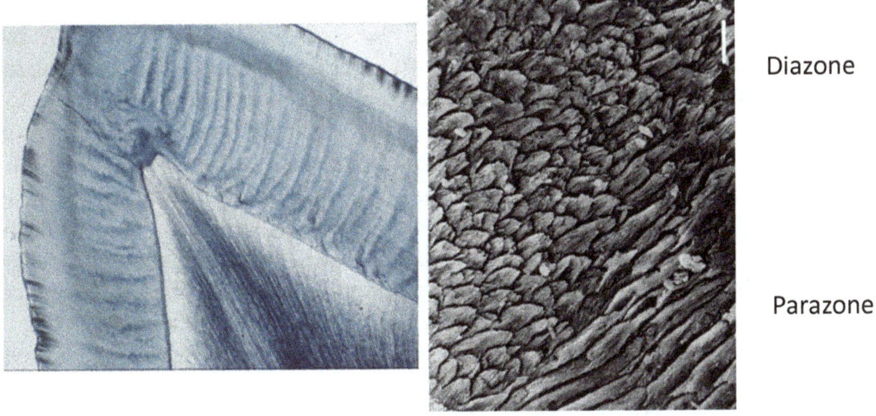

Fig. 1.19 Diazones and parazones form dark and lucent Hunter-Schreger bands (Goldberg personal collection)

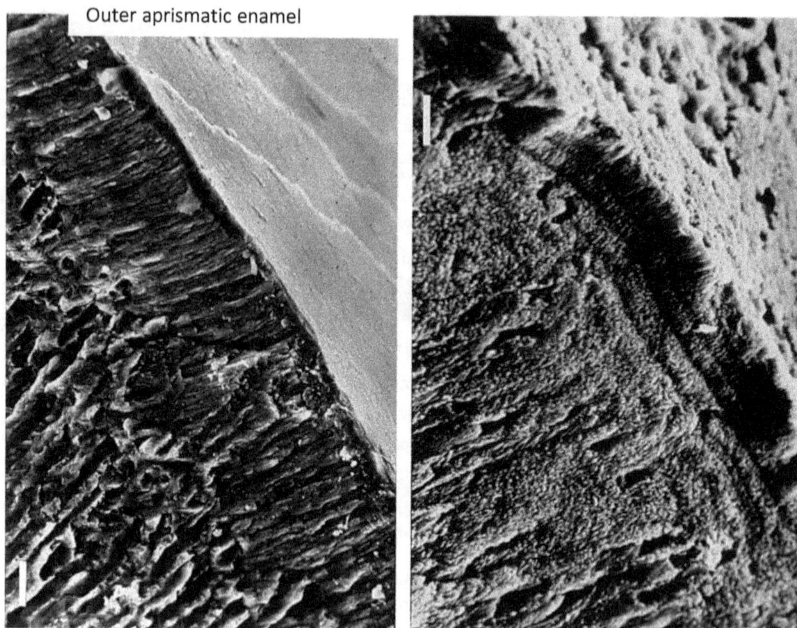

Fig. 1.20 Enamel surface with perikymata or/and lamellar structures. Goldberg – personal collection

(Fig. 1.20). The angle between prisms and the enamel surface is about 15°, and between the Striae of Retzius and the enamel surface is about 4°. Hypomineralized bands form them. Prismatic enamel is present, namely in most mammalians.

Group of crystallites is associated with prisms (enamel rods) (Fig. 1.17) measuring about 5 μm width, whereas the width of interprismatic enamel is about 0.5 μm. They merge more or less regularly. Each prism is composed of tightly packed small crystallites. Remnants of enamel proteins are located in the sheaths located at the surface of crystallites and in the organic part (enamel sheaths organic-rich) located between rods and interrods. Enamel rods constitute cylindric-like structures. Enamel sheath separates rods from interrods. They are forming a continuous network of honeycomb-like structures. The bulk of enamel is formed by rods and interrods structures, with crystallites forming a 60° angle between the two basic enamel structures. This is why preferential acidic dissolution occurs along the *c*-axis, and eliminates either rods or interrods, depending on the crystal orientation (Fig. 1.21). Therefore, the pattern of dissolution removes from enamel either rods or interrods or produces a surface apparently flat (Type 3 etching pattern Fig. 1.22) (Silverstone et al.).

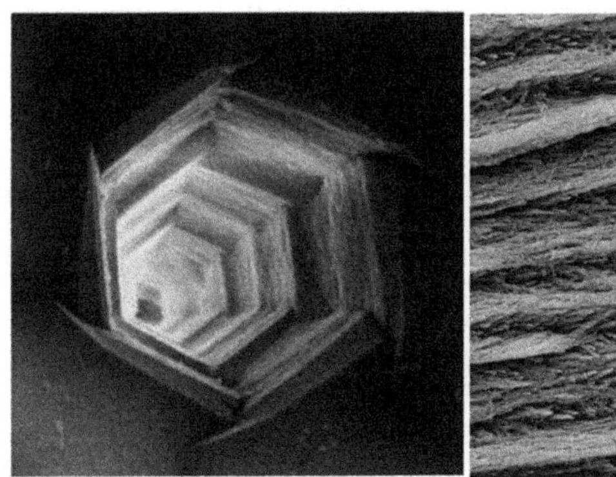

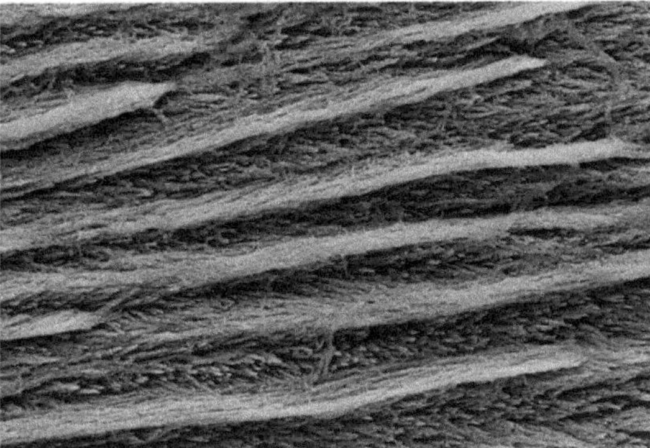

Fig. 1.21 Enamel crystal dissolution (link) and longitudinal section of enamel (rods and interrods)

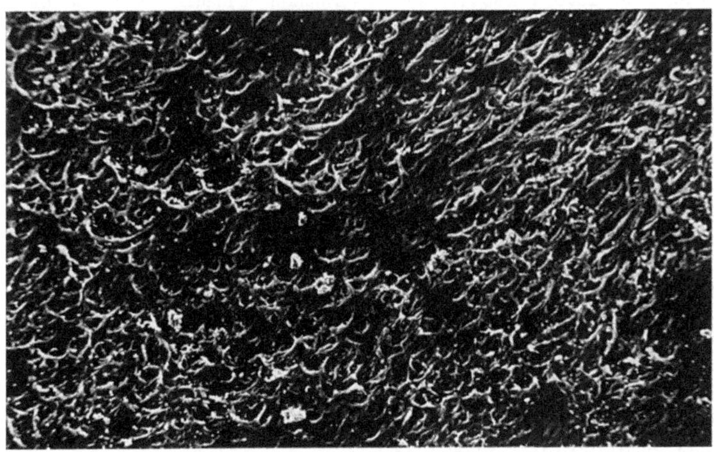

Fig. 1.22 Flat surface after etching (either fluorated enamel or aprismatic zone) (Type 3 etching pattern)

References

Anderson HC (1995) Molecular biology of matrix vesicles. Clin Orthop Relat Res:226–280

Anderson HC (2003) Matrix vesicles and calcification. Curr Rheumatol Rep 5:222–226. https://doi.org/10.1007/s11926-003-0071-z

Arends J, Jongebloed WL (1979) Ultrastructural studies of synthetic apatite crystalsl. J Dent Res 58 (special issue B):837–843

Bonewald LF (2017) The role of the osteocyte in bone and nonbone disease. Endocrinol Metab Clin North Am 46:1–18. https://doi.org/10.1016/j.ecl.2016.09.003

Boonrungsiman S, Gentleman E, Carzaniga R, Evans ND, McComb DW, Porter AE, Stevens MM (2012) The role of intracellular calcium phosphate in osteoblast-mediated bone apatite formation. Proc Natl Acad Sci U S A 109:14170–14175. https://doi.org/10.1073/pnas.1208916109

Boskey AL (2013) Bone composition: relationship to bone fragility and antiosteoporotic drug effects. Bonekey Rep 2:447. https://doi.org/10.1038/bonekey.2013.181

Boskey AL, Coleman R (2010) Aging and bone. J Dent Res 89:1333–1348. https://doi.org/10.1177/0022034510377791

Bosshardt DD, Nanci A (1997) Immunodetection of enamel- and cementum-related (bone) proteins at the enamel-free area and cervical portion of the tooth in rat molars. J Bone Miner Res 12:367–379. https://doi.org/10.1359/jbmr.1997.12.3.367

Bosshardt DD, Nanci A (2004) Hertwig's epithelial root sheath, enamel matrix proteins, and initiation of cementogenesis in porcine teeth. J Clin Periodontol 31:184–192. https://doi.org/10.1111/j.0303-6979.2004.00473.x

Bosshardt DD, Schroeder HE (1993) Attempts to label matrix synthesis of human root cementum in vitro. Cell Tissue Res 274:343–352. https://doi.org/10.1007/BF00318753

Bosshardt DD, Schroeder HE (1996) Cementogenesis reviewed: a comparison between human premolars and rodent molars. Anat Rec 245:267–292. https://doi.org/10.1002/(sici)1097-0185(199606)245:2<267::Aid-ar12>3.0.Co;2-n

Cate ART, Mills C, Solomon G (1971) The development of the periodontium. A transplantation and autoradiographic study. Anat Rec 170:365–379. https://doi.org/10.1002/ar.1091700312

Chaudhary SC et al (2016) Phosphate induces formation of matrix vesicles during odontoblast-initiated mineralization in vitro. Matrix Biol 52–54:284–300. https://doi.org/10.1016/j.matbio.2016.02.003

Chaussain-Miller C, Fioretti F, Goldberg M, Menashi S (2006) The role of matrix metalloproteinases (MMPs) in human caries. J Dent Res 85:22–32. https://doi.org/10.1177/154405910608500104

Chen J et al (2014) TGF-β1 and FGF2 stimulate the epithelial-mesenchymal transition of HERS cells through a MEK-dependent mechanism. J Cell Physiol 229:1647–1659. https://doi.org/10.1002/jcp.24610

Diekwisch TG (2001) The developmental biology of cementum. Int J Dev Biol 45:695–706

Foster BL et al (2014) Rare bone diseases and their dental, oral, and craniofacial manifestations. J Dent Res 93:7s–19s. https://doi.org/10.1177/0022034514529150

Garces-Ortiz M, Ledesma-Montes C, Reyes-Gasga J (2013) Presence of matrix vesicles in the body of odontoblasts and in the inner third of dentinal tissue: a scanning electron microscopic study. Med Oral Patol Oral Cir Bucal 18:e537–e541. https://doi.org/10.4317/medoral.18650

Gericke A, Qin C, Spevak L, Fujimoto Y, Butler WT, Sørensen ES, Boskey AL (2005) Importance of phosphorylation for osteopontin regulation of biomineralization. Calcif Tissue Int 77:45–54. https://doi.org/10.1007/s00223-004-1288-1

Goldberg M, Escaig F (1981) Odontoblastes: collagène dans la prédentine et la dentine d'incisive de rat- Etude par cryofracture. Biol Cell 40:203–216

Goldberg M, Septier D (1996) A comparative study of the transition between predentin and dentin, using various preparative procedures in the rat. Eur J Oral Sci 104:269–277. https://doi.org/10.1111/j.1600-0722.1996.tb00077.x

Goldberg M, Smith AJ (2004) Cells and extracellular matrices of dentin and pulp: a biological basis for repair and tissue engineering. Crit Rev Oral Biol Med 15:13–27. https://doi.org/10.1177/154411130401500103

Goldberg M, Septier D, Escaig-Haye F (1987) Glycoconjugates in dentinogenesis and dentin. Prog Histochem Cytochem 17(2):1–112. https://doi.org/10.1016/s0079-6336(87)80001-3

Goldberg M, Opsahl S, Aubin I, Septier D, Chaussain-Miller C, Boskey A, Guenet J-L (2008) Sphingomyelin degradation is a key factor in dentin and bone mineralization: lessons from the *fro/fro* mice. J Dent Res 87(1):9–13

Goldberg M, Kulkarni AB, Young M, Boskey A (2011) Dentin: structure, composition and mineralization. Front Biosci (Elite Ed) 3:711–735. https://doi.org/10.2741/e281

Hammarström L (1997) The role of enamel matrix proteins in the development of cementum and periodontal tissues. Ciba Found Symp 205:246–255; discussion 255–260

Hoshi K, Ozawa H (2000) Matrix vesicle calcification in bones of adult rats. Calcif Tissue Int 66:430–434. https://doi.org/10.1007/s002230010087

Huang X, Bringas P Jr, Slavkin HC, Chai Y (2009) Fate of HERS during tooth root development. Dev Biol 334:22–30. https://doi.org/10.1016/j.ydbio.2009.06.034

Kémoun P et al (2007) Human dental follicle cells acquire cementoblast features under stimulation by BMP-2/−7 and enamel matrix derivatives (EMD) in vitro. Cell Tissue Res 329:283–294. https://doi.org/10.1007/s00441-007-0397-3

Lorber M (1951) A study of the histochemical reactions of the dental cementum and alveolar bone. Anat Rec 111:129–144. https://doi.org/10.1002/ar.1091110202

Lester K, Boyde A (1968) The surface morphology of some crystalline components of dentine. In: Symons NBB (ed) Dentine and Pulp: their structure and reactions. E. & S. Livingstone Ltd., Edinburgh, pp 197–219

Li X, Zhang S, Zhang Z, Guo W, Chen G, Tian W (2019) Development of immortalized Hertwig's epithelial root sheath cell lines for cementum and dentin regeneration. Stem Cell Res Ther 10:3. https://doi.org/10.1186/s13287-018-1106-8

Murray PE, Stanley HR, Matthews JB, Sloan AJ, Smith AJ (2002) Age-related odontometric changes of human teeth. Oral Surg Oral Med Oral Pathol Oral Radiol Endodontol 93:474–482. https://doi.org/10.1067/moe.2002.120974

Nanci A (2007) Ten Cate's oral histology developmant, structure; and function. Mosby Elsevier, St Louis

Ruch JV (1998) Odontoblast commitment and differentiation. Biochem Cell Biol 76:923–938

Ruch J, Lesot H, Begue-Kirn C (1995) Odontoblast differentiation. Int J Dev Biol 39(1):51–68

Saito M et al (2005) Immortalization of cementoblast progenitor cells with Bmi-1 and TERT. J Bone Min Res 20:50–57. https://doi.org/10.1359/JBMR.041006

Sandhu SV, Gupta S, Bansal H, Singla K (2012) Collagen in health and disease. J Orofac Res 2:153–159

Schilke R, Lisson JA, Bauß O, Geurtsen W (2000) Comparison of the number and diameter of dentinal tubules in human and bovine dentine by scanning electron microscopic investigation. Arch Oral Biol 45:355–361. https://doi.org/10.1016/S0003-9969(00)00006-6

Schroeder L, Frank RM (1985) High-resolution transmission electron microscopy of adult human peritubular dentine. Cell Tissue Res 242:449–451. https://doi.org/10.1007/BF00214561

Sela J, Schwartz Z, Swain L, Amir D, Boyan B (1992) Extracellular matrix vesicles in endochondral bone development and in healing after injury. Cells Mater 2:6

Silverstone LM, Saxton CA, Dogon IL, Fejerskov O (1975) Variation in the pattern of acid etching of human dental enamel examined by scanning electron microscopy. Caries Res 9(5):373–387. https://doi.org/10.1159/000260179

Sonoyama W, Seo BM, Yamaza T, Shi S (2007) Human Hertwig's epithelial root sheath cells play crucial roles in cementum formation. J Dent Res 86:594–599. https://doi.org/10.1177/154405910708600703

Takano Y, Sakai H, Baba O, Terashima T (2000) Differential involvement of matrix vesicles during the initial and appositional mineralization processes in bone, dentin, and cementum. Bone 26:333–339. https://doi.org/10.1016/S8756-3282(00)00243-X

Thouverey C, Strzelecka-Kiliszek A, Balcerzak M, Buchet R, Pikula S (2009) Matrix vesicles originate from apical membrane microvilli of mineralizing osteoblast-like Saos-2 cells. J Cell Biochem 106:127–138. https://doi.org/10.1002/jcb.21992

Tsukasaki M, Takayanagi H (2019) Osteoimmunology: evolving concepts in bone–immune interactions in health and disease. Nat Rev Immunol 19:626–642. https://doi.org/10.1038/s41577-019-0178-8

Weiner S, Wagner HD (1998) The material bone: structure-mechanical function relations. Annu Rev Mater Sci 28:271–298. https://doi.org/10.1146/annurev.matsci.28.1.271
Weiner S, Traub W, Wagner HD (1999) Lamellar bone: structure-function relations. J Struct Biol 126:241–255. https://doi.org/10.1006/jsbi.1999.4107
Wuthier RE, Lipscomb GF (2011) Matrix vesicles: structure, composition, formation and function in calcification. Front Biosci (Landmark Ed) 16:2812–2902. https://doi.org/10.2741/3887
Wysolmerski JJ (2012) Osteocytic osteolysis: time for a second look? Bone Key Rep 1:229–229. https://doi.org/10.1038/bonekey.2012.229
Yamamoto T, Yamamoto T, Yamada T, Hasegawa T, Hongo H, Oda K, Amizuka N (2014) Hertwig's epithelial root sheath cell behavior during initial acellular cementogenesis in rat molars. Histochem Cell Biol 142:489–496. https://doi.org/10.1007/s00418-014-1230-1
Yamamoto T, Hasegawa T, Yamamoto T, Hongo H, Amizuka N (2016) Histology of human cementum: its structure, function, and development. Jpn Dent Sci Rev 52:63–74. https://doi.org/10.1016/j.jdsr.2016.04.002
Yuasa K et al (2004) Laminin alpha 2 is essential for odontoblast differentiation regulating dentin sialoprotein expression. J Biol Chem 279:10286–10292. https://doi.org/10.1074/jbc.M310013200
Zeichner-David M, Oishi K, Su Z, Zakartchenko V, Chen L-S, Arzate H, Bringas P Jr (2003) Role of Hertwig's epithelial root sheath cells in tooth root development. Dev Dyn 228:651–663. https://doi.org/10.1002/dvdy.10404

Chapter 2
Extracellular Matrix Proteins: Nomenclature and Functions in Biomineralization

Michel Goldberg, Pamela DenBesten, and Yukiko Nakano

Abstract Proteins in the extracellular matrix (ECM) of mineralized tissues have specific roles in promoting and/or inhibiting the mineralization processes of collagenous-rich (dentin, cementum, bone) or non-collagenous extracellular matrix components (enamel matrix proteins). These ECM proteins are comprised of two categories of molecules: (1) hydrophobic molecules that define the architecture of the system and the orientation of the crystal growth and (2) the charged molecules that play an active role in initiating, promoting, and regulating crystal formation usually present at the interface of the structural components and the nucleating apatite crystals.

2.1 ECM Proteins in Collagen-Derived Mineralized Tissues

ECM proteins in collagen-derived tissues are structural proteins, regulators for hydroxyapatite (HAp) crystal formation, and signaling molecules (Boskey 2013). They are categorized into collagens and non-collagenous proteins.

2.1.1 Collagens

In dentin, cementum, and bone, collagens are the major hydrophobic structural molecules to provide the structural support of the tissue and the site of nucleation

M. Goldberg (✉)
Faculté des siences fondamentales et médicales, Université Paris Cité, INSERM UMR-S 1124, Paris, France

P. DenBesten · Y. Nakano
Department of Orofacial Sciences, School of Dentistry, University of California San Francisco, San Francisco, CA, USA
e-mail: Pamela.DenBesten@ucsf.edu; Yukiko.Nakano@ucsf.edu

M. Goldberg, P. Den Besten (eds.), *Extracellular Matrix Biomineralization of Dental Tissue Structures*, Biology of Extracellular Matrix 10,
https://doi.org/10.1007/978-3-030-76283-4_2

and extension of the hydroxyapatite crystals. The weight percent composition of mineralized dentin is 70% mineral, closely associated with 20–30% organic matrix, and approximately 10% water. Among the organic matrix, 86 wt % is the type I collagen, with type I trimer (11 wt %), and type III, V, and VI as minor components.

Cementum is slightly less mineralized than root dentin, though the composition is similar to that of dentin and bone. The mineral phase accounts for 61 wt %, the organic matrix is about 27 wt %, and the water 12 wt %. Type I collagen comprises about 90 wt % of the organic matrix and type III collagen, which coats type I collagen fibrils, is 5 wt %. Type V and VI collagens have not been identified in cementum.

In alveolar bone, ECM includes approximately 60–70% mineral, 20% of collagen and non-collagenous extracellular matrix proteins, and 10–20% of water. Type I collagen comprises about 90 wt % of the organic matrix, and the remaining 10 % is non-collagenous proteins. The types of collagens in mineralized tissues and their primary functions are listed in Table 2.1.

2.1.2 Non-collagenous Proteins (NCPs)

The NCPs are identified as belonging to the SIBLING (small integrin-binding ligand, N-linked glycoprotein), the SLRPs (Small leucine-rich proteoglycans), glycosaminoglycans, GLA proteins (γ-carboxy-glutamic acid proteins), and CCN proteins (small secreted cysteine-rich protein). NCPs also include serum proteins (albumin, alpha2-HS glycoprotein/fetuin, immunoglobulins), osteocalcin, and non-collagenous structural proteins (including thrombospondins, fibronectin, and fibrillin).

Primary non-collagenous proteins that contribute to matrix mineralization are characterized by the capability of binding to Ca^{2+} ions. They include dentin sialophosphoprotein [DSPP, which is cleaved after secretion into dentin sialoprotein (DSP), dentin glycoprotein (DGP), and dentin phosphoprotein (DPP)], bone sialoprotein (BSP), osteocalcin (OCN), and osteopontin (OPN) (Goldberg et al. 2011; Goldberg and Smith 2004). These non-collagenous proteins and their role in matrix mineralization are discussed in greater detail in the subsequent chapters.

ECM mineralization is also regulated by enzymes regulating phosphate and pyrophosphate levels, including tissue nonspecific alkaline phosphatase (TNSALP), phosphoethanolamine (phosphocholine), phosphatase (PHOSPHO1), and ectonucleotide pyrophosphatase (phosphodiesterase 1, ENPP1). Matrix metalloproteinases (MMPs) also contribute to the mineralization process.

Growth factors are essential in stimulating cells within the dentine–pulp complex or bone to promote mineralization and regenerative processes (Giannobile 1996; Pitaru et al. 1992).

Furthermore, certain cytokines, such as interleukin-8 (IL-8), are expressed by odontoblasts. As with growth factors, following the release of these dentin matrix components, gradients are established within the dentin–pulp complex that are

Table 2.1 Classification of the different types of collagens

Collagen type	Classification	Distribution/remarks
I	Fibril-forming	Non-cartilaginous connective tissues—e.g., tendon, ligament, cornea, bone, annulus fibrosis, skin
II	Fibril-forming	Cartilage, vitreous humour, and nucleus pulposus
III	Fibril-forming	Co-distributes with collagen I, especially in embryonic skin and hollow organs
IV	Network-forming	Basement membranes
V	Fibril-forming	Co-distributes with collagen I, especially in embryonic tissues and in cornea
VI	Beaded-filament forming	Widespread, especially muscle
VII	Anchoring fibrils	Dermal-epidermal junction
VIII	Network-forming	Descemet's membrane
IX	FACIT	Co-distributes with collagen II, especially in cartilage and vitreous humour
X	Network-forming	Hypertrophic cartilage
XI	Fibril-forming	Co-distributes with collagen II
XII	FACIT	Found with collagen I
XIII	Transmembrane	Neuromuscular junctions, skin
XIV	FACIT	Found with collagen I
XV	Endostatins	Located between collagen fibrils that are close to basement membranes; found in the eye, muscle, and microvessels; a close structural homolog of collagen XVIII
XVI	FACIT	Integrated into collagen fibrils and fibrillin-1 microfibrils
XVII	Transmembrane	Also known as the bullous pemphigoid antigen 2/BP180; localized to epithelia; an epithelial adhesion molecule; ectodomain cleaved by ADAM proteinases
XVIII	Endostatins	Associated with basement membranes; endostatin is proteolytically released from the C-terminus of collagen XVIII; important for retinal vasculogenesis
XIX	FACIT	Rare; localized to basement membrane zones; contributes to muscle physiology and differentiation
XX	FACIT	Widespread distribution, most prevalent in corneal epithelium
XXI	FACIT	Widespread distribution
XXII	FACIT	Localized at tissue junctions—e.g., myotendinous junction, cartilage synovial fluid, hair follicle-dermis
XXIII	Transmembrane	Limited tissue distribution; exists as a transmembrane and shed form
XXIV	Fibril-forming	Shares sequence homology with the fibril-forming collagens; has minor interruptions in the triple helix; selective expression in developing cornea and bone

(continued)

Table 2.1 (continued)

Collagen type	Classification	Distribution/remarks
XXV	Transmembrane	CLAC-P—precursor protein for CLAC (collagenous Alzheimer amyloid plaque component)
XXVI	Beaded-filament forming	Also known as EMI domain-containing protein 2, protein Emu2, Emilin and multimerin domain-containing protein 2
XXVII	Fibril-forming	Shares sequence homology with the fibril-forming collagens; has minor interruptions in the triple helix; found in embryonic cartilage, developing dermis, cornea, inner limiting membrane of the retina, and major arteries of the heart; restricted to cartilage in adults; found in fibrillar-like assemblies
XXVIII	Beaded-filament forming	A component of the basement membrane around Schwann cells; a von Willebrand factor A domain-containing protein with numerous interruptions in the triple-helical domain
Ectodysplasin A	Transmembrane	Ectoderm
Gliomedin	Transmembrane	Myelinating Schwann cells

Modified from Kadler et al. (2007)

responsible for chemo-attraction, and the homing of cells involved in both the immune and the regenerative responses, i.e., stem/progenitor cells. Anti-inflammatory cytokines are also sequestrated within the dentin. Cariogenic bacterial acids can release these cytokines under caries conditions, and this immune response precedes tertiary dentin formation or remineralization of affected dentin (Takahashi and Nyvad 2016). In support of the dentine matrix components' immune regulatory potential, the key molecules to mineralization processes (e.g., DSP) can also promote the migration and pro-inflammatory activation of immune system cells when they are released from their matrix immobilized state (Smith et al. 2012).

2.1.2.1 The Small Integrin-Binding Ligand N-Linked Glycoproteins (SIBLINGs)

The SIBLING family of bone and dentin includes DSPP and its proteolytic products, DPP, DGP and DSP, DMP1, BSP, MEPE, and OPN. All these matrix proteins play a role in dentin mineralization and signaling molecules (Fisher and Fedarko 2003; MacDougall et al. 1992; von Marschall and Fisher 2010).

Dentin sialophosphoprotein (DSPP) is cleaved immediately after secretion into the ECM, becoming dentin sialoprotein (DSP) dentin and dentin phosphoprotein (DPP). C-terminal of DSP is further cleaved by MMP-2 and -20 to form dentin glycoprotein (DGP) (Yamakoshi et al. 2005b). DSP is suggested to regulate predentin thickness (Sreenath et al. 2003; Suzuki et al. 2009). The gene coding for DSPP is located on human chromosome 4 (4q21-q23), which also codes BSP, OPN, and DMP1.

Dentin phosphophoryn (DPP) is a cleavage product of DSPP, localized predominantly in dentinal tubules at the site of peritubular dentin. Physicochemical properties of DPP include a high composition of aspartic acid and serine (70–80% of the total amino acid residues), that it is highly phosphorylated (400/1000 residues phosphate or greater) usually at serine residues, is extremely anionic, and has a high affinitive for calcium ions. These properties suggest that DPP functions as a regulator of hydroxyapatite (HAp) crystal nucleation and growth in ECM. At high concentrations, DPP inhibits or slows HAp crystal growth, and at low concentrations, DPP promotes rapid nucleation of HAp in vitro (Boskey et al. 1990). Phosphorylation of the DPP has a crucial role in the initiation of dentin mineralization (Suzuki et al. 2009) and forming a framework for mineralization associated with type I collagen (Yamakoshi et al. 2005a). DPP is also located in the nucleus of developing odontoblasts and is suggested to have signaling functions during odontogenesis (Eapen et al. 2012).

Dentin Matrix Protein 1 (DMP1) is present in dentin and bone as an acidic phosphoprotein. Chondroitinase treatment demonstrates that the slower migration rate is due to the presence of GAGs, namely the C 4-sulfate. DMP1 is present predominantly at the mineralization front, dentinal tubules, and lacunae and canaliculi of osteocytes. Overexpressing DMP in mice demonstrated higher levels of mineralization (Bhatia et al. 2012).

DMP1 is proteolytically processed into the two forms of NH2-terminal (37 kDa) and COOH-terminal (37 kDa) fragments. The NH2 terminal fragment is composed predominantly of chondroitin 4-sulfate and takes a proteoglycan form of DMP1. This NH2-terminal property is given by the tetrasaccharide extended through the sequential addition of sugar precursors by specific glycosyltransferases and sulfotransferases located in the Golgi apparatus. Since both DMP1 and DSPP have the proteoglycan property with the similarity in the NH2-terminal fragments, they are suggested to share similar biological functions.

COOH-terminal has the nuclear localization signal, RGD sequence, and phosphorylation sites (Ravindran et al. 2008) thus suggested to regulate odontoblast differentiation and osteocyte maturation and phosphate metabolism (George et al. 1995). Also, DMP1 fragments are cleaved by MMP-2 and influence stem cell differentiation (Chaussain et al. 2009)

Osteopontin (OPN) is a highly phosphorylated sialoprotein member of the SIBLING family. It contributes to mineralized matrix formation, influences osteoclast attachment, and facilitates matrix mineralization, by influencing the shape and size of HAp crystals (Gericke et al. 2005). OPN is a highly conserved multifunctional phosphorylated glycoprotein expressed in many mineralized and soft tissues, including bone, dentin, elastin, muscle, tumors, and body fluids (milk, inner ear, and urine). In the cementum, OPN is mainly located in acellular cementum (McKee et al. 1996). During biomineralization, OPN is known to regulate mineralization by inhibiting HAp formation and growth (Addison et al. 2007). It is also known as secreted phosphoprotein-1 (SPP1), urinary stone phosphoprotein, uropontin, and early T cell activator (ETA-1). OPN contains an integrin receptor-binding arg-gly-asp (RGD) sequence suggesting its role in regulating intracellular signaling. Its mRNA expression is upregulated in response to 1, $25(OH)_2$ vitamin D_3, parathyroid hormone, and elevated phosphate and pyrophosphate levels (Addison et al. 2007; Gericke et al. 2005), indicating the role of OPN as ECM signaling mediator.

Matrix extracellular phosphoglycoprotein (MEPE) regulates bone and dentin mineralization by binding to hydroxyapatite via ASARM (aspartic acid-rich motif) domain located at its C-terminus (Addison et al. 2008; Rowe 2004; Rowe et al. 2000). MEPE is protected from cleavage by a zinc-dependent protein-protein specific interaction with PHEX (phosphate-ogies to endopeptidases located on the X-chromosome). In the familial X-linked hypophosphatemic rickets (XLH), mutations in *PHEX* result in the absence or the loss of function of PHEX and resulting in exposing unprotected MEPE, which is further cleaved and accumulated in the ECM. ASARM peptide is an acidic and stable peptide, and its pathologic accumulation partially explains the osteomalacia associated with XLH. MEPE deficiency results in accelerated osteoblast-mediated bone formation and mineralization without a change in phosphatemia. MEPE is also expressed in odontoblasts as a promotor of terminal differentiation, involved in dentin formation and mineralization. It was previously demonstrated an accumulation of dentin sialoprotein, dentin matrix protein, bone sialoprotein, osteopontin, MEPE, and osteocalcin in the unmineralized interglobular spaces of the XLH dentin (Salmon et al. 2013).

Osteocalcin (OCN) or bone gamma-carboxyglutamic acid-containing (gla) protein belongs to the SIBLING family that shares many common features, such as containing an RGD sequence, a cell-attachment site, and an acidic serine- and aspartate-rich motif (ASARM). OCN contributes as a mineralization inhibitor to the dentin extracellular matrix. Osteocalcin is expressed in cementoblasts.

2.1.2.2 Proteoglycans

Proteoglycans in collagen-derived mineralized tissues belong to the small leucine-rich interstitial family (SLRP family), including decorin, biglycan, lumican, fibromodulin, keratocan, PRRLP, osteoadherin, epiphycan, and osteoglycin (Embery et al. 2001; Goldberg et al. 2005, 2006; Goldberg and Takagi 1993; Haruyama et al. 2009; Milan et al. 2005). In dentin, decorin and biglycan are the major proteoglycans. C4S decorin is located in the extracellular matrix in cementum, while biglycan is immunolocalized in cementoblasts and pre-cement. Decorin links low-density lipoproteins with type I collagen. The dentin glycosaminoglycans (GAGs) bind to hydroxyapatite via their sulfate and hydroxyl groups and regulate crystal deposition and orientation. MMPs (MMP-2, MMP-9, MMP-3) are involved in proteoglycans degradation and activation forms.

Chondroitin sulfate (CS)/dermatan sulfate (DS) present in ECM as proteoglycans and a major member of the glycosaminoglycan (GAG) family. Cementum GAG constitutes 0.29 mg/mg of the dry weight formed by CS (chondroitin sulfate)/dermatan sulfate (DS) and hyaluronic acid with the proportion of CS as delta Di-OS 3%, delta Di-4S 87%, delta Di-6S 10%. All CS/DS presents in pericellular and extracellular cementum, and cementum extraction by guanidine/EDTA contains DS at 31%, CS at 53%, and HA at 16%. Further, after collagenase digestion, the extract comprises DS at 31% and CS at 69% (HA not determined).

2.1.2.3 Other Proteins Present in the ECM Implicated in Mineralization

Reelin is expressed by odontoblasts in the human tooth germ. Located between odontoblasts and nerve, ending a close contact was detected in this location between the two cells (Bleicher et al. 2001; Ganss et al. 1999).

Periostin is expressed by pre-odontoblasts. It is a negative regulator of mineralization, and its signaling functions are responsible for the ECM matrix organization.

Growth factors in dentin and bone include transforming growth factor-β (TGF-β1–3), bone morphogenetic proteins (BMPs), insulin-like growth factor-1 and 2 (IGF-1 and -2), fibroblast growth factor-2 (FGF-2), FGF-23, adrenomedullin, and several angiogenic growth factors, such as platelet-derived growth factor (PDGF) and vascular endothelial growth factor (VEGF) (Bosshardt and Schroeder 1996). In dentin–pulp complex and bone, TGF-β1 and bone morphogenetic protein-7 (BMP-7) can stimulate the regenerative process. Sclerostin is an osteoclast-derived bone morphogenetic protein antagonist. It inhibits the BMP6 and BMP7, but not BMP2 and BMP4 activity.

Two specific growth factors are also implicated in cementum formation: the cementum-derived attachment protein (CAP) and the purified cementum-derived growth factor (CGF, 23 kDa), which is involved in cell proliferation and differentiation (Pitaru et al. 1992). Cementum-derived attachment protein (CAP) promotes migration and adhesion of fibroblasts. It has a RGD sequence, without cross-over with fibronectin, osteopontin, and BSP. It binds to cementum with a high affinity for HAp. It also stimulates PGs synthesis and alkaline phosphatase expression in a dose-dependent manner (Giannobile 1996). In cementum, fibronectin, tenascin, vitronectin are widely distributed in cementum ECM.

Tissue nonspecific alkaline phosphatase (TNSALP) plays critical roles in the mineralization of skeletal tissue (Fedde et al. 1999; Waymire et al. 1995). TNSALP deficient mice suggest its role in acellular cementum formation (Beertsen et al. 1999), as well as in the mineralization of molar root dentin and cementum (Anderson et al. 2004; McKee et al. 2011). The physiologic substrate and exact role(s) of the TNSALP are yet to be fully understood. Nevertheless, previous reports suggest hydrolysis of the mineralization inhibitor (pyrophosphate/PPi) is one of the significant roles of TNSALP in mineralization of type 1 collagen-driven tissue like bones (Addison et al. 2007; Murshed et al. 2005).

Metalloproteinases (MMPs) are a family of 24 secreted and membrane-bound zinc endopeptidases in humans. They can degrade all ECM components, and their specific tissue inhibitors (TIMPs) regulate bone maturation (including the remodeling of normal bone and its turnover). In the tooth, two gelatinases, MMP-2 and MMP-9, contribute to dentin mineralization (Bourd-Boittin et al. 2005). They are synthesized and secreted as pro-enzymes. MMP-2 is activated and migrates at 66 kDa. MMP-9 is derived from the zymogen proMMP-9 migrating at 92 kDa. MT1-MMP (a membrane-type MMP) acts on tooth eruption, root growth, and periodontal remodeling. Additionally, in bone, MMP-13 plays a role in bone remodeling (Tang et al. 2012) and ADAMTS18 regulates bone mass density (Xiong et al. 2009).

Matrix Gla protein (MGP) is a potent inhibitor of mineralization. Rodent models of ablated *Mgp* expression, and nonfunctionalized MGP resulting from inhibiting post-translational gamma-carboxylation of MGP, have identified MGP as the essential inhibitor of mineralization (Price et al. 1983).

PHOSPHO1 (Phospho ethanolamine/phosphocholine phosphatase 1) is an essential contributor to skeletal mineralization. PHOSPHO1 is an intravesicular phosphatase releasing phosphate from membrane-associated phosphoethanolamine and phosphocholine as part of an initial step in the process of MV-mediated mineralization. Observations on skeletal MV function in PHOSPHO1 knockout mice and on the known importance of MVs suggest its role in initiating mantle dentin and bone mineralization (McKee et al. 2013; Roberts et al. 2007).

PHEX (phosphate regulating endopeptidase homolog X-linked) is associated with the familial X-linked hypophosphatemic rickets (XLH), osteomalacia, and impaired renal phosphate reabsorption. PHEX is involved in either activation or degradation of factor(s) that directly or indirectly regulate(s) renal phosphate reabsorption and bone mineralization (Bai et al. 2002).

Runt-related transcription factor 2 (RUNX2)/Core binding factor 1 (Cbfa 1) is a transcription factor critical for bone and dental mineralization (Komori et al. 1997). In odontoblasts and osteoblasts, RUNX2 is essential to regulate the expression of downstream molecules like OSX, DMP1, DLX3, and DSPP to regulate cell differentiation and ECM formation and mineralization (Chen et al. 2005, 2009; Ducy 2000; Narayanan et al. 2006; Yang et al. 2017). Growth factors such as BMPs and WNTs stimulate RUNX2 expression in osteoblasts and odontoblasts (Gaur et al. 2005; Kao et al. 2020).

Cathepsins are lysosomal cysteine proteinases. They have the capacity to degrade ECM proteins such as albumin, immunoglobulins, and transferrin.

Asporin binds type-1 collagen and is inhibited by asporin fragment LRR(leucine-rich repeat) and by full-length decorin, but not by biglycan (Kalamajski et al. 2009).

2.1.2.4 Lipids

The total lipids extracted from calcified tissues contained cholesterol, cholesterol esters, mono-di-triglycerides, free fatty acids, and various phospholipids. The average total lipids were 20.28/100 g dentin, 1.97 mg/100 g enamel, and 2.97 mg/100 g human bone (Dirksen and Marinetti 1970; Fincham et al. 1972; Goldberg and Boskey 1996; Goldberg et al. 1999). Sixty percent were extracted from the original tissue, the remainder being extracted from decalcified tissue, except cardiolipids extracted from calcified dentin (Prout et al. 1973). Before demineralization, a group of phospholipids was extracted by lipid solvents and was associated with cell membrane. The other group of phospholipids was associated with the extracellular matrix and linked to the mineral phase. Goldberg and Septier (2002) showed that the blood-serum-labeled [^{3}H]choline was detected as early as 30 min before any labeling was seen in odontoblasts and predentin. Hence, the blood-serum-labeled phospholipids pass between odontoblasts and predentin and diffuse in dentin prior to any cellular uptake and phospholipid synthesis (Fincham et al. 1972). Deletion of the gene encoding sphingomyelin phosphodiesterase 3 (Smpd3) leads to a syndrome of severe dentinogenesis imperfecta (Aubin et al. 2005). This provides insight into human pathologies and the role of lipids in enamel and dentin biomineralization.

2.2 ECM Proteins in Non-collagen-Derived Mineralized Tissue/Enamel

The extracellular matrix that directs hydroxyapatite (HAp) formation in enamel is uniquely different from that of collagen-derived tissues. Enamel mineralization is regulated by various organic molecules and, importantly, that their efficacy as inhibitors of mineralization may be modulated through their degradation. Based on the data presented by Margolis et al. (Margolis et al. 2006), appropriate expression of amelogenin, enamelin, ameloblastin, and enamelysin, and the processing of these matrix proteins by proteinases and subsequent removal of protein fragments from the mineralizing matrix is essential for proper enamel mineral formation (see Chaps. 10 and 11 for more details.)

During enamel matrix formation, the hydrophobic alternatively spliced amelogenin proteins form a scaffold. Simultaneously, crystal nucleation and growth are directed by charged molecules, including enamelin, ameloblastin present in much smaller quantities within the matrix. Modulation of these matrix proteins by extracellular matrix proteinases, such as matrix metalloproteinase 20 (MMP-20) and serine proteinase kallikrein 4 (KLK4), regulates the structure and growth of the HAp crystals. The initial enamel crystal is considered to be HAp or octacalcium phosphate (OCP). The c-axis of these crystals always coincides with their long axis whereas, a and b crystallographic axis coincides with two other (thickness and width) morphological axis of the crystal. As HAp crystals develop, most of the structural proteins are removed in the maturation process. Consequently, fully formed enamel is approximately 90% mineral, with the remaining 10% made up of water and peptides remaining in the matrix. Free water is mostly located in tiny intercrystallite spaces.

In the forming enamel, ECM includes a mixture of enamel structural proteins (amelogenins, ameloblastin, enamelin, amelotin, and odontogenic ameloblast-associated gene/ODAM), enzymes [metalloproteinases (MMP-2, -3, 7, and 10), enamelysin (MMP-20), KLK4, chymotrypsin C, cathepsin C, adaptor protein complex-2, Lamp 1, Lamp2, CD63], phospholipids and type XVII collagen. The structural proteins are composed predominantly of amelogenin and its cleavage products, making up over 90% of the enamel matrix. These mostly 20- to 25-kDa proteins are primarily hydrophobic, rich in proline (25%), glutamine (14%), leucine (9%), and histidine (7%), which account for more than 50% of their amino acids. Between prisms and interprism of forming enamel, a structural discontinuity is seen as a prism sheath, where most of the organic components are recognized to accumulate ECM proteins.

2.2.1 Composition

Amelogenins exist with multiple molecular mass (18–25 kDa) resulting from the varients of alternative splicing product and cleavage by proteases in the matrix. The major amelogenin protein is translated from an mRNA near the full length of the coding sequence and consists of 174 amino acids in humans (H174) and 180 amino acids in mice (M180). This form of amelogenin is divided into three domains (Fig. 2.1): Domain A (amino acid 1–42) is a hydrophobic N-terminal domain and rich in Tyr and therefore is called TRAP (Tyrosine-rich amelogenin peptide). This domain is essential for amelogenin-to-amelogenin interactions. Domain B (amino acid 157–173) is a C-terminal region and hydrophilic and almost conserved across mammalian species. This hydrophilic domain provides the amelogenin capability of binding to the HAp crystals. Domain B is cleaved a short time after amelogenin's secretion into the extracellular space. The central region (domain C) is hydrophobic and composed of Xxxx-Yyy-Pro-repeat motif. When amelogenin takes nanosphere

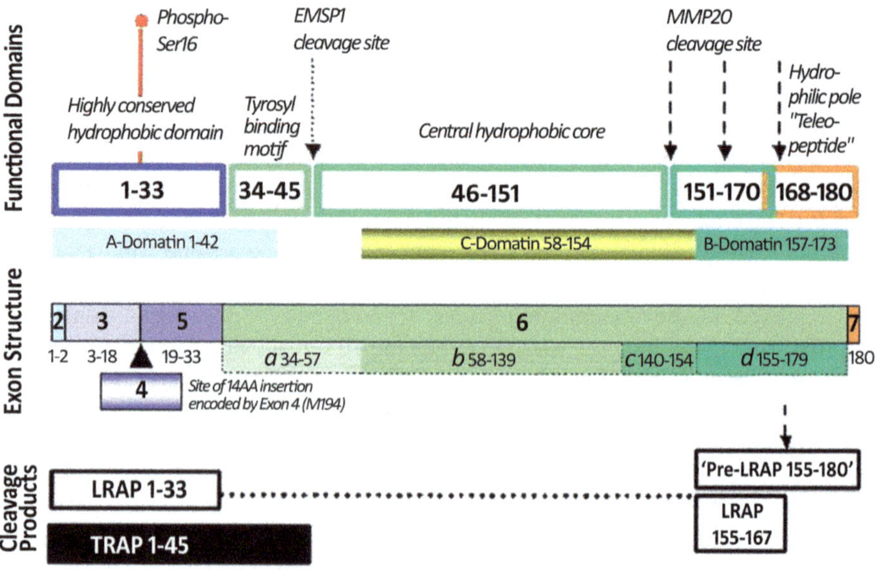

Fig. 2.1 Amelogenin architecture. Modified from Bansal et al. (2012)

form, it forms the nanospheres' dense central area (Bansal et al. 2012). The self-assembly of amelogenin depends on the variants of the amino acid sequence. The observation that a truncated amelogenin without domain "B" will no longer interact with domain "A" explains matrix disassembly following proteolytic cleavage at the carboxyl end of amelogenin by enamel protease (Fincham et al. 1995). A shorter form of amelogenin with 59 amino acid residues is translated from an mRNA variant lacking the majority of exon 6, called leucine-rich amelogenin peptide (LRAP). It is suggested to function as a signaling molecule during enamel formation (Gibson et al. 2009; Le et al. 2007).

Amelotin (AMTN) is localized in the basement membrane of maturation stage ameloblasts and junctional epithelium (Somogyi-Ganss et al. 2012). The three proteins co-localized and form supramolecular aggregates. Amelotin is a secreted enamel protein predominantly expressed during the maturation stage of enamel formation. It accumulates in a basal lamina-like structure at the interface between ameloblasts and enamel minerals and it co-localizes with the enamel protein odontogenic ameloblast-associated protein (ODAM). All the genes coding for enamel proteins are situated on the so-called secretory calcium-binding phosphoprotein proline-glutamine rich 1 (SCPPPQ1) gene cluster. They mediate the attachment of epithelial cells to tooth surfaces (Moffatt et al. 2006).

Ameloblastin has a molecular mass of 65 kDa and is also called **sheathelin** or **amelin**. It plays a role in crystal growth. It is a cell adhesion molecule that facilitates the attachment of ameloblasts to the enamel matrix (Mazumder et al. 2016). Ameloblastin is cleaved to include products that have an apparent molecular weight of 13, 15, 27, and 29 kDa, that concentrate in the sheath space separating rod and interrod enamel. C-terminal cleavage. It appears that 13 and 15 kDa are the calcium-binding domain of sheathelin (Hu et al. 1997).

Enamelin (ENAM) gene maps to chromosome 4q21. In full length, it is a glycoprotein with a molecular mass of ~186 kDa (Dohi et al. 1998). Once it is secreted into the enamel matrix, it is rapidly cleaved by MMP-20. The primary cleavage product is 32 kDa and present throughout the forming enamel matrix (Uchida et al. 1991). In matured enamel, it is detected as the trace constituents. Thus, the majority of ENAM is removed or further degraded by proteases before maturation, suggesting its role in the secretory stage of the enamel matrix. ENAM gene mutations in humans and deletion of ENAM gene in mice result in severe enamel hypoplasia (Hu et al. 2014). ENAM cooperatively functions with amelogenin (Deutsch et al. 1995).

Enamel proteinases are believed to play two crucial roles during enamel formation. First, during the secretory stage, they process enamel matrix proteins into a large number of stable cleavage products. Both the intact enamel proteins and their proteolytic cleavage products are believed to play active roles during amelogenesis (Den Besten et al. 1998). The second function of enamel proteinases is to degrade the enamel matrix during the transition to maturation stages, which allows the enamel layer to attain its high degree of mineralization. Matrix metalloproteinase 20 (MMP-20, enamelysin 41–45 kDa) appears to act early during enamel development to process the structural enamel proteins so that the cleavage products will localize to specific areas of the enamel matrix (Bartlett et al. 1998). Subsequently, the serine proteinase kallikrein 4 (KLK4, also designated as enamel matrix serine proteinase 1/EMSP1, 31–34 kDa) appears to act later during enamel development to degrade the organic matrix (Hu et al. 2002). It is optimally active around pH 6 against enamel protein and facilitates the enamel matrix removal while the matrix undergoes pH modulation during the maturation stage (see Chap. 11 for detail). Enamel proteinases also include a serine protease caldecrin/chymotrypsin C, which increases in the maturation stage (Lacruz et al. 2011).

2.3 Conclusion

Dental tissues are either collagen-rich structures including dentins (primary, secondary, and tertiary dentin) with some differences between the crown and the roots, a variety of cementum and alveolar bone. Whereas dental tissues may include non-collagenous proteins. Enamel is formed by amelogenin, ameloblastin, amelotin, enamelin, tuft proteins, and sheathelin. Dentin and enamel are also regulated by proteases (MMPs, Klk4, serine proteases) and others complex molecules, either during the secretory or the maturation stages.

Most of the extracellular matrix proteins are implicated in the mineralization and a few are inhibitors of mineralization. It is also clear that enamel mineralization implies non-collagenous components (ECM). In addition, some molecules are acting as promoters or inhibitors of mineralization. In addition to collagens, the ECM non-collagenous molecules will be reviewed in the different chapters of this book.

References

Addison WN, Azari F, Sørensen ES, Kaartinen MT, McKee MD (2007) Pyrophosphate inhibits mineralization of osteoblast cultures by binding to mineral, up-regulating osteopontin, and inhibiting alkaline phosphatase activity. J Biol Chem 282:15872–15883. https://doi.org/10.1074/jbc.M701116200

Addison WN, Nakano Y, Loisel T, Crine P, McKee MD (2008) MEPE-ASARM peptides control extracellular matrix mineralization by binding to hydroxyapatite: an inhibition regulated by PHEX cleavage of ASARM. J Bone Miner Res 23:1638–1649. https://doi.org/10.1359/jbmr.080601

Almushayt A, Narayanan K, Zaki AE, George A (2006) Dentin matrix protein 1 induces cytodifferentiation of dental pulp stem cells into odontoblasts. Gene Ther 13:611–620. https://doi.org/10.1038/sj.gt.3302687

Anderson HC et al (2004) Impaired calcification around matrix vesicles of growth plate and bone in alkaline phosphatase-deficient mice. Am J Pathol 164:841–847. https://doi.org/10.1016/s0002-9440(10)63172-0

Aubin I, Adams CP, Opsahl S, Septier D, Bishop CE, Auge N et al (2005) A deletion in the gene encoding sphingomyelin phosphodiesterase 3 -*Smpd* 3- results in osteogenesis and dentinogenesis imperfecta in the mouse. Nat Genet 37:803–805

Bai X, Miao D, Panda D, Grady S, McKee MD, Goltzman D, Karaplis AC (2002) Partial rescue of the Hyp phenotype by osteoblast-targeted PHEX (phosphate-regulating gene with homologies to endopeptidases on the X chromosome) expression. Mol Endocrinol 16:2913–2925. https://doi.org/10.1210/me.2002-0113

Bansal AK, Shetty DC, Bindal R, Pathak A (2012) Amelogenin: a novel protein with diverse applications in genetic and molecular profiling. J Oral Maxillofac Pathol 16:395–399. https://doi.org/10.4103/0973-029X.102495

Bartlett JD, Ryu OH, Xue J, Simmer JP, Margolis HC (1998) Enamelysin mRNA displays a developmental defined pattern of expression and encodes a protein which degrades amelogenin. Connect Tissue Res 39:101–109. https://doi.org/10.3109/03008209809023916

Beertsen W, VandenBos T, Everts V (1999) Root development in mice lacking functional tissue non-specific alkaline phosphatase gene: inhibition of acellular cementum formation. J Dent Res 78:1221–1229. https://doi.org/10.1177/00220345990780060501

Bhatia A et al (2012) Overexpression of DMP1 accelerates mineralization and alters cortical bone biomechanical properties in vivo. J Mech Behav Biomed Mater 5:1–8. https://doi.org/10.1016/j.jmbbm.2011.08.026

Bleicher F, Couble ML, Buchaille R, Farges JC, Magloire H (2001) New genes involved in odontoblast differentiation. Adv Dent Res 15:30–33. https://doi.org/10.1177/08959374010150010701

Boskey AL (2013) Bone composition: relationship to bone fragility and antiosteoporotic drug effects. Bonekey Rep 2:447. https://doi.org/10.1038/bonekey.2013.181

Boskey AL, Maresca M, Doty S, Sabsay B, Veis A (1990) Concentration-dependent effects of dentin phosphophoryn in the regulation of in vitro hydroxyapatite formation and growth. Bone Miner 11:55–65. https://doi.org/10.1016/0169-6009(90)90015-8

Bosshardt DD, Schroeder HE (1996) Cementogenesis reviewed: a comparison between human premolars and rodent molars. Anat Rec 245:267–292. https://doi.org/10.1002/(sici)1097-0185(199606)245:2<267::Aid-ar12>3.0.Co;2-n

Bourd-Boittin K, Fridman R, Fanchon S, Septier D, Goldberg M, Menashi S (2005) Matrix metalloproteinase inhibition impairs the processing, formation and mineralization of dental tissues during mouse molar development. Exp Cell Res 304:493–505. https://doi.org/10.1016/j.yexcr.2004.11.024

Chaussain C et al (2009) MMP2-cleavage of DMP1 generates a bioactive peptide promoting differentiation of dental pulp stem/progenitor cell. Eur Cell Mater 18:84–95. https://doi.org/10.22203/ecm.v018a08

Chen S et al (2005) Differential regulation of dentin sialophosphoprotein expression by Runx2 during odontoblast cytodifferentiation. J Biol Chem 280:29717–29727. https://doi.org/10.1074/jbc.M502929200

Chen S et al (2009) Runx2, osx, and dspp in tooth development. J Dent Res 88:904–909. https://doi.org/10.1177/0022034509342873

Den Besten PK, Punzi JS, Li W (1998) Purification and sequencing of a 21 kDa and 25 kDa bovine enamel metalloproteinase. Eur J Oral Sci 106(Suppl 1):345–349. https://doi.org/10.1111/j.1600-0722.1998.tb02196.x

Deutsch D, Palmon A, Dafni L, Catalano-Sherman J, Young MF, Fisher LW (1995) The enamelin (tuftelin) gene. Int J Dev Biol 39:135–143

Dirksen TR, Marinetti GV (1970) Lipids of bovine enamel and dentin and human bone. Calcif Tissue Res 6:1–10

Dohi N et al (1998) Immunocytochemical and immunochemical study of enamelins, using antibodies against porcine 89-kDa enamelin and its N-terminal synthetic peptide, in porcine tooth germs. Cell Tissue Res 293:313–325. https://doi.org/10.1007/s004410051123

Ducy P (2000) Cbfa1: a molecular switch in osteoblast biology. Dev Dyn 219:461–471. https://doi.org/10.1002/1097-0177(2000)9999:9999<::Aid-dvdy1074>3.0.Co;2-c

Eapen A, Ramachandran A, George A (2012) Dentin phosphoprotein (DPP) activates integrin-mediated anchorage-dependent signals in undifferentiated mesenchymal cells. J Biol Chem 287:5211–5224. https://doi.org/10.1074/jbc.M111.290080
Embery G, Hall R, Waddington R, Septier D, Goldberg M (2001) Proteoglycans in dentinogenesis. Crit Rev Oral Biol Med 12:331–349. https://doi.org/10.1177/10454411010120040401
Fedde KN et al (1999) Alkaline phosphatase knock-out mice recapitulate the metabolic and skeletal defects of infantile hypophosphatasia. J Bone Miner Res 14:2015–2026. https://doi.org/10.1359/jbmr.1999.14.12.2015
Fincham AG, Burkland GA, Shapiro IM (1972) Lipophilia of enamel matrix—a chemical investigation of the neutral lipids and lipophilic proteins of enamel. Calcif Tissue Res 9:247–259
Fincham AG, Moradian-Oldak J, Diekwisch TG, Lyaruu DM, Wright JT, Bringas P Jr, Slavkin HC (1995) Evidence for amelogenin "nanospheres" as functional components of secretory-stage enamel matrix. J Struct Biol 115:50–59. https://doi.org/10.1006/jsbi.1995.1029
Fisher LW, Fedarko NS (2003) Six genes expressed in bones and teeth encode the current members of the SIBLING family of proteins. Connect Tissue Res 44(Suppl 1):33–40
Ganss B, Kim RH, Sodek J (1999) Bone sialoprotein. Crit Rev Oral Biol Med 10:79–98. https://doi.org/10.1177/10454411990100010401
Gaur T et al (2005) Canonical WNT signaling promotes osteogenesis by directly stimulating Runx2 gene expression. J Biol Chem 280:33132–33140. https://doi.org/10.1074/jbc.M500608200
George A, Silberstein R, Vais A (1995) In situ hybridization shows DMP1 (AG1) to be a developmentally regulated dentin-specific protein produces by mature odontoblasts. Connect Tissue Res 33(1–3):67–72
Gericke A, Qin C, Spevak L, Fujimoto Y, Butler WT, Sørensen ES, Boskey AL (2005) Importance of phosphorylation for osteopontin regulation of biomineralization. Calcif Tissue Int 77:45–54. https://doi.org/10.1007/s00223-004-1288-1
Giannobile WV (1996) Periodontal tissue engineering by growth factors. Bone 19:23s–37s. https://doi.org/10.1016/s8756-3282(96)00127-5
Gibson CW et al (2009) The leucine-rich amelogenin peptide alters the amelogenin null enamel phenotype. Cells Tissues Organs 189:169–174. https://doi.org/10.1159/000151384
Goldberg M, Boskey AL (1996) Lipids and biomineralizations. Prog Histochem Cytochem 31:1–187
Goldberg M, Septier D (2002) Phospholipids in amelogenesis and dentinogenesis. Crit Rev Oral Biol Med 13(3):276–290
Goldberg M, Smith AJ (2004) Cells and extracellular matrices of dentin and pulp: a biological basis for repair and tissue engineering. Crit Rev Oral Biol Med 15:13–27. https://doi.org/10.1177/154411130401500103
Goldberg M, Takagi M (1993) Dentine proteoglycans: composition, ultrastructure and functions. Histochem J 25:781–806. https://doi.org/10.1007/BF02388111
Goldberg M, Lécolle S, Vermelin L, Benghezal A, Septier D, Godeau G (1999) [^{3}H]choline uptake and turnover into membrane and extracellular matrix phospholipids, visualized by radioautography in rat incisor dentin and enamel. Calcif Tissue Int 65:66–72
Goldberg M, Septier D, Rapoport O, Iozzo RV, Young MF, Ameye LG (2005) Targeted disruption of two small leucine-rich proteoglycans, biglycan and decorin, excerpts divergent effects on enamel and dentin formation. Calcif Tissue Int 77:297–310. https://doi.org/10.1007/s00223-005-0026-7
Goldberg M, Septier D, Oldberg Å, Young MF, Ameye LG (2006) Fibromodulin-deficient mice display impaired collagen fibrillogenesis in predentin as well as altered dentin mineralization and enamel formation. J Histochem Cytochem 54:525–537. https://doi.org/10.1369/jhc.5A6650.2005
Goldberg M, Kulkarni AB, Young M, Boskey A (2011) Dentin: structure, composition and mineralization. Front Biosci (Elite Ed) 3:711–735. https://doi.org/10.2741/e281
Haruyama N et al (2009) Genetic evidence for key roles of decorin and biglycan in dentin mineralization. Matrix Biol 28:129–136. https://doi.org/10.1016/j.matbio.2009.01.005

Hu CC et al (1997) Sheathlin: cloning, cDNA/polypeptide sequences, and immunolocalization of porcine enamel sheath proteins. J Dent Res 76:648–657. https://doi.org/10.1177/00220345970760020501

Hu JC-C, Sun X, Zhang C, Liu S, Bartlett JD, Simmer JP (2002) Enamelysin and kallikrein-4 mRNA expression in developing mouse molars. Eur J Oral Sci 110:307–315. https://doi.org/10.1034/j.1600-0722.2002.21301.x

Hu JCC et al (2014) Enamelin is critical for ameloblast integrity and enamel ultrastructure formation. PLoS One 9:e89303–e89303. https://doi.org/10.1371/journal.pone.0089303

Kadler KE, Baldock C, Bella J, Boot-Handford RP (2007) Collagens at a glance. J Cell Sci 120:1955–1958. https://doi.org/10.1242/jcs.03453

Kalamajski S, Aspberg A, Lindblom K, Heinegård D, Oldberg A (2009) Asporin competes with decorin for collagen binding, binds calcium and promotes osteoblast collagen mineralization. Biochem J 423:53–59. https://doi.org/10.1042/bj20090542

Kao YH et al (2020) Fluoride alters signaling pathways associated with the initiation of dentin mineralization in enamel fluorosis susceptible mice. Biol Trace Elem Res. https://doi.org/10.1007/s12011-020-02434-y

Komori T et al (1997) Targeted disruption of *Cbfa1* results in a complete lack of bone formation owing to maturational arrest of osteoblasts. Cell 89:755–764. https://doi.org/10.1016/S0092-8674(00)80258-5

Lacruz RS et al (2011) Chymotrypsin C (caldecrin) is associated with enamel development. J Dent Res 90:1228–1233. https://doi.org/10.1177/0022034511418231

Le TQ, Zhang Y, Li W, Denbesten PK (2007) The effect of LRAP on enamel organ epithelial cell differentiation. J Dent Res 86:1095–1099. https://doi.org/10.1177/154405910708601114

Lu Y et al (2011) The biological function of DMP-1 in osteocyte maturation is mediated by its 57-kDa C-terminal fragment. J Bone Miner Res 26:331–340. https://doi.org/10.1002/jbmr.226

MacDougall M, Slavkin HC, Zeichner-David M (1992) Characteristics of phosphorylated and non-phosphorylated dentine phosphoprotein. Biochem J 287(Pt 2):651–655. https://doi.org/10.1042/bj2870651

Margolis HC, Beniash E, Fowler CE (2006) Role of macromolecular assembly of enamel matrix proteins in enamel formation. J Dent Res 85:775–793. https://doi.org/10.1177/154405910608500902

Martin A, David V, Li H, Dai B, Feng JQ, Quarles LD (2012) Overexpression of the DMP1 C-terminal fragment stimulates FGF23 and exacerbates the hypophosphatemic rickets phenotype in Hyp mice. Mol Endocrinol 26:1883–1895. https://doi.org/10.1210/me.2012-1062

Mazumder P, Prajapati S, Bapat R, Moradian-Oldak J (2016) Amelogenin-ameloblastin spatial interaction around maturing enamel rods. J Dent Res 95:1042–1048. https://doi.org/10.1177/0022034516645389

McKee MD, Zalzal S, Nanci A (1996) Extracellular matrix in tooth cementum and mantle dentin: localization of osteopontin and other noncollagenous proteins, plasma proteins, and glycoconjugates by electron microscopy. Anat Rec 245:293–312. https://doi.org/10.1002/(sici)1097-0185(199606)245:2<293::Aid-ar13>3.0.Co;2-k

McKee MD et al (2011) Enzyme replacement therapy prevents dental defects in a model of hypophosphatasia. J Dent Res 90:470–476. https://doi.org/10.1177/0022034510393517

McKee MD, Yadav MC, Foster BL, Somerman MJ, Farquharson C, Millán JL (2013) Compounded PHOSPHO1/ALPL deficiencies reduce dentin mineralization. J Dent Res 92:721–727. https://doi.org/10.1177/0022034513490958

Milan AM, Sugars RV, Embery G, Waddington RJ (2005) Modulation of collagen fibrillogenesis by dentinal proteoglycans. Calcif Tissue Int 76:127–135. https://doi.org/10.1007/s00223-004-0033-0

Moffatt P, Smith CE, St-Arnaud R, Simmons D, Wright JT, Nanci A (2006) Cloning of rat amelotin and localization of the protein to the basal lamina of maturation stage ameloblasts and junctional epithelium. Biochem J 399:37–46. https://doi.org/10.1042/bj20060662

Murshed M, Harmey D, Millán JL, McKee MD, Karsenty G (2005) Unique coexpression in osteoblasts of broadly expressed genes accounts for the spatial restriction of ECM mineralization to bone. Genes Dev 19:1093–1104. https://doi.org/10.1101/gad.1276205

Narayanan K, Gajjeraman S, Ramachandran A, Hao J, George A (2006) Dentin matrix protein 1 regulates dentin sialophosphoprotein gene transcription during early odontoblast differentiation. J Biol Chem 281:19064–19071. https://doi.org/10.1074/jbc.M600714200

Pitaru S, Savion N, Hekmati H, Olsen S, Narayanan SA (1992) Binding of a cementum attachment protein to extracellular matrix components and to dental surfaces. J Periodontal Res 27:640–646. https://doi.org/10.1111/j.1600-0765.1992.tb01748.x

Price PA, Urist MR, Otawara Y (1983) Matrix Gla protein, a new gamma-carboxyglutamic acid-containing protein which is associated with the organic matrix of bone. Biochem Biophys Res Commun 117:765–771. https://doi.org/10.1016/0006-291x(83)91663-7

Prout RES, Odutuga AA, Tring FC (1973) Lipid analyses of rat enamel and dentin. Arch Oral Biol 18(3):373–380

Ravindran S, Narayanan K, Eapen AS, Hao J, Ramachandran A, Blond S, George A (2008) Endoplasmic reticulum chaperone protein GRP-78 mediates endocytosis of dentin matrix protein 1. J Biol Chem 283:29658–29670. https://doi.org/10.1074/jbc.M800786200

Roberts S, Narisawa S, Harmey D, Millan JL, Farquharson C (2007) Functional involvement of PHOSPHO1 in matrix vesicle-mediated skeletal mineralization. J Bone Miner Res 22:617–627. https://doi.org/10.1359/Jbmr.070108

Rowe PS (2004) The wrickkened pathways of FGF23, MEPE and PHEX. Crit Rev Oral Biol Med 15:264–281. https://doi.org/10.1177/154411130401500503

Rowe PS, de Zoysa PA, Dong R, Wang HR, White KE, Econs MJ, Oudet CL (2000) MEPE, a new gene expressed in bone marrow and tumors causing osteomalacia. Genomics 67:54–68. https://doi.org/10.1006/geno.2000.6235

Salmon B et al (2013) MEPE-derived ASARM peptide inhibits odontogenic differentiation of dental pulp stem cells and impairs mineralization in tooth models of X-linked hypophosphatemia. PLoS One 8:e56749. https://doi.org/10.1371/journal.pone.0056749

Smith AJ, Scheven BA, Takahashi Y, Ferracane JL, Shelton RM, Cooper PR (2012) Dentine as a bioactive extracellular matrix. Arch Oral Biol 57:109–121. https://doi.org/10.1016/j.archoralbio.2011.07.008

Somogyi-Ganss E, Nakayama Y, Iwasaki K, Nakano Y, Stolf D, McKee MD, Ganss B (2012) Comparative temporospatial expression profiling of murine amelotin protein during amelogenesis. Cells Tissues Organs 195:535–549. https://doi.org/10.1159/000329255

Sreenath T et al (2003) Dentin sialophosphoprotein knockout mouse teeth display widened predentin zone and develop defective dentin mineralization similar to human dentinogenesis imperfecta type III. J Biol Chem 278:24874–24880. https://doi.org/10.1074/jbc.M303908200

Suzuki S et al (2009) Dentin sialoprotein and dentin phosphoprotein have distinct roles in dentin mineralization. Matrix Biol 28:221–229. https://doi.org/10.1016/j.matbio.2009.03.006

Takahashi N, Nyvad B (2016) Ecological hypothesis of dentin and root caries. Caries Res 50:422–431. https://doi.org/10.1159/000447309

Tang SY, Herber R-P, Ho SP, Alliston T (2012) Matrix metalloproteinase–13 is required for osteocytic perilacunar remodeling and maintains bone fracture resistance. J Bone Miner Res 27:1936–1950. https://doi.org/10.1002/jbmr.1646

Uchida T, Tanabe T, Fukae M, Shimizu M (1991) Immunocytochemical and immunochemical detection of a 32 kDa nonamelogenin and related proteins in porcine tooth germs. Arch Histol Cytol 54:527–538. https://doi.org/10.1679/aohc.54.527

von Marschall Z, Fisher LW (2010) Dentin sialophosphoprotein (DSPP) is cleaved into its two natural dentin matrix products by three isoforms of bone morphogenetic protein-1 (BMP1). Matrix Biol 29:295–303. https://doi.org/10.1016/j.matbio.2010.01.002

Waymire KG, Mahuren JD, Jaje JM, Guilarte TR, Coburn SP, MacGregor GR (1995) Mice lacking tissue non-specific alkaline phosphatase die from seizures due to defective metabolism of vitamin B-6. Nat Genet 11:45–51. https://doi.org/10.1038/ng0995-45

Xiong DH et al (2009) Genome-wide association and follow-up replication studies identified ADAMTS18 and TGFBR3 as bone mass candidate genes in different ethnic groups. Am J Hum Genet 84:388–398. https://doi.org/10.1016/j.ajhg.2009.01.025

Yamakoshi Y, Hu JCC, Fukae M, Iwata T, Kim J-W, Zhang H, Simmer JP (2005a) Porcine dentin sialoprotein is a proteoglycan with glycosaminoglycan chains containing chondroitin 6-sulfate. J Biol Chem 280:1552–1560. https://doi.org/10.1074/jbc.M409606200

Yamakoshi Y, Hu JCC, Fukae M, Zhang H, Simmer JP (2005b) Dentin glycoprotein: the protein in the middle of the dentin sialophosphoprotein chimera. J Biol Chem 280:17472–17479. https://doi.org/10.1074/jbc.M413220200

Yang G, Yuan G, MacDougall M, Zhi C, Chen S (2017) BMP-2 induced Dspp transcription is mediated by Dlx3/Osx signaling pathway in odontoblasts. Sci Rep 7:10775. https://doi.org/10.1038/s41598-017-10908-8

Part II
Collagen-Derived Extracellular Matrix Components: Dentin, Bone and Cementum

Chapter 3
Collagenous Mineralized Tissues: Composition, Structure, and Biomineralization

Elia Beniash

Abstract Mineralized tissues play multiple essential roles in the body, including locomotion, mechanical protection and support of the soft tissues, mastication; sense of gravity and linear acceleration, mineral ion depot, and others. The vast majority of mineralized tissues in the human body, such as bone, dentin, and cementum, belong to the family of collagenous mineralized tissues (CMTs). Two major components of these tissues are collagen type I fibrils and nonstoichiometric carbonated apatite nanocrystals which comprise the mineralized collagen fibrils—the basic building blocks of CMTs. The mineralized collagen fibrils organize into a variety of structural patterns at several hierarchical levels from nano- to macroscale. In addition to the two major components, CMTs contain a number of noncollagenous extracellular matrix components, including proteins and proteoglycans, carbohydrates, and lipids. These noncollagenous biomolecules play a role in the regulation of the mineralization process of CMTs. CMTs are biogenic hierarchical nanocomposite materials and understanding of their formation and function requires interdisciplinary efforts of researchers from different fields of study including chemists, material scientists, and biologists. Not only these studies help us to understand how mineralized tissues form and function, they also provide inspiration for the design of novel materials using principles learned in nature. This chapter provides a general overview of the composition, structural organization, and biomineralization of CMTs with the emphasis on recent advances in the field.

E. Beniash (✉)
Department of Oral and Craniofacial Sciences, University of Pittsburgh School of Dental Medicine, Pittsburgh, PA, USA
e-mail: ebeniash@pitt.edu

M. Goldberg, P. Den Besten (eds.), *Extracellular Matrix Biomineralization of Dental Tissue Structures*, Biology of Extracellular Matrix 10,
https://doi.org/10.1007/978-3-030-76283-4_3

3.1 Composition of CMTs

CMTs comprise three major constituencies—mineral, organic matrix, and water. They contain between 50 and 70% mineral by weight, with cementum being the least mineralized (~50%) and dentin the most mineralized (~70%). The organic content of CMTs varies between 20 and 40% by weight, while water content is between 5 and 10% (Clarke 2008). Differences in the mineral content of CMTs directly affect the mechanical properties of the tissues, and the tissues containing more mineral are stiffer (Currey 1999, 2013). Similarly, dehydration of CMTs increases their stiffness and brittleness (Nyman et al. 2006; Bajaj et al. 2006). Collagen type I is the major organic component, comprising roughly 90% of the matrix, with other organic constituencies, including noncollagenous matrix proteins (NCPs), proteoglycans, and lipids together accounting for the rest. The organic components of CMTs contribute to fracture toughness and resilience, and are potentially involved in self-healing of the microdamages (Fantner et al. 2005, 2007; Gupta et al. 2006; Currey 2003; Fratzl, 2008; Wang et al. 2020).

3.1.1 Mineral Phases in CMTs

3.1.1.1 Nano-crystalline Nonstoichiometric Apatite: The Major Mineral Phase of CMTs

The major mineral phase in CMTs comprises tiny crystallites of nonstoichiometric hydroxyapatite (Rey et al. 2009; Wopenka and Pasteris 2005; Von Euw et al. 2019). These mineral platelets, 30–100 nm across and 2–8 nm thick (Fratzl et al. 1996; Ziv and Weiner 1994; Turunen et al. 2016), are possibly the smallest biogenic crystals known (Lowenstam and Weiner 1989). The large variations in the crystallite dimensions reported are likely due to the age, tissue location, and the characterization technique used (Ziv and Weiner 1994; Turunen et al. 2016; Roschger et al. 2001; Kerschnitzki et al. 2013; Pabisch et al. 2013; Boskey and Coleman 2010). The small dimensions of the crystallites provide unique advantages, such as high strength, tolerance to flaws (Gupta et al. 2006; Gao et al. 2003), and reduced solubility (Tang et al. 2004; Ebacher and Wang 2009). Another unique feature of the crystallites is their extremely high surface to bulk ratio (Padilla et al. 2008), resulting in greater intermolecular interactions between the crystallites and the environment, specifically the matrix macromolecules and water, which likely contributes to the unique mechanical properties of CMTs at the nanoscale (Gupta et al. 2006; Almer and Stock 2005; Wang et al. 2013; Stock 2015). Despite their minute dimensions, the CMT mineral particles demonstrate complex organization with the bulk of the particles made of low crystallinity nonstoichiometric apatitic phase coated with a 0.5–1.0 nm thick surface hydrated amorphous layer (Von Euw et al. 2019; Jäger et al. 2006; Wang et al. 2013; Von Euw et al. 2017). Two models of the distribution of the hydrated layer around the crystalline core have emerged. The core/shell model envisages that the hydrated amorphous layer envelopes the core crystallite (Von Euw et al. 2019; Jäger et al. 2006; Wang et al. 2013; Von Euw et al. 2017), while in the core/crown model the amorphous layer is only present around the edges of the crystallites (Bertolotti et al. 2021).

The apatitic mineral in the CMTs is poorly crystalline (Farlay et al. 2010; Legros et al. 1987); it is Ca^{2+} (Wopenka and Pasteris 2005; Von Euw et al. 2019; Legros et al. 1987) and OH^{-} (Cho et al. 2003; Rey et al. 1995; Pasteris et al. 2004; Termine and Lundy 1973) deficient and contains significant fractions of CO_3^{2-} (Legros et al. 1987; Paschalis et al. 1996; Miller et al. 2001) and HPO_4^{2-} (Pleshko et al. 1991; Rey et al. 1990; Wu et al. 1994). Its carbonate fraction varies between 3 and 5% depending on the type of the tissue, location, and age (Legros et al. 1987). Carbonate can substitute either hydroxyl (A-substitution) or phosphate (B-substitution) (Pleshko et al. 1991; LeGeros et al. 1969) in the crystal lattice. A number of cations such as Mg^{2+}, Sr^{2+}, and Na^{+} substitute Ca^{2+} which leads to calcium deficiency of the mineral phase (Wopenka and Pasteris 2005; Von Euw et al. 2019). Remarkably, bone mineral particles contain only a fifth of OH^{-} ions present in the stoichiometric hydroxyapatite (Cho et al. 2003). Hydrogen phosphate is present primarily in the disorganized amorphous surface layer of the crystallites (Von Euw et al. 2019) and it is more prominent in the newly formed mineral (Legros et al. 1987).

It is abundantly clear from the data presented above that although the major mineral phase of CMTs exhibits certain structural and compositional characteristics of hydroxyapatite it is qualitatively and quantitatively different. These differences can be due to the nanoscopic dimensions and the high surface to bulk ratio of these mineral particles. They can also be a result of biological control, although the later possibility is less likely since the apatitic minerals of similar size and morphology synthesized abiotically demonstrate properties very similar to those of the mineral particles from CMTs (Bertolotti et al. 2021; Pasteris et al. 2004; Pleshko et al. 1991). Hence, this mineral phase should be called nano-apatite and not hydroxyapatite as it is commonly referred in the literature.

3.1.1.2 Minor Mineral Phases in CMTs

The presence of mineral phases other than nano-apatite in the CMTs was a subject of heated debates among researchers in the second part of the twentieth century (Glimcher et al. 1981; Boskey 1997; Grynpas and Omelon 2007; Termine and Posner 1966; Grynpas et al. 1984; Posner et al. 1980; Betts et al. 1975). Two models have been developed over the years however none of them till recently could be fully corroborated. According to one, proposed by Posner and colleagues the nano-apatite crystallites develop via transient mineral precursor phases, namely amorphous calcium phosphate (ACP) (Boskey 1997; Betts et al. 1975). This model was based on the studies of calcium phosphate mineralization in vitro at ambient conditions from supersaturated solutions. They have shown that ACP forms first and then transforms into a poorly crystalline apatitic phase, which structurally resembles bone mineral (Termine and Posner 1966; Posner et al. 1980; Betts et al. 1975). It was proposed that ACP consists of randomly organized spherical clusters 0.95 nm in diameter (later named Posner clusters), with a formula $Ca_9(PO_4)_6$ (Posner et al. 1980; Betts et al. 1975). These clusters fit well into hydroxyapatite unit cell (Posner et al. 1980; Posner and Betts 1975), suggesting that a solid phase transition can potentially occur by rearranging the clusters. This model was further supported by the observations of amorphous mineral in mitochondria and intracellular secretory vesicles of mineralizing cells (Betts et al. 1975; Lehninger 1970) and electron microscopy studies suggested the presence of amorphous mineral phase in the newly formed bone (Robinson and Watson 1955; Landis and Glimcher 1978). Alternative model postulates that the mineral phase in CMTs is a poorly crystalline apatite, which over time matures and becomes more crystalline (Glimcher et al. 1981) and that no other mineral phase is present in CMTs. This model is based on the radial distribution function (RDF) analysis of X-ray diffraction patterns (Glimcher et al. 1981; Grynpas and Omelon 2007; Grynpas et al. 1984). Unfortunately, these earlier studies were significantly limited by inadequate characterization methods and the evidence in support of either model was not strong enough.

The question regarding the transient mineral phases was only resolved with the advent of modern characterization techniques which helped to overcome the limitations of earlier studies. One significant limitation was the need to dehydrate the samples which can lead to transformation of transient mineral phases. Another limitation was the lack of spatial resolution of Fourier Transform Infrared (FTIR) spectroscopy and X-ray crystallography, the two main methods used by early researchers, while the use of transmission electron microscopy (TEM) for these studies was significantly limited by the need for extensive processing of the TEM samples (Landis and Glimcher 1978; Landis et al. 1977). A number of new methodologies, such as cryo electron microscopy (CryoEM) and Raman microspectroscopy, became widely available to the research community which allowed to address the question regarding the presence of mineral phases other than nano-apatite in CMTs with much higher accuracy and fidelity. In 2006 Crane et al. (Crane et al. 2006) published a seminal work, where they identified ACP and OCP in calvaria sutures of mice, which lead them to a conclusion that bone mineralization proceeds through transient metastable phases. The greatest advantage of Raman microspectroscopy is that it can be performed on fully hydrated live samples with 0.5μm resolution. This makes Raman microspectroscopy the method of choice for studies of biomineralization in living tissues (Mandair and Morris 2015; Feng et al. 2017; Akiva et al. 2016). In 2008 Mahamid et al. have identified ACP in forming bone of zebrafish using cryoEM (Mahamid et al. 2008) and in a follow-up study they have mapped transition of ACP to a crystalline apatitic phase in the same system using a combination of synchrotron diffraction and cryo scanning electron microscopy (cryoSEM) techniques (Mahamid et al. 2010). Later the presence of disordered calcium phosphate phase was confirmed in osteoblasts and the newly deposited bone matrix using a variety of characterization techniques including electron diffraction of the cryosectioned mouse calvaria (Mahamid et al. 2011). Taken together these recent studies clearly demonstrate the presence of transient

ACP at the sites of bone formation. The evidence on the presence of other minerals is not as clear, although the data suggest that the biomineralization process involves other transient mineral phases (Akiva et al. 2016; Thomas et al. 2020).

3.1.2 Organic Components of CMTs

3.1.2.1 Collagen Type I: The Major Component of CMTs

Collagen type I is the most abundant and probably the most studied structural protein in the human body (Kadler et al. 1996; Shoulders and Raines 2009). Individual molecule of collagen consists of three polypeptide chains (two identical α1[I] and one α2[I]) organized into a coiled coil motif, called triple helix, in which individual chains adopt a left-handed polyproline type II helical structure (Kramer et al. 1999). Each polypeptide chain contains more than 1000 amino acids with glycine, proline, and hydroxyproline constituting three most abundant amino acids in collagen molecule. The bulk of collagen sequence, forming the triple helix, consists of amino acid repeats Gly-Xxx-Yyy, where glycine always occupies the first position, with proline accounting for 28% of amino acids in second and hydroxyproline for 38% of amino acids in the third position (Ramshaw et al. 1998). Collagen is synthesized and secreted as a proprotein procollagen, containing N- and C-terminal propeptides, which are cleaved extracellularly by procollagenases (Kadler et al. 1996). After cleavage collagen assumes its active form—tropocollagen, which consists of the main triple helical domain with short mainly unstructured telopeptides on both ends of the molecule. The tropocollagen molecules are rigid rods which are 300 nm long and 1.5 nm wide.

Tropocollagen molecules spontaneously self-assemble into fibrils at physiological temperature and pH. In TEM stained collagen fibrils have a characteristic 67 nm periodic pattern (D-spacing) of alternating bands of high and low electron density. To explain this characteristic of the collagen fibrils in 1963 Hodge and Petruska came up with the first structural model of collagen fibril (Kadler et al. 1996; Shoulders and Raines 2009; Hodge and Petruska 1963). This two-dimensional model proposed that the tropocollagen molecules align along the fibrillar axis in linear arrays with all their N-termini pointing in the same direction with a gap of 36 nm between C-terminus of one tropocollagen molecule and N-terminus of another. The neighboring arrays of the tropocollagen molecules are staggered in such a way that in the equatorial direction each fifth triple helix is in the same position as first one. This model explains the axial periodic appearance of collagen fibrils in 2D projections of TEM micrographs, with less electron dense regions containing gaps and denser bands corresponding to overlaps. However, this model is not sufficient to explain the three-dimensional structure of collagen fibrils. Numerous models of collagen fibril organization in three dimensions have been proposed over the years (Hulmes and Miller 1979; Hulmes et al. 1995; Prockop and Fertala 1998), but they could not satisfactory describe the organization of collagen fibrils, till more recent structural studies of fully hydrated collagen fibrils using synchrotron light source by Orgel and colleagues, which lead to the breakthrough in our understanding of the fibrillar structure of collagen (Orgel et al. 2001, 2006). This model proposes that five collagen triple-helices organize into quasi-hexagonal super-twisted microfibrils, which interdigitate with the neighboring microfibrils. This structure is further stabilized by cross-links between the molecules in individual microfibers as well as between different microfibrils. Very recently the model by Orgel et al. was further refined with the focus on the structural organization of the gap region (Xu et al. 2020). They determined that since one out of five triple-helices in the microfibrils is missing in the gap region due to the quarter-staggering, each unit cell of the collagen fibril in the gap contains a 2–3 nm channel which is aligned with the fibril axis.

3.1.2.2 Noncollagenous Macromolecules

Noncollagenous macromolecules of CMTs include proteins, proteoglycans, and glycosaminoglycans (GAGs). The discovery of these macromolecules began in the mid-twentieth century when it was shown that collagen is the major component of both nonmineralized and mineralized connective tissues, and this finding led to a question regarding the reason why some collagenous tissues mineralize while others do not (Veis 2004). Early studies of CMTs revealed that the main characteristic distinguishing them from nonmineralized connective tissues is that, in addition to collagen, CMTs contain a fraction of unusually acidic proteins and glycoproteins, collectively known as noncollagenous proteins (NCPs), with a high content of acidic amino acids such as aspartic and glutamic acid, phosphoserine and γ-carboxyglutamic acid (Veis and Perry 1967; Richardson et al. 1978; Butler et al. 1983; Veis et al. 1972; Stetlerstevenson and Veis 1983; Gundberg et al. 1984; Qin et al. 2004). It was suggested that these proteins might play a role in the regulation of mineralization and a number of studies supported this notion (Boskey 1989; Boskey et al. 1990; Hunter and Goldberg 1993; Gorski 2011). The major progress in the studies of NCPs was driven by the development of molecular biology and genetics tools which allowed cloning and sequencing of these proteins, production of recombinant proteins, and development of genetically modified animal models, although despite this progress many fundamental questions remain. For the current status of our knowledge about NCPs and two other major classes of macromolecules, proteoglycans and GAGs I will refer the readers to several other chapters in this volume, which explicitly focus on the noncollagenous macromolecules in CMTs (Chaps. 2, 5, 6, 7, and 9).

3.1.2.3 Small Molecules

There is a number of small molecules identified in CMTs, although our knowledge of these molecules is not as extensive as of collagen and NCPs.

Lipids are one of the groups of small molecules found in CMTs. Lipids are the major component of matrix vesicles—membrane delineated compartments which bud from the cell membranes and become embedded in the collagen matrix at early mineralization sites (During et al. 2015; Anderson 1995; Golub 2009). Phospholipids play a role in Ca2+ transport in matrix vesicles and in the mineral nucleation sites (Golub 2009; Wu et al. 1997; Kirsch et al. 1997). A number of small phosphate containing molecules such as pyrophosphate, AMP, ATP, and beta-GP can act as sources of inorganic phosphate or inhibitors of mineralization (Garimella et al. 2006). Finally, citrate was recently identified on the surface of bone crystallites and its role in the regulation of the nanocrystals morphology, mineral–mineral and mineral–collagen interactions have been proposed (Davies et al. 2014; Costello et al. 2014; Shao et al. 2018).

3.2 Structural Organization of CMTs

CMTs are nanocomposites, organized at several levels of hierarchy from atomic and nanoscale to the macroscale (Weiner and Wagner 1998; Fratzl and Weinkamer 2007; Beniash 2011). The basic blocks of all CMTs are mineralized collagen fibrils (Weiner and Wagner 1998). These fibrils organize into arrays forming basic structural patterns, such as lamellae in bones and cementum, which serve as principal elements of more complex structural arrangements (Yamamoto et al. 2010; Reznikov et al. 2014a, b) optimized for the maximum mechanical performance (Roschger et al. 2001; Ebacher and Wang 2009; Koester et al. 2008; Ebacher et al. 2007; Weinkamer and Fratzl 2011; Nalla et al. 2003). The structural organization of CMTs at the nanoscale is primarily driven by an interplay between the macromolecular assemblies and forming mineral (Cantaert et al. 2013; Nudelman et al. 2013; Deshpande et al. 2011), while at the higher levels of hierarchy it is genetically (Karsenty et al. 2009; Long and Ornitz 2013) and environmentally controlled (Weinkamer and Fratzl 2011; Barak et al. 2011) through coordinated movements and directional matrix deposition by groups of mineralizing cells and tissue resorption by specialized cells, i.e., osteoclasts in bone. It is important to emphasize here that the basic structural organization at the nano- and mesoscale is similar among CMTs while there is a significant increase in structural diversity at the higher levels of hierarchy. We therefore will focus primarily on the common structural characteristics of CMTs, i.e., at the level of nano- and mesoscale.

The structural organization of the mineralized collagen fibrils has been extensively studied since 1950s and yet we do not have a complete and clear understanding of it today. The decades of studies of this basic structure produced a series of more and more sophisticated models, reflecting advances in characterization techniques. One of the first TEM studies of bone by Robertson and Watson in 1952 (Robinson and Watson 1952) identified the main characteristics of the mineralized collagen fibrils, which are still true today. They concluded that the mineralized collagen fibrils contain plate-shaped apatitic crystals organized with their crystallographic *c*-axes along the long axis of the fibril. They also noticed that the crystals in the collagen fibrils are axially arranged in periodic bands at 63 nm intervals (Robinson and Watson 1952). With the development of the triple staggered collagen fibril model by Hodge and Petruska (1963) the periodic nature of the mineral distribution was explained by the crystals being limited to the gap regions of the collagen fibrils. For many years this was a consensus model of bone and other CMTs, although a number of questions remained unresolved, including the organization of the crystallites inside the fibrils, the presence of extrafibrillar mineral, and the fact that this model underestimated the amount of mineral present (Weiner and Traub 1992; Katz and Li 1973; Lees et al. 1984). Weiner and Traub have conducted a TEM study of isolated dehydrated mineralized collagen fibrils from the turkey tendon, where they have shown that although the mineral platelets in the fibril congregate in the gap regions, creating the striated appearance, they are longer than the gap length, i.e., extend into overlaps. They have also demonstrated that the crystallites form stacks which organize into parallel arrays spanning the length of the fibril (Weiner and Traub 1986) with the crystallographic c-axes aligned with the long axis of the fibril. In the follow-up study they later confirmed these findings by cryoEM of vitrified samples of the fibrils to eliminate the possibility that ordering of the platelets was an artifact of drying (Traub et al. 1989). Around the same time, Arsenault has shown that mineralization begins in the gap regions and the mineral crystals expand into the overlaps in more mature mineralizing fibrils (Arsenault 1989; Larry Arsenault 1991). This model was further supported by a series of electron tomography studies by Landis and colleagues (Fratzl et al. 1996; Landis et al. 1993; Landis 1995). Although the presence of extrafibrillar mineral has been previously suggested by neutron diffraction and other indirect methods (Katz and Li 1973; Bonar et al. 1985), in the 1990s a number of direct observations of the extrafibrillar mineral by TEM have been reported (Fratzl et al. 1996; Su et al. 2003; Landis and Song 1991; Prostak and Lees 1996). Later AFM studies have confirmed the presence extrafibrillar mineral in CMTs (Sasaki et al. 2002; Hassenkam et al. 2004). These studies produced a consensus "deck of cards" model according to which the bulk of the mineral is present intrafibrillary in a form of stacks of plate-shaped crystallites organized into parallel arrays, with their *c*-axes co-aligned with the long axes of the fibrils, with a smaller fraction of mineral particles incrusting the surfaces of the fibrils. These highly anisotropic fibrils organize into arrays, which in turn form higher order structures, such as lamellae and osteons. This concept provided the structural basis for our current understanding of the mechanical performance of CMTs at the nano- and microscales (Gupta et al. 2006; Fantner et al. 2005, 2007; Gao et al. 2003).

Over the last decade, a number of studies have challenged the "deck of cards" model. Focused ion beam (FIB) milling technology has allowed scientists to obtain ultrathin sections from mature bone tissues without introducing mechanical artifacts inherent to ultramicrotomy (McNally et al. 2012). In a series of studies of ion beam milled samples of bones, Schwarcz et al. have shown that the majority of mineral particles reside outside of the fibrils in the form stacks of 200 nm long, 60 nm wide, and 5 nm thick plates surrounding the collagen fibrils (McNally et al. 2012, 2013;

Schwarcz et al. 2014, 2017). The darkfield analysis of the FIB milled sections revealed that these plates are not single crystals but rather assemblies of smaller plate-like crystallites (Schwarcz et al. 2014). They have calculated that this extrafibrillar mineral accounts for roughly 70% of total mineral in the bone tissue, while other 30% correspond to the intrafibrillar mineral which is localized in the gap regions of the fibrils (McNally et al. 2012). This model is in good agreement with estimates of extra- vs intrafibrillar mineral based on the neutron scattering data (Lees et al. 1984; Lees 2003). The authors further conducted finite element analysis comparing their structural model, with and without noncollagenous matrix surrounding the platelets, with the consensus "deck of cards" model of interfibrillar staggered crystallites (Jäger and Fratzl 2000) and found that the model with the extrafibrillar plates surrounded by the noncollagenous matrix is closer to the experimental data than the consensus model (Schwarcz et al. 2017).

More recently Reznikov et al. systemically analyzed FIB milled sections of bone using high-resolution TEM and electron tomography (Reznikov et al. 2018). They have prepared sections in three specific projections—longitudinal (along the fibril axis), intermediate (45° from the fibril axis), and transverse (perpendicular to the fibril axis). These data were combined to build a new model of the mineralized collagen fibrils' arrays. This model proposes that there is a continuum of mineral structures spanning the intra- and extrafibrillar spaces, with the intrafibrillar mineral being in the form of acicular (needle-shaped) slightly curved particles 5 nm across and 50–100 nm long. These needles exit the fibrils where they merge into platelets 20 nm across and 50–100 nm long. The platelets form stacks which in turn form larger aggregates incasing multiple fibrils. This model combines the features of the "deck of cards" model (Weiner and Traub 1992) with the extrafibrillar mineralization model proposed by the McMaster University group (McNally et al. 2012).

Yet in another electron tomography study using FIB milled samples, recently published by Somerdijk and colleagues, paints a very different picture (Xu et al. 2020). Based on the results of their tomographic reconstruction the bulk of the mineral resides inside of collagen fibrils in a form of plate-shaped crystallites (3 nm × 20 nm × 65 nm) aligned with their longest dimension corresponding to the crystallographic *c*-axis along the fibril axis, with only a few mineral particles detected outside of the fibrils. At the same time in contrast to the "deck of cards" model, the crystallites were essentially randomly oriented in two other dimensions, i.e., the mineral crystallites were oriented inside the fibril uniaxially. Stacks containing two to eight crystallites were also observed, but these stacks were not organized into arrays over the length of the fibril. Interestingly, the authors described finger-like protrusions at the ends of the platelets, similar to those observed by Reznikov et al. (2018) but overall this uniaxial model differs drastically not only from the older "deck of cards" model but also from the more recent studies conducted using FIB milled samples on advanced TEM microscopes.

It is quite remarkable that after 70 years of studies of the mineralized collagen fibrils in CMTs and dozens of publications on the subject there is still no clear understanding of their structural organization. There are only a few attributes which seem to be undisputed and consistent throughout the literature, namely the mineral in

CMTs is in the form of 3–6 nm thick plates arranged with their *c*-axes along the axis of the fibril—the facts reported in the first publication on this subject by Robinson and Watson in 1952 (Robinson and Watson 1952). This is even more peculiar, considering that researchers using the same tissue, sample preparation methods, and characterization tools arrive at very different conclusions (Xu et al. 2020; Reznikov et al. 2018).

3.3 Biomineralization of CMTs

Initial mineralization events in CMTs universally involve matrix vesicles (MVs)—small membrane-bound vesicles, which bud from the plasma membrane of mineralized cells into the newly deposited nonmineralized collagenous matrix (Anderson 1995; Golub 2009; Cui et al. 2016; Hasegawa et al. 2017; Goldberg et al. 2011). MVs have a unique composition distinct from that of the cytoplasm and the plasma membrane. Their membranes are enriched in phospholipids and they contain phosphatases such as TNAP and Phospho-1, integrins, annexins, and metalloproteinases (Golub 2009). The phosphatases are responsible for hydrolysis of pyrophosphate, which plays a dual role in the initiation of mineralization. Pyrophosphate is a potent mineralization inhibitor and its hydrolysis removes the inhibitory effects, while at the same time it increases phosphate concentration and hence the driving force toward mineralization. Phosphate and Ca^{2+} are transported into the MVs where the mineral nucleates on the inner surface of the MV membrane containing phospholipids and annexin compexes (During et al. 2015; Golub 2009; Wu et al. 1997). It is not clear if the first mineral phase formed in MVs is ACP, although some studies indicate that the early mineral is amorphous (Wu et al. 1997; Chaudhary et al. 2016; Cruz et al. 2020; Sauer and Wuthier 1988). Eventually, mineral crystals appear on the inner membrane surface of the MVs and then more crystals grow inside the vesicles. At a certain point, they puncture the membrane and become exposed to the matrix environment and spark the mineralization process in the collagen fibrils around them (Hasegawa et al. 2017; Stratmann et al. 1996; Takano et al. 2000), probably through heterogeneous nucleation of mineral crystallites on the surface of the exposed MV crystals. Once the initial mineralized layer is formed, the mineralizing cells continue to deposit nonmineralized matrix such as osteoid in bone and predentin in dentin which mineralizes a few microns away from the cell bodies at the mineralization front (Goldberg et al. 2011; Anderson 1989). The nonmineralized collagenous matrices undergo maturation including changes in their structural organization and composition prior to mineralization (Embery et al. 2001; Beniash et al. 2000; Lormée et al. 1996; Bronckers et al. 1985; Weinstock and Leblond 1973; Septier et al. 2001; Kazama et al. 1992).

One of the important questions in the field, which still remains unclear is the transport of phosphate and calcium to the mineralization sites in CMTs (Gay et al. 2000). A number of recent studies using advanced TEM techniques have shed some light on this process. Mahamid et al. performed cryoEM and cryoSEM studies of vitrified samples of forming zebrafish caudal bones (Mahamid et al. 2008, 2010) and calvaria from mouse embryos (Mahamid et al. 2011). The authors observed intracellular membrane-bound granules of disordered calcium phosphate which was secreted into the extracellular matrix via exocytosis where it eventually crystallized. In another study, Boonrungsiman et al. analyzed freeze substituted samples of osteoblast tissue cultures (Boonrungsiman et al. 2012). They also observed calcium phosphate-containing vesicles inside the cells and amorphous mineral aggregates in the extracellular matrix adjacent to the cells. They further confirmed the presence of calcium phosphate mineral in mitochondria and demonstrated that the mineral-containing vesicles are associated with mitochondria, the observation that suggests a role of mitochondria in sequestering and storage of calcium and phosphate for mineralization.

A number of recent studies identified mineral aggregates, both membrane delineated and free in the blood, which rapidly migrate to the sites of active bone growth via intercellular spaces (Haimov et al. 2020; Akiva et al. 2015, 2019; Kerschnitzki et al. 2016a, b). These recent findings taken together imply that the mineral is transported to the mineralization sites in a solid phase instead of an ionic form, both intercellularly and intracellularly. This mode of mineral transport is highly efficient since mineral ions in a solid phase are more concentrated than in a solution and are bound, which reduces their activity. The later point is especially important for the intracellular mineral transport since Ca^{2+} at high concentrations is highly cytotoxic and its concentration in the cytoplasm is maintained at the nM level (Bagur and Hajnóczky 2017). Interestingly, no classical ion transport pathways, via transmembrane ion pumps and channels, and at the levels which would account to the high rate of mineral deposition, have been identified in bone tissue so far (Schlesinger et al. 2020), which supports the solid phase transport model.

3.4 Conclusion

Despite the decades of studies, which revealed an enormous complexity and unique properties of CMTs, many fundamental questions posed at the inception of the field remain unanswered. The progress in our understanding of these materials depends on the development of new research methodologies and on interdisciplinary collaborations of scientists from diverse fields, such as physical chemistry and crystallography, material sciences, biophysics, biochemistry, molecular, and cell biology. Recent technological developments in many areas, including single cell transcriptomics (Krivanek et al. 2020), super-resolution light microscopy (Sigal et al. 2018), microspectroscopy (Byrne et al. 2018), electron microscopy (Courtland 2018; Callaway 2015), and others will help to address the enigmatic questions which evaded scientists for decades. At this point in time, we are standing at a threshold of great discoveries, and it is an exciting time for young investigators entering the field.

References

Akiva A, Malkinson G, Masic A, Kerschnitzki M, Bennet M, Fratzl P, Addadi L, Weiner S, Yaniv K (2015) On the pathway of mineral deposition in larval zebrafish caudal fin bone. Bone 75:192–200. https://doi.org/10.1016/j.bone.2015.02.020

Akiva A, Kerschnitzki M, Pinkas I, Wagermaier W, Yaniv K, Fratzl P, Addadi L, Weiner S (2016) Mineral formation in the larval zebrafish tail bone occurs via an acidic disordered calcium phosphate phase. J Am Chem Soc 138:14481–14487. https://doi.org/10.1021/jacs.6b09442

Akiva A, Nelkenbaum O, Schertel A, Yaniv K, Weiner S, Addadi L (2019) Intercellular pathways from the vasculature to the forming bone in the zebrafish larval caudal fin: Possible role in bone formation. J Struct Biol 206:139–148. https://doi.org/10.1016/j.jsb.2019.02.011

Almer JD, Stock SR (2005) Internal strains and stresses measured in cortical bone via high-energy X-ray diffraction. J Struc Biol 152:14–27. https://doi.org/10.1016/j.jsb.2005.08.003

Anderson HC (1989) Mechanism of mineral formation in bone. Lab Invest 60:320–330

Anderson HC (1995) Molecular-biology of matrix vesicles. Clin Orthop Relat Res 266–280

Arsenault AL (1989) A comparative electron microscopic study of apatite crystals in collagen fibrils of rat bone, dentin and calcified turkey leg tendons. Bone Miner 6:165–177. https://doi.org/10.1016/0169-6009(89)90048-2

Bagur R, Hajnóczky G (2017) Intracellular Ca(2+) sensing: its role in calcium homeostasis and signaling. Mol Cell 66:780–788. https://doi.org/10.1016/j.molcel.2017.05.028

Bajaj D, Sundaram N, Nazari A, Arola D (2006) Age, dehydration and fatigue crack growth in dentin. Biomaterials 27:2507–2517. https://doi.org/10.1016/j.biomaterials.2005.11.035

Barak MM, Lieberman DE, Hublin J-J (2011) A Wolff in sheep's clothing: Trabecular bone adaptation in response to changes in joint loading orientation. Bone 49:1141–1151. https://doi.org/10.1016/j.bone.2011.08.020

Beniash E (2011) Biominerals—hierarchical nanocomposites: the example of bone. Wiley Interdiscip Rev Nanomed Nanobiotechnol 3:47–69. https://doi.org/10.1002/wnan.105

Beniash E, Traub W, Veis A, Weiner S (2000) A transmission electron microscope study using vitrified ice sections of predentin: structural changes in the dentin collagenous matrix prior to mineralization. J Struct Biol 132:212–225. https://doi.org/10.1006/jsbi.2000.4320

Bertolotti F, Carmona FJ, Dal Sasso G, Ramírez-Rodríguez GB, López JMD, Pedersen JS, Ferri F, Masciocchi N, Guagliardi A (2021) On the amorphous layer in bone mineral and biomimetic apatite: a combined small- and wide-angle X-ray scattering analysis. Acta Biomater 120:167–180

Betts F, Blumenthal NC, Posner AS, Becker GL, Lehninger AL (1975) Atomic structure of intracellular amorphous calcium phosphate deposits. Proc Natl Acad Sci USA 72:2088–2090. https://doi.org/10.1073/pnas.72.6.2088

Bonar LC, Lees S, Mook HA (1985) Neutron diffraction studies of collagen in fully mineralized bone. J Mol Biol 181:265–270. https://doi.org/10.1016/0022-2836(85)90090-7

Boonrungsiman S, Gentleman E, Carzaniga R, Evans ND, McComb DW, Porter AE, Stevens MM (2012) The role of intracellular calcium phosphate in osteoblast-mediated bone apatite formation. Proc Natl Acad Sci USA 109:14170–14175. https://doi.org/10.1073/pnas.1208916109

Boskey AL (1989) Noncollagenous matrix proteins and their role in mineralization. Bone Miner 6:111–123. https://doi.org/10.1016/0169-6009(89)90044-5

Boskey AL (1997) Amorphous calcium phosphate: the contention of bone. J Dent Res 76:1433–1436. https://doi.org/10.1177/00220345970760080501

Boskey AL, Coleman R (2010) Aging and bone. J Dent Res 89:1333–1348. https://doi.org/10.1177/0022034510377791

Boskey AL, Maresca M, Doty S, Sabsay B, Veis A (1990) Concentration-dependent effects of dentin phosphophoryn in the regulation of in vitro hydroxyapatite formation and growth. Bone Miner 11:55–65. https://doi.org/10.1016/0169-6009(90)90015-8

Bronckers AL, Gay S, Dimuzio MT, Butler WT (1985) Immunolocalization of gamma-carboxyglutamic acid-containing proteins in developing molar tooth germs of the rat. Coll Relat Res 5:17–22. https://doi.org/10.1016/s0174-173x(85)80044-3
Butler WT, Bhown M, Dimuzio MT, Cothran WC, Linde A (1983) Multiple forms of rat dentin phosphoproteins. Arch Biochem Biophys 225:178–186. https://doi.org/10.1016/0003-9861(83)90021-8
Byrne HJ, Bonnier F, Casey A, Maher M, McIntyre J, Efeoglu E, Farhane Z (2018) Advancing Raman microspectroscopy for cellular and subcellular analysis: towards in vitro high-content spectralomic analysis. Appl Opt 57:E11–E19
Callaway E (2015) The revolution will not be crystallized: a new method sweeps through structural biology. Nature 525:172–174. https://doi.org/10.1038/525172a
Cantaert B, Beniash E, Meldrum FC (2013) Nanoscale confinement controls the crystallization of calcium phosphate: relevance to bone formation. Chem Eur J 19:14918–14924. https://doi.org/10.1002/chem.201302835
Chaudhary SC, Kuzynski M, Bottini M, Beniash E, Dokland T, Mobley CG, Yadav MC, Poliard A, Kellermann O, Millán JL, Napierala D (2016) Phosphate induces formation of matrix vesicles during odontoblast-initiated mineralization in vitro. Matrix Biol 52–54:284–300. https://doi.org/10.1016/j.matbio.2016.02.003
Cho GY, Wu YT, Ackerman JL (2003) Detection of hydroxyl ions in bone mineral by solid-state NMR spectroscopy. Science 300:1123–1127. https://doi.org/10.1126/science.1078470
Clarke B (2008) Normal bone anatomy and physiology. Clin J Am Soc Nephrol CJASN 3(Suppl 3): S131–S139. https://doi.org/10.2215/CJN.04151206
Costello LC, Chellaiah M, Zou J, Franklin RB, Reynolds MA (2014) The status of citrate in the hydroxyapatite/collagen complex of bone; and Its role in bone formation. J Regen Med Tissue Eng 3:4. https://doi.org/10.7243/2050-1218-3-4
Courtland R (2018) The microscope revolution that's sweeping through materials science. Nature 563:462–464. https://doi.org/10.1038/d41586-018-07448-0
Crane NJ, Popescu V, Morris MD, Steenhuis P, Ignelzi MA (2006) Raman spectroscopic evidence for octacalcium phosphate and other transient mineral species deposited during intramembranous mineralization. Bone 39:434–442. https://doi.org/10.1016/j.bone.2006.02.059
Cruz MAE, Ferreira CR, Tovani CB, de Oliveira FA, Bolean M, Caseli L, Mebarek S, Millán JL, Buchet R, Bottini M, Ciancaglini P, Paula Ramos A (2020) Phosphatidylserine controls calcium phosphate nucleation and growth on lipid monolayers: A physicochemical understanding of matrix vesicle-driven biomineralization. J Struct Biol 212:107607. https://doi.org/10.1016/j.jsb.2020.107607
Cui L, Houston DA, Farquharson C, MacRae VE (2016) Characterisation of matrix vesicles in skeletal and soft tissue mineralisation. Bone 87:147–158. https://doi.org/10.1016/j.bone.2016.04.007
Currey JD (1999) The design of mineralised hard tissues for their mechanical functions. J Exp Biol 202:3285–3294
Currey JD (2003) Role of collagen and other organics in the mechanical properties of bone. Osteoporosis Int 14:29–36. https://doi.org/10.1007/s00198-003-1470-8
Currey JD (2013) Bones: structure and mechanics. Princeton University Press, Princeton
Davies E, Müller KH, Wong WC, Pickard CJ, Reid DG, Skepper JN, Duer MJ (2014) Citrate bridges between mineral platelets in bone. Proc Natl Acad Sci USA 111:E1354–E1363. https://doi.org/10.1073/pnas.1315080111
Deshpande AS, Fang P-A, Zhang X, Jayaraman T, Sfeir C, Beniash E (2011) Primary structure and phosphorylation of dentin matrix protein 1 (DMP1) and dentin phosphophoryn (DPP) uniquely determine their role in biomineralization. Biomacromolecules 12:2933–2945. https://doi.org/10.1021/bm2005214
During A, Penel G, Hardouin P (2015) Understanding the local actions of lipids in bone physiology. Prog Lipid Res 59:126–146. https://doi.org/10.1016/j.plipres.2015.06.002

Ebacher V, Wang R (2009) A unique microcracking process associated with the inelastic deformation of haversian bone. Adv Funct Mater 19:57–66. https://doi.org/10.1002/adfm.200801234

Ebacher V, Tang C, McKay H, Oxland TR, Guy P, Wang R (2007) Strain redistribution and cracking behavior of human bone during bending. Bone 40:1265–1275. https://doi.org/10.1016/j.bone.2006.12.065

Embery G, Hall R, Waddington R, Septier D, Goldberg M (2001) Proteoglycans in dentinogenesis. Crit Rev Oral Biol Med 12:331–349. https://doi.org/10.1177/10454411010120040401

Fantner GE, Hassenkam T, Kindt JH, Weaver JC, Birkedal H, Pechenik L, Cutroni JA, Cidade GAG, Stucky GD, Morse DE, Hansma PK (2005) Sacrificial bonds and hidden length dissipate energy as mineralized fibrils separate during bone fracture. Nat Mater 4:612–616. https://doi.org/10.1038/nmat1428

Fantner GE, Adams J, Turner P, Thurner PJ, Fisher LW, Hansma PK (2007) Nanoscale Ion mediated networks in bone: osteopontin can repeatedly dissipate large amounts of energy. Nano Lett 7:2491–2498. https://doi.org/10.1021/nl0712769

Farlay D, Panczer G, Rey C, Delmas PD, Boivin G (2010) Mineral maturity and crystallinity index are distinct characteristics of bone mineral. J Bone Miner Metab 28:433–445. https://doi.org/10.1007/s00774-009-0146-7

Feng G, Ochoa M, Maher JR, Awad HA, Berger AJ (2017) Sensitivity of spatially offset Raman spectroscopy (SORS) to subcortical bone tissue. J Biophotonics 10:990–996. https://doi.org/10.1002/jbio.201600317

Fratzl P (2008) Collagen: structure and mechanics, an introduction. In: Fratzl P (ed) Collagen: structure and mechanics. Springer, New York, pp 1–13

Fratzl P, Weinkamer R (2007) Nature's hierarchical materials. Prog Mater Sci 52:1263–1334. https://doi.org/10.1016/j.pmatsci.2007.06.001

Fratzl P, Paris O, Klaushofer K, Landis W (1996) Bone mineralization in an osteogenesis imperfecta mouse model studied by small-angle x-ray scattering. J Clin Invest 97:396–402

Gao H, Ji B, Jäger IL, Arzt E, Fratzl P (2003) Materials become insensitive to flaws at nanoscale: Lessons from nature. Proc Natl Acad Sci USA 100:5597–5600. https://doi.org/10.1073/pnas.0631609100

Garimella R, Bi XH, Anderson HC, Camacho NP (2006) Nature of phosphate substrate as a major determinant of mineral type formed in matrix vesicle-mediated in vitro mineralization: An FTIR imaging study. Bone 38:811–817. https://doi.org/10.1016/j.bone.2005.11.027

Gay CV, Gilman VR, Sugiyama T (2000) Perspectives on osteoblast and osteoclast function. Poult Sci 79:1005–1008. https://doi.org/10.1093/ps/79.7.1005

Glimcher MJ, Bonar LC, Grynpas MD, Landis WJ, Roufosse AH (1981) Recent studies of bone mineral: Is the amorphous calcium phosphate theory valid? J Cryst Growth 53:100–119. https://doi.org/10.1016/0022-0248(81)90058-0

Goldberg M, Kulkarni AB, Young M, Boskey A (2011) Dentin: structure, composition and mineralization. Front Biosci (Elite Ed) 3:711–735. https://doi.org/10.2741/e281

Golub EE (2009) Role of matrix vesicles in biomineralization. Biochim Biophys Acta Gen Subj 1790:1592–1598. https://doi.org/10.1016/j.bbagen.2009.09.006

Gorski JP (2011) Biomineralization of bone: a fresh view of the roles of non-collagenous proteins. Front Biosci (Landmark ed) 16:2598

Grynpas MD, Omelon S (2007) Transient precursor strategy or very small biological apatite crystals? Bone 41:162–164. https://doi.org/10.1016/j.bone.2007.04.176

Grynpas MD, Bonar LC, Glimcher MJ (1984) Failure to detect an amorphous calcium-phosphate solid phase in bone mineral: A radial distribution function study. Calcif Tissue Int 36:291–301. https://doi.org/10.1007/BF02405333

Gundberg CM, Hauschka PV, Lian JB, Gallop PM (1984) Osteocalcin—isolation, characterization, and detection. Methods Enzymol 107:516–544

Gupta HS, Seto J, Wagermaier W, Zaslansky P, Boesecke P, Fratzl P (2006) Cooperative deformation of mineral and collagen in bone at the nanoscale. Proc Natl Acad Sci USA 103:17741–17746. https://doi.org/10.1073/pnas.0604237103

Haimov H, Shimoni E, Brumfeld V, Shemesh M, Varsano N, Addadi L, Weiner S (2020) Mineralization pathways in the active murine epiphyseal growth plate. Bone 130:115086. https://doi.org/10.1016/j.bone.2019.115086

Hasegawa T, Yamamoto T, Tsuchiya E, Hongo H, Tsuboi K, Kudo A, Abe M, Yoshida T, Nagai T, Khadiza N, Yokoyama A, Oda K, Ozawa H, de Freitas PHL, Li M, Amizuka N (2017) Ultrastructural and biochemical aspects of matrix vesicle-mediated mineralization. Jpn Dent Sci Rev 53:34–45. https://doi.org/10.1016/j.jdsr.2016.09.002

Hassenkam T, Fantner GE, Cutroni JA, Weaver JC, Morse DE, Hansma PK (2004) High-resolution AFM imaging of intact and fractured trabecular bone. Bone 35:4–10. https://doi.org/10.1016/j.bone.2004.02.024

Hodge AJ, Petruska JA (1963) Recent studies with the electron microscope on ordered aggregates of the tropocollagen macromolecule. In: Ramachandran GN (ed) Aspects of protein structure. Academic, New York, pp 289–300

Hulmes DJS, Miller A (1979) Quasi-hexagonal molecular packing in collagen fibrils. Nature 282:878–880. https://doi.org/10.1038/282878a0

Hulmes DJ, Wess TJ, Prockop DJ, Fratzl P (1995) Radial packing, order, and disorder in collagen fibrils. Biophys J 68:1661–1670. https://doi.org/10.1016/S0006-3495(95)80391-7

Hunter GK, Goldberg HA (1993) Nucleation of hydroxyapatite by bone sialoprotein. Proc Natl Acad Sci USA 90:8562–8565. https://doi.org/10.1073/pnas.90.18.8562

Jäger I, Fratzl P (2000) Mineralized collagen fibrils: a mechanical model with a staggered arrangement of mineral particles. Biophys J 79:1737–1746. https://doi.org/10.1016/S0006-3495(00)76426-5

Jäger C, Welzel T, Meyer-Zaika W, Epple M (2006) A solid-state NMR investigation of the structure of nanocrystalline hydroxyapatite. Magn Reson Chem 44:573–580

Kadler KE, Holmes DF, Trotter JA, Chapman JA (1996) Collagen fibril formation. Biochem J 316 (Pt 1):1–11. https://doi.org/10.1042/bj3160001

Karsenty G, Kronenberg HM, Settembre C (2009) Genetic control of bone formation. Annu Rev Cell Dev Biol 25:629–648. https://doi.org/10.1146/annurev.cellbio.042308.113308

Katz EP, Li S-T (1973) Structure and function of bone collagen fibrils. J Mol Biol 80:1–15. https://doi.org/10.1016/0022-2836(73)90230-1

Kazama T, Takagi M, Ishii T, Toda Y (1992) Immunoelectron microscopic studies of glycosaminoglycans in the metaphyseal bone trabeculae of growing rats. Histochem J 24:747–755. https://doi.org/10.1007/bf01460827

Kerschnitzki M, Kollmannsberger P, Burghammer M, Duda GN, Weinkamer R, Wagermaier W, Fratzl P (2013) Architecture of the osteocyte network correlates with bone material quality. J Bone Miner Res 28:1837–1845. https://doi.org/10.1002/jbmr.1927

Kerschnitzki M, Akiva A, Ben Shoham A, Asscher Y, Wagermaier W, Fratzl P, Addadi L, Weiner S (2016a) Bone mineralization pathways during the rapid growth of embryonic chicken long bones. J Struct Biol 195:82–92. https://doi.org/10.1016/j.jsb.2016.04.011

Kerschnitzki M, Akiva A, Shoham AB, Koifman N, Shimoni E, Rechav K, Arraf AA, Schultheiss TM, Talmon Y, Zelzer E, Weiner S, Addadi L (2016b) Transport of membrane-bound mineral particles in blood vessels during chicken embryonic bone development. Bone 83:65–72. https://doi.org/10.1016/j.bone.2015.10.009

Kirsch T, Nah HD, Demuth DR, Harrison G, Golub EE, Adams SL, Pacifici M (1997) Annexin V-mediated calcium flux across membranes is dependent on the lipid composition: implications for cartilage mineralization. Biochemistry 36:3359–3367. https://doi.org/10.1021/bi9626867

Koester KJ, Ager JW, Ritchie RO (2008) The true toughness of human cortical bone measured with realistically short cracks. Nat Mater 7:672–677. https://doi.org/10.1038/nmat2221

Kramer RZ, Bella J, Mayville P, Brodsky B, Berman HM (1999) Sequence dependent conformational variations of collagen triple-helical structure. Nat Struct Biol 6:454–457. https://doi.org/10.1038/8259

Krivanek J, Soldatov RA, Kastriti ME, Chontorotzea T, Herdina AN, Petersen J, Szarowska B, Landova M, Matejova VK, Holla LI (2020) Dental cell type atlas reveals stem and differentiated cell types in mouse and human teeth. Nat Comm 11:1–18
Landis WJ (1995) The strength of a calcified tissue depends in part on the molecular structure and organization of its constituent mineral crystals in their organic matrix. Bone 16:533–544
Landis WJ, Glimcher MJ (1978) Electron diffraction and electron probe microanalysis of the mineral phase of bone tissue prepared by anhydrous techniques. J Ultrastruct Res 63:188–223. https://doi.org/10.1016/S0022-5320(78)80074-4
Landis WJ, Song MJ (1991) Early mineral deposition in calcifying tendon characterized by high voltage electron microscopy and three-dimensional graphic imaging. J Struct Biol 107:116–127. https://doi.org/10.1016/1047-8477(91)90015-o
Landis WJ, Paine MC, Glimcher MJ (1977) Electron microscopic observations of bone tissue prepared anhydrously in organic solvents. J Ultrastruct Res 59:1–30. https://doi.org/10.1016/S0022-5320(77)80025-7
Landis W, Song M, Leith A, McEwen L, McEwen B (1993) Mineral and organic matrix interaction in normally calcifying tendon visualized in three dimensions by high-voltage electron microscopic tomography and graphic image reconstruction. J Struct Biol 110:39–54
Larry Arsenault A (1991) Image analysis of mineralized and non-mineralized type I collagen fibrils. J Electron Microsc Tech 18:262–268. https://doi.org/10.1002/jemt.1060180308
Lees S (2003) Mineralization of type I collagen. Biophys J 85:204–207. https://doi.org/10.1016/S0006-3495(03)74466-X
Lees S, Bonar LC, Mook HA (1984) A study of dense mineralized tissue by neutron diffraction. Int J Biol Macromol 6:321–326. https://doi.org/10.1016/0141-8130(84)90017-5
LeGeros RZ, Trautz OR, Klein E, LeGeros JP (1969) Two types of carbonate substitution in the apatite structure. Experientia 25:5–7. https://doi.org/10.1007/bf01903856
Legros R, Balmain N, Bonel G (1987) Age-related changes in mineral of rat and bovine cortical bone. Calcif Tissue Int 41:137–144. https://doi.org/10.1007/BF02563793
Lehninger AL (1970) Mitochondria and calcium ion transport. Biochem J 119:129–138
Long F, Ornitz DM (2013) Development of the endochondral skeleton. Cold Spring Harb Perspect Biol 5:a008334
Lormée P, Septier D, Lécolle S, Baudoin C, Goldberg M (1996) Dual incorporation of (35S)sulfate into dentin proteoglycans acting as mineralization promotors in rat molars and predentin proteoglycans. Calcif Tissue Int 58:368–375. https://doi.org/10.1007/bf02509387
Lowenstam HA, Weiner S (1989) On biomineralization. Oxford University Press on Demand, Oxford
Mahamid J, Sharir A, Addadi L, Weiner S (2008) Amorphous calcium phosphate is a major component of the forming fin bones of zebrafish: Indications for an amorphous precursor phase. Proc Natl Acad Sci USA 105:12748–12753. https://doi.org/10.1073/pnas.0803354105
Mahamid J, Aichmayer B, Shimoni E, Ziblat R, Li C, Siegel S, Paris O, Fratzl P, Weiner S, Addadi L (2010) Mapping amorphous calcium phosphate transformation into crystalline mineral from the cell to the bone in zebrafish fin rays. Proc Natl Acad Sci USA 107:6316–6321
Mahamid J, Sharir A, Gur D, Zelzer E, Addadi L, Weiner S (2011) Bone mineralization proceeds through intracellular calcium phosphate loaded vesicles: A cryo-electron microscopy study. J Struct Biol 174:527–535. https://doi.org/10.1016/j.jsb.2011.03.014
Mandair GS, Morris MD (2015) Contributions of Raman spectroscopy to the understanding of bone strength. Bonekey Rep 4:620–620. https://doi.org/10.1038/bonekey.2014.115
McNally EA, Schwarcz HP, Botton GA, Arsenault AL (2012) A model for the ultrastructure of bone based on electron microscopy of ion-milled sections. PLoS One 7:e29258. https://doi.org/10.1371/journal.pone.0029258
McNally E, Nan F, Botton GA, Schwarcz HP (2013) Scanning transmission electron microscopic tomography of cortical bone using Z-contrast imaging. Micron 49:46–53. https://doi.org/10.1016/j.micron.2013.03.002

Miller LM, Vairavamurthy V, Chance MR, Mendelsohn R, Paschalis EP, Betts F, Boskey AL (2001) In situ analysis of mineral content and crystallinity in bone using infrared microspectroscopy of the v(4) PO43-vibration. Biochem Biophys Acta Gen Subj 1527:11–19. https://doi.org/10.1016/s0304-4165(01)00093-9

Nalla RK, Kinney JH, Ritchie RO (2003) Effect of orientation on the in vitro fracture toughness of dentin: the role of toughening mechanisms. Biomaterials 24:3955–3968. https://doi.org/10.1016/S0142-9612(03)00278-3

Nudelman F, Lausch AJ, Sommerdijk NAJM, Sone ED (2013) In vitro models of collagen biomineralization. J Struct Biol 183:258–269. https://doi.org/10.1016/j.jsb.2013.04.003

Nyman JS, Roy A, Shen XM, Acuna RL, Tyler JH, Wang XD (2006) The influence of water removal on the strength and toughness of cortical bone. J Biomech 39:931–938. https://doi.org/10.1016/j.jbiomech.2005.01.012

Orgel JPRO, Miller A, Irving TC, Fischetti RF, Hammersley AP, Wess TJ (2001) The in situ supermolecular structure of Type I collagen. Structure 9:1061–1069. https://doi.org/10.1016/S0969-2126(01)00669-4

Orgel JPRO, Irving TC, Miller A, Wess TJ (2006) Microfibrillar structure of type I collagen in situ. Proc Natl Acad Sci USA 103:9001–9005. https://doi.org/10.1073/pnas.0502718103

Pabisch S, Wagermaier W, Zander T, Li C, Fratzl P (2013) In: De Yoreo JJ (ed) Methods in enzymology (vol 532). Academic, London, pp 391–413

Padilla S, Izquierdo-Barba I, Vallet-Regi M (2008) High specific surface area in nanometric carbonated hydroxyapatite. Chem Mater 20:5942–5944. https://doi.org/10.1021/cm801626k

Paschalis EP, DiCarlo E, Betts F, Sherman P, Mendelsohn R, Boskey AL (1996) FTIR microspectroscopic analysis of human osteonal bone. Calcif Tissue Int 59:480–487. https://doi.org/10.1007/bf00369214

Pasteris JD, Wopenka B, Freeman JJ, Rogers K, Valsami-Jones E, van der Houwen JAM, Silva MJ (2004) Lack of OH in nanocrystalline apatite as a function of degree of atomic order: implications for bone and biomaterials. Biomaterials 25:229–238. https://doi.org/10.1016/s0142-9612(03)00487-3

Pleshko N, Boskey A, Mendelsohn R (1991) Novel infrared spectroscopic method for the determination of crystallinity of hydroxyapatite minerals. Biophys J 60:786–793. https://doi.org/10.1016/s0006-3495(91)82113-0

Posner AS, Betts F (1975) Synthetic amorphous calcium phosphate and its relation to bone mineral structure. Acc Chem Res 8:273–281

Posner AS, Betts F, Blumenthal NC (1980) Formation and structure of synthetic and bone hydroxyapatites. Prog Cryst Growth Char 3:49–64

Prockop DJ, Fertala A (1998) The collagen fibril: the almost crystalline structure. J Struct Biol 122:111–118

Prostak KS, Lees S (1996) Visualization of crystal-matrix structure. In situ demineralization of mineralized turkey leg tendon and bone. Calcif Tissue Int 59:474–479. https://doi.org/10.1007/bf00369213

Qin C, Baba O, Butler W (2004) Post-translational modifications of sibling proteins and their roles in osteogenesis and dentinogenesis. Crit Rev Oral Biol Med 15:126–136

Ramshaw JA, Shah NK, Brodsky B (1998) Gly-XY tripeptide frequencies in collagen: a context for host–guest triple-helical peptides. J Struct Biol 122:86–91

Rey C, Shimizu M, Collins B, Glimcher MJ (1990) Resolution-enhanced fourier transform infrared spectroscopy study of the environment of phosphate ions in the early deposits of a solid phase of calcium-phosphate in bone and enamel, and their evolution with age. I: Investigations in thev 4 PO 4 domain. Calcif Tissue Int 46:384–394

Rey C, Miquel JL, Facchini L, Legrand AP, Glimcher MJ (1995) Hydroxyl groups in bone mineral. Bone 16:583–586. https://doi.org/10.1016/8756-3282(95)00101-i

Rey C, Combes C, Drouet C, Glimcher MJ (2009) Bone mineral: update on chemical composition and structure. Osteoporosis Int 20:1013–1021. https://doi.org/10.1007/s00198-009-0860-y

Reznikov N, Shahar R, Weiner S (2014a) Bone hierarchical structure in three dimensions. Acta Biomater 10:3815–3826. https://doi.org/10.1016/j.actbio.2014.05.024
Reznikov N, Shahar R, Weiner S (2014b) Three-dimensional structure of human lamellar bone: the presence of two different materials and new insights into the hierarchical organization. Bone 59:93–104. https://doi.org/10.1016/j.bone.2013.10.023
Reznikov N, Bilton M, Lari L, Stevens MM, Kröger R (2018) Fractal-like hierarchical organization of bone begins at the nanoscale. Science 360:eaao2189. https://doi.org/10.1126/science.aao2189
Richardson WS, Munksgaard EC, Butler WT (1978) Rat incisor phosphoprotein—nature of phosphate and quantitation of phosphoserine. J Biol Chem 253:8042–8046
Robinson RA, Watson ML (1952) Collagen-crystal relationships in bone as seen in the electron microscope. Anat Rec 114:383–409. https://doi.org/10.1002/ar.1091140302
Robinson RA, Watson ML (1955) Crystal-collagen relationships in bone as observed in the electron microscope. III. Crystal and collagen morphology as a function of age. Ann N Y Acad Sci 60:596–630. https://doi.org/10.1111/j.1749-6632.1955.tb40054.x
Roschger P, Grabner BM, Rinnerthaler S, Tesch W, Kneissel M, Berzlanovich A, Klaushofer K, Fratzl P (2001) Structural development of the mineralized tissue in the human L4 vertebral body. J Struct Biol 136:126–136. https://doi.org/10.1006/jsbi.2001.4427
Sasaki N, Tagami A, Goto T, Taniguchi M, Nakata M, Hikichi K (2002) Atomic force microscopic studies on the structure of bovine femoral cortical bone at the collagen fibril-mineral level. J Mater Sci Mater Med 13:333–337. https://doi.org/10.1023/a:1014079421895
Sauer GR, Wuthier RE (1988) Fourier transform infrared characterization of mineral phases formed during induction of mineralization by collagenase-released matrix vesicles in vitro. J Biol Chem 263:13718–13724
Schlesinger PH, Braddock DT, Larrouture QC, Ray EC, Riazanski V, Nelson DJ, Tourkova IL, Blair HC (2020) Phylogeny and chemistry of biological mineral transport. Bone 141:115621. https://doi.org/10.1016/j.bone.2020.115621
Schwarcz HP, McNally EA, Botton GA (2014) Dark-field transmission electron microscopy of cortical bone reveals details of extrafibrillar crystals. J Struct Biol 188:240–248. https://doi.org/10.1016/j.jsb.2014.10.005
Schwarcz HP, Abueidda D, Jasiuk I (2017) The ultrastructure of bone and its relevance to mechanical properties. Front Phys 5:39. https://doi.org/10.3389/fphy.2017.00039
Septier D, Hall RC, Embery G, Goldberg M (2001) Immunoelectron microscopic visualization of pro- and secreted forms of decorin and biglycan in the predentin and during dentin formation in the rat incisor. Calcif Tissue Int 69:38–45. https://doi.org/10.1007/s002230020047
Shao C, Zhao R, Jiang S, Yao S, Wu Z, Jin B, Yang Y, Pan H, Tang R (2018) Citrate improves collagen mineralization via interface wetting: a physicochemical understanding of biomineralization control. Adv Mater 30:1704876. https://doi.org/10.1002/adma.201704876
Shoulders MD, Raines RT (2009) Collagen structure and stability. Annu Rev Biochem 78:929–958. https://doi.org/10.1146/annurev.biochem.77.032207.120833
Sigal YM, Zhou R, Zhuang X (2018) Visualizing and discovering cellular structures with super-resolution microscopy. Science 361:880–887
Stetlerstevenson WG, Veis A (1983) Bovine dentin phosphophoryn—composition and molecular-weight. Biochemistry 22:4326–4335. https://doi.org/10.1021/bi00287a025
Stock SR (2015) The mineral-collagen interface in bone. Calcif Tissue Int 97:262–280. https://doi.org/10.1007/s00223-015-9984-6
Stratmann U, Schaarschmidt K, Wiesmann HP, Plate U, Höhling HJ (1996) Mineralization during matrix-vesicle-mediated mantle dentine formation in molars of albino rats: a microanalytical and ultrastructural study. Cell Tissue Res 284:223–230. https://doi.org/10.1007/s004410050582
Su X, Sun K, Cui FZ, Landis WJ (2003) Organization of apatite crystals in human woven bone. Bone 32:150–162. https://doi.org/10.1016/s8756-3282(02)00945-6

Takano Y, Sakai H, Baba O, Terashima T (2000) Differential involvement of matrix vesicles during the initial and appositional mineralization processes in bone, dentin, and cementum. Bone 26:333–339. https://doi.org/10.1016/s8756-3282(00)00243-x

Tang RK, Wang LJ, Orme CA, Bonstein T, Bush PJ, Nancollas GH (2004) Dissolution at the nanoscale: Self-preservation of biominerals. Ang Chem Int Ed 43:2697–2701. https://doi.org/10.1002/anie.200353652

Termine JD, Lundy DR (1973) Hydroxide and carbonate in rat bone mineral and its synthetic analogues. Calcif Tissue Res 13:73–82. https://doi.org/10.1007/BF02015398

Termine JD, Posner AS (1966) Infrared analysis of rat bone: age dependency of amorphous and crystalline mineral fractions. Science 153:1523–1525. https://doi.org/10.1126/science.153.3743.1523

Thomas J, Worch H, Kruppke B, Gemming T (2020) Contribution to understand the biomineralization of bones. J Bone Miner Metab 38:456–468. https://doi.org/10.1007/s00774-020-01083-4

Traub W, Arad T, Weiner S (1989) Three-dimensional ordered distribution of crystals in turkey tendon collagen fibers. Proc Natl Acad Sci USA 86:9822–9826

Turunen MJ, Kaspersen JD, Olsson U, Guizar-Sicairos M, Bech M, Schaff F, Tägil M, Jurvelin JS, Isaksson H (2016) Bone mineral crystal size and organization vary across mature rat bone cortex. J Struct Biol 195:337–344. https://doi.org/10.1016/j.jsb.2016.07.005

Veis A (2004) Biomineralization: on the trail of the phosphate. Part II: phosphophoryn, the DMPs, and more. J Dent Res 83:6–10. https://doi.org/10.1177/154405910408300102

Veis A, Perry A (1967) The phosphoprotein of the dentin matrix. Biochemistry 6:2409–2416. https://doi.org/10.1021/bi00860a017

Veis A, Spector AR, Zamoscianyk H (1972) Isolation of an EDTA-soluble phosphoprotein from mineralizing bovine dentin. Biochim Biophys Acta 257:404. https://doi.org/10.1016/0005-2795(72)90293-0

Von Euw S, Ajili W, Delices A, Laurent G, Babonneau F, Nassif N, Azaïs T (2017) Amorphous surface layer versus transient amorphous precursor phase in bone—a case study investigated by solid-state NMR spectroscopy. Acta Biomater 59:351–360

Von Euw S, Wang Y, Laurent G, Drouet C, Babonneau F, Nassif N, Azaïs T (2019) Bone mineral: new insights into its chemical composition. Sci Rep 9:8456. https://doi.org/10.1038/s41598-019-44620-6

Wang Y, Von Euw S, Fernandes FM, Cassaignon S, Selmane M, Laurent G, Pehau-Arnaudet G, Coelho C, Bonhomme-Coury L, Giraud-Guille MM, Babonneau F, Azais T, Nassif N (2013) Water-mediated structuring of bone apatite. Nat Mater 12:1144–1153. https://doi.org/10.1038/nmat3787

Wang Y, Morsali R, Dai Z, Minary-Jolandan M, Qian D (2020) Computational nanomechanics of noncollagenous interfibrillar interface in bone. ACS Appl Mater Interfaces 12:25363–25373. https://doi.org/10.1021/acsami.0c01613

Weiner S, Traub W (1986) Organization of hydroxyapatite crystals within collagen fibrils. FEBS Lett 206:262–266

Weiner S, Traub W (1992) Bone structure: from angstroms to microns. FASEB J 6:879–885

Weiner S, Wagner HD (1998) The material bone: structure-mechanical function relations. Annu Rev Mater Sci 28:271–298. https://doi.org/10.1146/annurev.matsci.28.1.271

Weinkamer R, Fratzl P (2011) Mechanical adaptation of biological materials — The examples of bone and wood. Mater Sci Eng C 31:1164–1173. https://doi.org/10.1016/j.msec.2010.12.002

Weinstock M, Leblond CP (1973) Radioautographic visualization of the deposition of a phosphoprotein at the mineralization front in the dentin of the rat incisor. J Cell Biol 56:838–845. https://doi.org/10.1083/jcb.56.3.838

Wopenka B, Pasteris JD (2005) A mineralogical perspective on the apatite in bone. Mater Sci Eng C 25:131–143. https://doi.org/10.1016/j.msec.2005.01.008

Wu Y, Glimcher MJ, Rey C, Ackerman JL (1994) A unique protonated phosphate group in bone mineral not present in synthetic calcium phosphates: identification by phosphorus-31 Solid State NMR spectroscopy. J Mol Biol 244:423–435. https://doi.org/10.1006/jmbi.1994.1740

Wu LN, Genge BR, Dunkelberger DG, LeGeros RZ, Concannon B, Wuthier RE (1997) Physico-chemical characterization of the nucleational core of matrix vesicles. J Biol Chem 272:4404–4411. https://doi.org/10.1074/jbc.272.7.4404

Xu Y, Nudelman F, Eren ED, Wirix MJM, Cantaert B, Nijhuis WH, Hermida-Merino D, Portale G, Bomans PHH, Ottmann C, Friedrich H, Bras W, Akiva A, Orgel JPRO, Meldrum FC, Sommerdijk N (2020) Intermolecular channels direct crystal orientation in mineralized collagen. Nat Comm 11:5068. https://doi.org/10.1038/s41467-020-18846-2

Yamamoto T, Li M, Liu Z, Guo Y, Hasegawa T, Masuki H, Suzuki R, Amizuka N (2010) Histological review of the human cellular cementum with special reference to an alternating lamellar pattern. Odontology 98:102–109. https://doi.org/10.1007/s10266-010-0134-3

Ziv V, Weiner S (1994) Bone crystal sizes: a comparison of transmission electron microscopic and X-ray diffraction line width broadening techniques. Connect Tissue Res 30:165–175. https://doi.org/10.3109/03008209409061969

Chapter 4
Non-collagenous ECM Matrix Components Growth Factors and Cytokines Involved in Matrix Mineralization

Annette Merkel, Elizabeth Guirado, Karthikeyan Narayanan, Amudha Ganapathy, and Anne George

Abstract Biomineralization is a widespread process in living organisms to form minerals to harden tissues via hierarchically structured organic–inorganic composites. These processes are very diverse and complex. However, the site-specific location, size, and morphology of the crystals formed are specifically controlled by various cellular activities that vary between organisms. In vertebrates, bone, cartilage, and teeth are calcified tissues that are essential for the functional and structural integrity of the organism. The composition, crystal morphology, and materials properties of these structural elements are intriguing to multidisciplinary scientists. Further, the simplicity of biomineralization process with low energy costs, high quality, and formation under ambient conditions are impossible to reproduce in a bio-free system. Studies have implicated that the organic component of the mineralized tissues which include collagen and non-collagenous proteins (NCPs) play a critical role in mineral nucleation and growth. In this review, we highlight recent developments in dentin biomineralization process with respect to matrix proteins and growth factors.

4.1 Introduction

Biomineralization, the generation of hard tissues is a fascinating process by which living organisms produce hard tissues composed of biominerals (Veis and Dorvee 2013; Alvares and Veis 2012; Boskey 1990, 1998, 2003a, 2013; Estroff and Cohen 2011). The control of biomineralization, the deposition of mineral crystals on a

Annette Merkel and Elizabeth Guirado contributed equally with all other contributors.

A. Merkel · E. Guirado · K. Narayanan · A. Ganapathy · A. George (✉)
Brodie Tooth Development Genetics & Regenerative Medicine Research Laboratory, Department of Oral Biology, University of Illinois at Chicago, Chicago, IL, USA
e-mail: anneg@uic.edu

M. Goldberg, P. Den Besten (eds.), *Extracellular Matrix Biomineralization of Dental Tissue Structures*, Biology of Extracellular Matrix 10,
https://doi.org/10.1007/978-3-030-76283-4_4

biologically formed matrix is crucial to the function of the mineralized tissue (Boskey 1992, 2003b; Halvorsen et al. 2001; Tian et al. 2020; Ma et al. 2020; Paine and Snead 2005). Disruptions in this process are detrimental to human health (Mckee et al. 2013; Pohjolainen et al. 2008; Koehne et al. 2013; Opsahl Vital et al. 2012; Pragnère et al. 2021). Biominerals have complex hierarchical structures, possess exceptional properties such as mechanical, electrical, or magnetic properties which are achieved under the direct control of biomolecules (Saunders et al. 2011; Gorski 2011). Bone and dentin formation are classified as "matrix-mediated biomineralization" as the organisms control mineral deposition (Feng 2011; Young et al. 1992). Specifically, the organism controls the nature, orientation, size, and shape of the mineral by creating closed compartments or defined channels in which the mineral crystal forms (Veis and Dorvee 2013; Dorvee and Veis 2013). The consensus is that both matrix macromolecules and tissue architecture formed by the compartment are of vital importance in dictating specific sites for localizing mineral formation, nucleation, crystal growth, crystal orientation, crystal morphology, and size regulation. Thus, minerals of biogenic origin are created at near ambient temperatures and pressures with the matrix dictating the formation of unique patterns.

Dentin mineralization is an example of matrix-mediated mineralization (Beniash et al. 2011; Foster et al. 2013; Goldberg et al. 2011; George and Veis 2008). Odontoblasts, which are the principal cells, are responsible for the synthesis and secretion of both the scaffolding proteins and specialized proteins that are involved in crystal nucleation and growth (Zurick et al. 2013; Goldberg et al. 2008; He et al. 2003).

4.2 Extracellular Matrix Components

4.2.1 Collagen

Type I collagen is the principal structural protein in bone and dentin. The self-assembled collagen matrix forms the scaffold for mineral deposition (George and Veis 1991; Helseth and Veis 1981). In earlier studies, Type I collagen was considered as a passive template deposited by the odontoblasts as it did not have the capacity to initiate mineral nucleation. Several recent studies have shown that type I collagen forms a dynamic and interactive scaffold for ordered mineral deposition (Nudelman et al. 2010; Yao et al. 2019). In dentin, the odontoblasts deposit collagen fibrils with a 67 nm periodicity. The ~300 nm rod-like collagen molecules self-assemble in a staggered manner leaving gaps between the molecular ends. The gaps are registered in packed fibril-producing channels that run transverse to the fibril axis; their reduced density produces cross-striations in the fibrils. Due to the specific binding of the non-collagenous proteins in the gap regions of the collagen fibrils, the mineral crystals begin to grow in the gap spaces. The initial precipitated amorphous nanoclusters of Ca and P transform to needle-like crystals of hydroxyapatite which eventually coalesce to form plate-like crystals and fill the gap space. The unique feature of the deposited crystals is that their *c*-axis align with the long axis of the collagen fibrils. Thus, the alignment of the collagen fibrils is a requirement to promote long-range order (Duer and Veis 2013; Yao et al. 2019).

4.2.2 Non-collagenous Proteins (NCPs)

NCPs synthesized by the odontoblasts are responsible for mineral nucleation, crystal growth, hydroxyapatite crystallinity, inhibit nonspecific mineral deposition, self-assembly of collagen fibrils, and coordinate cell–matrix interactions (Kim et al. 2004; Boskey 1989; George et al. 1993).

Primary Structure of NCPs The characteristic feature of the NCPs involved in mineralization is that they are acidic and negatively charged. The negative charge of the NCPs is contributed by both the high number of negatively charged amino acid residues such as aspartic acid and glutamic acid and by the post-translational phosphorylation modification of amino acids such as serines, threonines, and tyrosines. Such negatively charged proteins/domains facilitate avid binding to Ca^{2+} and PO_4^{3-} ions which are supersaturated in the ECM milieu. The binding of Ca^{2+} and PO_4^{3-} by collagen-immobilized NCPs is necessary to form the initial mineral nidus. The negatively charged amino acids can shape crystal morphology and promote the transformation of amorphous calcium phosphate to crystalline hydroxyapatite. Some of the proteins synthesized by the odontoblasts function as crystal inhibitors and prevent the growth by binding to the surface of the nascent mineral nuclei (Giachelli 2005).

Structure-Based Classification of NCPs Based on the structural analysis, several of these proteins have been classified as intrinsically disordered proteins (IDPs) or have regions that do not fold into a 3D structure, and these are referred to as intrinsically disordered protein regions (IDRPs) (Boskey and Villarreal-Ramirez 2016; Tavafoghi and Cerruti 2016). Recent reports suggest that a large fraction of the human proteome and all organisms comprises proteins that under physiological conditions lack ordered 3D structures (Veis and Dorvee 2013; Alvares and Veis 2012; Boskey 1990, 1998, 2003a, 2013; Estroff and Cohen 2011). IDPs are characterized by conformational flexibility as their structures are variable and do not fold into the conventional secondary structures seen in structural proteins. Such structural plasticity enables them to engage in a wide-array of biological functions which cannot be performed by structured proteins (Kalmar et al. 2012). However, these proteins can assume secondary structures when they are phosphorylated or interact with their binding partners. The intrinsic disorder is a unique structural feature that enables these proteins to participate in many signaling events.

IDPs are often hydrophilic, have amino acid domains containing repeat sequences which are either positively or negatively charged domains (Boskey and Villarreal-Ramirez 2016). Many of the noncollagenous proteins of bone and teeth have no defined structure and perhaps this conformational flexibility is highly beneficial for binding to Ca^{2+} and initiating the process of mineral nucleation and subsequent formation of amorphous calcium phosphate mineral (Wald et al. 2017). The importance of IDPs were demonstrated when dysregulated proteins engage in unwanted interactions leading to the development of several pathological conditions such as neurodegeneration, cardiovascular, amyloidosis and genetic diseases.

4.3 Noncollagenous Proteins in the Dentin Matrix That Regulate Mineralization

4.3.1 Introduction

Dentin is a composite of organic macromolecules and hydroxyapatite minerals (Zhang et al. 2014). It is formed by odontoblast cells as the foundation for the overlaying enamel and cementum layers that face the oral cavity environment.

Dentinogenesis is classified into three phases, primary, secondary, and tertiary dentin formation.

Primary dentinogenesis precedes birth and begins at around 14 weeks of gestation, when odontoblast differentiation begins. A 5–30 micron thick layer of mantle dentin is deposited first at the dentino-enamel junction, followed by circumpulpal dentin (Kawashima and Okiji 2016). Primary dentinogenesis continues until tooth eruption. After eruption into the oral cavity and the establishment of functional contacts between teeth are completed, any further dentin deposition is termed secondary dentinogenesis. Tertiary dentinogenesis has been used to describe the reparative process of dentin deposition following insult to the pulp, such as caries or restorative procedures. Pre-dentin, a layer 10–30 microns thick, lies in direct contact with the odontoblasts at the periphery of the pulp chamber and represents an unmineralized dentin matrix (Linde and Goldberg 1993).

The mechanical properties of dentin depend on its mineral content and ultrastructure (Angker et al. 2004). Although dentin's hardness and elastic modulus remain fairly constant, its fracture resistance decreases with increasing mineral deposition (Zhang et al. 2014).

Dentin mineralization is a meticulously controlled process dependent on the organic components of the dentin matrix. This dentin matrix is composed of type I collagen (86%), as well as types III, V, and VI collagens, and a variety of non-collagenous proteins (Fig. 4.1). The small integrin-binding ligand, N-linked glycoprotein (SIBLING) family of non-collagenous proteins is the focus of this chapter. They represent a group of dentin matrix proteins known to regulate the biomineralization process.

SIBLING protein genes are contained within a contiguous region of chromosome 4 (Fisher and Fedarko 2003). They include: dentin matrix protein 1 (DMP1), osteopontin (OPN), bone sialoprotein (BSP), matrix extracellular

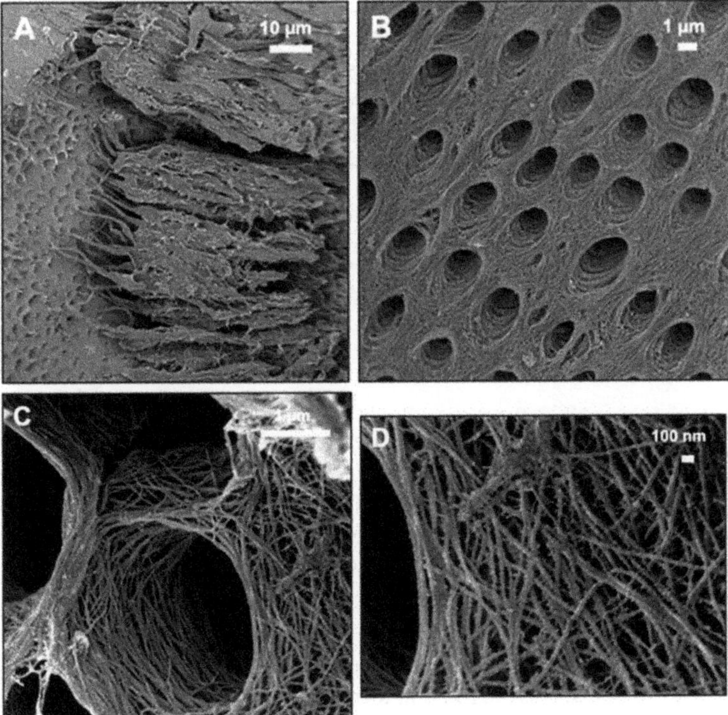

Fig. 4.1 Scanning Electron Microscopy Images of dentin. Decalcified maxillary first molars from mice were freeze-fractured in liquid nitrogen, dehydrated in hexamethyldisilazane, coated with gold/palladium (10 nm), and imaged using field-emission scanning electron microscopy, 3.0 kV, WD 6.0 mm. (**a**) An odontoblast layer lines the periphery of the pulp chamber, projecting their cellular processes (odontoblastic processes) through the dentinal tubules. Scale = 10 um, ×1000. (**b**) During dentinogenesis, odontoblasts deposit a scaffold of unmineralized dentin matrix that is optimized for nucleating minerals and forming the mineralized tissue that is dentin. This dentin matrix is composed primarily of type I collagen. Scale = 1 um, ×4500. (**c**) A magnified view of the dentinal tubule wall reveals the characteristic type 1 collagen 67 nm D-band periodicity. The collagen density appears highest closest to the tubule wall. Scale = 1 um, ×19,000 (**d**) Further magnification reveals non-collagenous cross-linkers within the dentin matrix. These non-collagenous components, to which the SIBLING family belongs, are thought to mediate the mineralization of the dentin matrix. Scale = 100 nm, ×40,000

phosphoglycoprotein (MEPE), and dentin sialophosphoprotein (DSPP), the latter being a compound protein consisting of dentin sialoprotein (DSP) and dentin phosphoprotein (DPP).

In general, SIBLING proteins are extended, flexible proteins in solution thought to orchestrate mineral formation. Though they have poor homologies at the amino acid level, they all contain conserved regions such as the integrin-binding RGD motif, the NXS/T motif for N-linked oligosaccharides, and multiple casein kinase II-type phosphorylation sites (Fisher and Fedarko 2003).

Several human diseases have been attributed to the dysfunction of these SIBLING proteins, emphasizing their importance to dentin formation. The phosphorylation status of SIBLING proteins affects their physiological role in mineralized tissues (Gericke et al. 2005). In fact, defects in global phosphorylation of the SIBLINGs account for the phenotype of Raine syndrome (Whyte et al. 2017). An understanding of the biology of these proteins is thus paramount to the understanding of dentin formation.

4.3.2 Dentin Sialoprotein (DSP) and Dentin Phosphoprotein (DPP)

The dentin sialophosphoprotein (DSPP) gene encodes the major SIBLING protein in the dentin matrix, dentin sialoprotein, and dentin phosphoprotein (Veis et al. 1972). Despite its early discovery, cloning of the DSPP gene lagged partly due to the gene's many domains (George et al. 1999). Further complexity to the gene arises from DSPP's posttranslational cleavage into two distinct proteins, the carbohydrate-rich dentin sialoprotein (DSP) and highly phosphorylated dentin phosphoprotein (DPP/DMP2) (Qin et al. 2004).

Murine models of dentinogenesis reveal that the DSPP gene is actively transcribed by odontoblasts before mineral formation and sustained through all phases of primary and secondary dentinogenesis (D'souza et al. 1997). DPP is secreted and appears rapidly at the mineralization front (Weinstock and Leblond 1973), while DSP is found predominantly in highly mineralized peritubular dentin (Hao et al. 2009). The DSPP gene has been mapped to the long arm of human chromosome four, 4q21, implicating it in the pathophysiology of dentinogenesis imperfecta type II (Takagi et al. 1983; MacDougall et al. 1997; George et al. 1999).

4.3.3 Dentin Matrix Protein 1 (DMP1, AG1)

Dentin matrix protein 1 (DMP1) was the first serine-rich, acidic protein to have been cloned and sequenced (George et al. 1999), different isoforms of whom were identified concurrently (MacDougall et al. 1998). It is an extremely hydrophilic protein containing a single RGD integrin-binding motif and N-glycosylation sequences (George et al. 1993).

The protein is cleaved into N-terminal (37 kDa) and C-terminal fragments (57 kDa), the latter of which is highly phosphorylated. Sequence analysis predicts that casein kinases I and II may be responsible for this phosphorylation. Within the dentin matrix, the phosphorylated form of DMP1 is found only at low levels. Flexibility at the C-terminus may help facilitate collagen binding in the dentin matrix, as well as mineral nucleation. In vitro, recombinant DMP1 is able to nucleate nanocrystalline hydroxyapatite ribbons (George et al. 2018).

Murine models of dentinogenesis reveal that the DMP1 gene is transcribed in newly differentiated odontoblasts after polarization is achieved, increases after birth in secretory stage odontoblasts, and becomes downregulated in mature odontoblasts (George et al. 1995; D'souza et al. 1997).

In-situ localization and chromosomal mapping localize the DMP1 gene to the 4q21 locus of the human chromosome. DMP1 was once thought to be a candidate gene for dentinogenesis imperfecta type II (George et al. 1994; Hirst et al. 1997); however, today it is known to be involved in the pathophysiology of recessive hypophosphatemic rickets (Lorenz-Depiereux et al. 2006).

4.3.4 *Matrix Extracellular Phosphoglycoprotein (MEPE) (Osteocyte/Osteoblast Factor 45 (OF45), Osteoregulin)*

Matrix extracellular phosphoglycoprotein (MEPE) was originally cloned from oncogenic hypophosphatemic osteomalacia (OHO) tumors (Rowe et al. 2000). Unlike other SIBLING proteins, MEPE is strongly basic (pI = 9.2) (Fisher and Fedarko 2003).

Murine models of dentinogenesis have revealed that MEPE is found in polarized odontoblast cells, both young and mature. In the largely non-mineralized pre-dentin layer, MEPE is particularly concentrated at the dentin mineralization front and its expression is sustained throughout root formation (Gullard et al. 2016).

Murine models reveal that the gene has distinct functions in the tooth organ versus the bone. MEPE knockout models display hypermineralized molars with thicker pre-dentin, dentin, and enamel layers, concurrently with hypomineralized cranial bones (Gullard et al. 2016).

MEPE-derived acidic serine- and aspartate-rich motif (ASARM) peptides have been implicated in the physiological role of the protein in tooth and bone mineralization (Martin et al. 2008; David and Quarles 2010). In vitro studies have revealed that MEPE-derived ASARM peptides inhibit odontogenic differentiation and matrix mineralization (Salmon et al. 2013), however, more research is necessary to determine the role of this protein and its peptide products in dentinogenesis.

Genetic disorders directly resulting from MEPE dysfunction have not been reported. However, abnormal MEPE accumulation has been reported in the teeth of X-linked Hypophosphatemic (XLH) patients (Salmon et al. 2014).

4.3.5 *Osteopontin (OPN) (Bone Sialoprotein 1 (BSP1)/ Secreted Phosphoprotein 1 (SPP1))*

Osteopontin (OPN) was first isolated from bovine bone (Franzén and Heinegård 1985). The protein is not tooth-specific, being transcribed by both odontoblasts and osteoblasts; and, has been proposed as a marker of odontoblasts differentiation (Surdilovic et al. 2018).

Dentin and bone protein extracts reveal that OPN levels in the dentin matrix are less than one-seventieth of that in bone (Qin et al. 2001).

Murine models of dentinogenesis have revealed strong OPN expression by alveolar bone osteoblasts and transient expression by cementoblasts. Similar to murine models of MEPE-knockout, OPN-knockout mice present with significantly increased dentin volume and mineral density. OPN transcription has not been identified in odontoblasts, pulp or periodontal ligament cells (Foster et al. 2018). Despite this, OPN protein has been localized to pre-dentin and to clusters of odontoblasts adjacent to areas devoid of dentin, but not dentin, pre-odontoblasts, or enamel organ (Mark et al. 1988). Like MEPE, X-linked hypophosphatemic (XLH) teeth present with prominent and abnormal OPN localization (Salmon et al. 2014).

4.3.6 *Bone Sialoprotein (BSP2)*

Bone sialoprotein (BSP) expression is restricted to bone and cementum. Competition ELISA of 4 M guanidine HCl extracts of fetal calf dentin has revealed small amounts (0.4%) of BSP2 protein (Fisher et al. 1983). Other studies failed to detect such miniscule levels in bovine dentin (Franzén and Heinegård 1985). Murine models of dentinogenesis have failed to detect BSP transcription in the rat tooth follicle until day 21 of gestation, particularly in odontoblasts adjacent to coronal dentin (Chen et al. 1992). Insights into BSP and its importance can be found in the following reviews (Ganss et al. 1999; Fisher et al. 2001).

4.3.7 *Type II TGF-β Receptor Interacting Protein-1 (TRIP1)*

TRIP1 is a member of a family of structurally conserved proteins, the WD-40 repeat proteins (Chen et al. 1995). The WD-40 proteins contain 4 or more copies of a conserved Trp-Asp motif, the so-called WD-40 repeat, which forms a scaffold for binding other proteins (Neer et al. 1994). TRIP-1 has been identified as a phosphorylation target of the TGFβR-II kinase during in vitro studies and as a functional component of eukaryotic translation initiator factor 3 (eiF3) multi-protein complexes (Hershey et al. 1996; Asano et al. 1997). WD40-repeat proteins are one of the largest protein families that often serve as platforms to assemble functional complexes and thus play vital roles in many biological processes (Neer et al. 1994). The name WD40 was derived from the conserved WD dipeptide and the conserved length of approximately 40 amino acids in a single repeat (Neer and Smith 1996). A typical WD40 domain consists of 6–8 structurally conserved WD40 repeats, each of which contains four-stranded anti-β-sheet which then folds into a β-propeller structure often comprising seven blades. These structures are stabilized by hydrophobic interactions.

Eukaryotic initiation factor subunit I (EIF3i) also called p36 or TRIP-1 is a part of the translation initiation complex and acts as a modulator of TGF-β signaling. It is a subunit of the eIF3 complex and is also termed eIF3i. eIF3 is the largest of the eIFs (Behlke et al. 1986) and plays an essential role in the initiation of eukaryotic translation. The presence of TRIP-1 from yeast to mammals indicates an evolutionarily conserved function in eukaryotes (Wrana et al. 1992) Recently, we have identified TRIP1 as a noncollagenous protein in the mineralized matrices of bone and dentin (Chen and George 2018). A signal peptide that is necessary to exit the cell is absent in TRIP1. However, expression was observed in the matrix and in the secretome of various cell lines.

The immunohistochemical analysis demonstrates the localization of TRIP-1 in the tooth at all stages of development (Ramachandran et al. 2012). During tooth development, reciprocal interactions and controlled programming take place between the oral epithelium and the neural-crest-derived mesenchyme. Signals transmitted by the epithelial cells promote condensation of the mesenchymal cells called the dental papilla. At the bud stage of tooth development, TRIP-1 was expressed in both the epithelial cells of the oral ectoderm and specifically in the dental papilla. With further development, the dental papilla differentiates to form the odontoblast cells that will initially produce the uncalcified predentin matrix and subsequently mineralized dentin, while the dental lamina differentiates to form the ameloblasts that will produce enamel. Expression of TRIP1 was observed in both the ameloblasts and odontoblasts until day three in the molars. At later developmental stages, TRIP-1 is expressed only by the odontoblasts. As the incisors are continuously erupting, expression of TRIP-1 was seen in both odontoblasts and ameloblasts. Undifferentiated pulp cells show lower expression levels of TRIP-1 during early stages and higher expression levels with development. These observations suggest that TRIP-1 might play a regulatory role during the differentiation of ameloblasts and odontoblasts and might participate in the epithelial–mesenchymal cross-talk during development. The expression of TRIP-1 in the dentin matrix is interesting. On day three, the odontoblasts are fully polarized and secrete type I collagen and several noncollagenous proteins necessary for matrix mineralization. During this process, TRIP-1 is localized at the mineralization front where the first crystals of calcium phosphate are deposited (Ramachandran et al. 2012). This suggests that TRIP-1 might play a role in matrix mineralization. This function was demonstrated by in vitro nucleation assay, which demonstrated that TRIP-1 could bind calcium and initiate the nucleation process.

To investigate the role of TRIP1 in biomineralization, demineralized dentin wafers which contained intact collagenous matrix were used to demonstrate their ability to nucleate calcium phosphate. Sparse mineral deposits were observed at 7 days within the collagen gap and overlap zones of the deproteinized and demineralized dentin wafer adsorbed with TRIP-1, while enhanced deposits were observed at the end of 14 days (Ramachandran et al. 2018). Interestingly, at low concentrations of TRIP1, self-assembly of the protein into fibrillar structures was clearly evident. With increasing TRIP-1 concentrations, nanosized calcium phosphate deposits were embedded within the protein meshwork. These particles coalesced to form agglomerates of HAP with higher concentrations of TRIP-1. TEM imaging showed the characteristic diffraction rings corresponding to the crystallographic planes of hydroxyapatite (Ramachandran et al. 2018). It is possible that the fibrillary supramolecular structure of TRIP1 may function as a nucleating template and higher amounts of TRIP-1 would bind more calcium phosphate nanoclusters and lower the interfacial energy for nucleation and promote hydroxyapatite formation.

Interaction of TRIP1 with Collagen The possibility of rTRIP1 to nucleate calcium phosphate suggested that TRIP1 could bind to type I collagen. Surface plasmon resonance analysis suggested that TRIP-1 bound to type I collagen in a dose-dependent manner with a K_D a measure the affinity between the two molecules of 48.5 μM and with the fast association and dissociation rates (Ramachandran et al. 2016). This indicated that TRIP1 binds with Type I collagen and this binding affinity is weaker than the binding of DMP1 and collagen. Immunogold labeling also suggested that TRIP1 could bind to monomeric collagen aggregates. These observations suggest that TRIP1 secreted to the matrix could bind both collagen and calcium ions to initiate matrix mineralization.

4.4 Glucose Regulatory Protein-78 (GRP78)

The alveolar bone and dentin require precise coordination of proteins and molecules to form a mineralized matrix. Dentin Matrix Protein-1 (DMP1), a SIBLING family protein, plays a major role in forming the matrix (Sun et al. 2011). Although DMP1 has been shown to function in the extracellular matrix of bone and dentin, the role of DMP1 inside the cell also has implications in osteoblast and odontoblast differentiation (Jacob et al. 2014). DMP1 is subject to posttranslational processing where it can function within the nucleus as a transcription factor due to its NLS signal, NES signal, and N-terminal domain (Narayanan et al. 2003). Its function as a transcription factor is still being explored, but DMP1 has been shown to activate the transcription of osteoblast- and odontoblast-specific genes like alkaline phosphatase and osteocalcin (Narayanan et al. 2003). When DMP1 was inactivated or repressed in embryonic mesenchymal stem cells, the levels of matrix mineralization and osteoblasts specific genes were decreased (Narayanan et al. 2003). The translocation of DMP1 from the extracellular matrix and to the nucleus has recently been described

(Merkel et al. 2019). Through mass spectrometry, DMP1 has been shown to interact with Glucose Regulated Protein-78 (GRP78) at the membrane (Ravindran et al. 2012). The coordinated efforts between DMP1 and GRP78 demonstrate that these two proteins are important and necessary in biomineralization.

Glucose-regulated protein-78 (GRP78/BiP) is a member of the heat shock family of proteins essential in the cell to help regulate protein folding and the unfolded protein response. Structurally, GRP78 is a 78 kilodalton protein that is subject to post-translational modifications. The protein is highly conserved with an N-terminal binding domain (NBD) and a C-terminal substrate-binding domain that are connected by a linker structure (SBD (Ni et al. 2011). GRP78 also has an ATPase function that controls the on and off state through conformational changes to allow substrate binding. Often referred to as the master regulator of the endoplasmic reticulum, GRP78 is one of the key chaperone proteins that enables proper folding of proteins that enter the cell for those proteins to reach their desired destinations (Pfaffenbach and Lee 2011). GRP78 is a multifunctional and dynamic protein that functions throughout the cell to ultimately maintain homeostasis.

Although GRP78 is known for its function within the endoplasmic reticulum lumen, the protein has functions outside of the ER, namely the cell surface. Under conditions of stress within the cell, GRP78 has the ability to relocate throughout the cell in order to act in different compacities. Cell surface GRP78 has been shown in many studies to function as a receptor and signaling molecule due to a KDEL motif that allows its cellular relocation upon stress signals (Tsai and Lee 2018). In recent studies, GRP78 on the cell surface promotes stemness or promoting stem cell self-renewal properties in breast cancer cells and pluripotent stem cells. Additionally, cell surface GRP78 has been found to promote pluripotency of stem cells in human and embryonic stem cells (Conner et al. 2020). Viruses like the Japanese Encephalitis virus, Coxsackie virus, and the newly emerged SARS-CoV-2, utilize cell surface GRP78 for entry into the host cell (Allam et al. 2020; Ni et al. 2011). The cell surface function of GRP78 suggests its importance in molecular signaling, endocytosis, and stem cell maintenance.

Another main function of GRP78 is to be the main signaling molecule in the Unfolded Protein Response. The Unfolded Protein Response (UPR) is a mechanism for the cell to respond to stress conditions and reduce the changes of cell apoptosis (Hetz 2012). Conditions of excess proteins, proteins improperly folded, and excess calcium are examples of conditions that activate the UPR (Bahar et al. 2016). Under cell and endoplasmic reticulum stress, GRP78 changes conformation to release three cascades of signaling molecules to help return to a homeostatic state. The three arms of the Unfolded Protein Response include the PERK (protein kinase RNA-like endoplasmic reticulum kinase), ATF6 (Activating Transcription Factor 4), and IRE1 (inositol requiring enzyme 1) cascades. With the activation of the UPR, the dissociation of GRP78 from PERK results in the auto-phosphorylation of the protein to thus phosphorylate eIF2 leading to cell cycle arrest or transcription of cytoprotective genes. In the ATF6 pathway, the transcription factor is cleaved in the Golgi apparatus for transport to the nucleus to allow transcription of cytoprotective genes, chaperones, and stress response genes. Lastly, the IRE-1 pathway works by splicing the transcription factor XBP1 to allow for the transcription and translation of ER stress response genes as well as ER genes associated with degradation (Walter and Ron 2011). Although all three of these pathways lead to the transcription and translation of genes to help restore homeostasis, another function of the UPR stress cascades is to provide outlets for apoptosis if cellular stress and damage exceed restorability. The PERK and IRE-1 pathways allow for apoptosis to occur through the CHOP pathway leading to cell death through BCL2 (Hu et al. 2019). The UPR balance provides a check and balances system for the cell to maintain homeostasis during cellular stress.

The process of mineralization is necessary yet requires numerous proteins and calcium that can activate the UPR stress response (Bahar et al. 2016). As previously mentioned, DMP1 functions in the nucleus and the influx of calcium into the nucleus from the cytoplasm. This triggers the phosphorylation of DMP1 and allows DMP1 to translocate from the extracellular matrix into the cell (Narayanan et al. 2003). Calcium signaling is a key player in the activation of the ER stress response and the UPR in the cell. The excess of calcium in the cell can result in GRP78 activation and translocation to the membrane (Bahar et al. 2016). Previous papers have described the interactions of GRP78 and DMP1 at the plasma membrane through mass spectrometry and immunolocalization in both stem cells, specifically T4-4cells, and wild-type mouse models (Ravindran et al. 2008). Additionally, membrane GRP78 has been isolated after cell treatment with DMP1 through a pull-down assay and Western Blotting (Merkel et al. 2019). Entry into cells can result through phagocytosis, endocytosis, or pinocytosis. Clathrin-mediated and caveolin-mediated endocytosis are two of the main pathways that proteins and extracellular material can enter the cell (Nabi and Le 2003). Inhibition of the clathrin pathway by the inhibitor Pitstop demonstrated no difference in DMP1 levels intracellularly (Dutta et al. 2012). However, inhibition of the caveolin pathway by β cyclodextrin has shown a blockage of DMP1 levels intracellularly. Intracellular trafficking within the cell is routed by Rab GTPases, which are proteins of the Ras superfamily that regulate membrane trafficking, subcellular vesicle trafficking, and vesicle coordination within the cell (Stenmark 2009). Rab5 and EEA1 proteins are associated with the early endosome after endocytosis. An increase in Rab5 proteins occurs with an increase in caveolin-1 proteins, the caveolae proteins, suggesting these two proteins work together during caveolin-mediated endocytosis (Hagiwara et al. 2009). Studies have shown a DMP1 associated with Rab5 and EEA1 within human periodontal ligament stem cells and wild-type mouse immunohistochemistry (Fig. 4.2) (Merkel et al. 2019). Next, Rab7, a late endosomal marker, coordinates with Rab5 to transfer DMP1 from the early to the late endosome. DMP1 can enter the nucleus after late endosomal vesicle trafficking due to its NLS signal. Lastly, Rab11, a recycling vesicle marker, has been shown to transport DMP1 back from the nucleus to the plasma membrane from which the protein can function in the extracellular matrix (Merkel et al. 2019).

Overall, the coordination of DMP1 and GRP78 depends on the activation of the stress response due to the excess calcium and protein folding occurring during osteoblast and odontoblast differentiation. Ultimately, these two proteins work

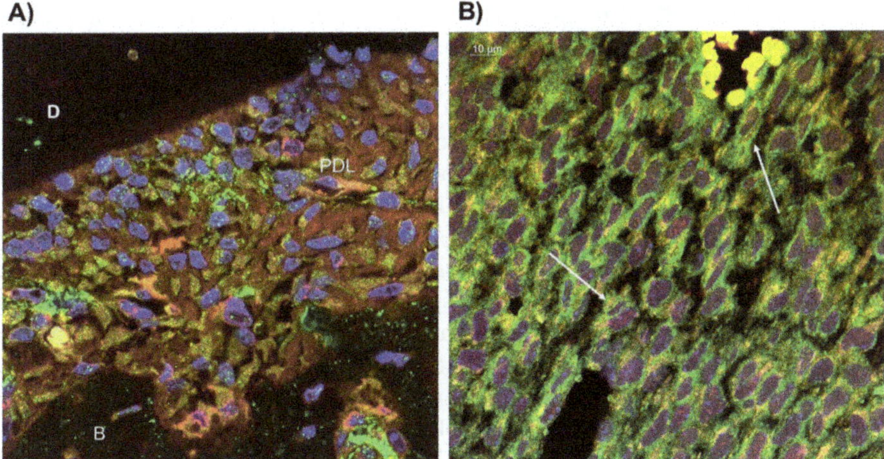

Fig. 4.2 Interaction of GRP78 with DMP1 and Rab5: (**a**) Immunolocalization of GRP78 (FITC) and DMP1 (TRITC) and DAPI in one month mouse mandible sections: Colocalization is seen as yellow. *D* dentin, *PDL* periodontal ligament, *B* bone. Bar represents 20 μm. (**b**) Localization of Rab5 (FITC), GRP78 (TRITC), and DAPI in 2 month DMP1-overexpressing mouse pulp cells. The colocalization of the two proteins is indicated as yellow. Arrows show co-localization along the membrane of 2 proteins. Bar represents 10 μm

synergistically to further aid in the mineralization of bone and dentin. GRP78 has been isolated from the secretome of primary mouse calvarial cells suggesting a role not only in the cell with DMP1 but also in the extracellular matrix. Additionally, in vitro nucleation studies have shown calcium phosphate formation from GRP78 protein on type 1 collagen matrices (Ravindran et al. 2012). DMP1 and other SIBLING proteins have been targets in research for tissue and bone engineering; however, literature has not shown efforts to use GRP78 in tissue engineering. GRP78 stands to be a multifunctional and important protein in our cells that goes beyond modulating the ER response. Harnessing and understanding this relationship between the stress response protein, GRP78, and the extracellular matrix protein, DMP1, can lead to novel mechanisms of tissue engineering and an enhancement in biomineralization of the bone and dentin.

4.5 Growth Factors and Cytokines

Odontoblast differentiation and formation of the dentin matrix is regulated by signaling factors, and in this review, our focus is on bone morphogenetic proteins (BMPs) and transforming growth factor-b (TGFβ). The TGF beta superfamily of growth factors are present in relative abundance in a mineralized matrix and stimulate proliferation and differentiation of cell types associated with mineralized tissue formation both in vitro and in vivo.

Bone Morphogenetic Proteins (BMPs) In the 1960s, Marshall Urist, an orthopedic surgeon identified that protein complex in the demineralized rabbit bone able to act as osteoinductive in vitro and in vivo and named this factor as bone morphogenetic protein (BMP) (Nogami and Urist 1970; Urist and Peltier 2002; Urist and Strates 1971). BMPs family comprises more than 30 members. Molecular cloning of these members identified them as members of the TGFβ superfamily (Table 4.1). Though the secretion of BMPs was initially identified in osteoprogenitor cells, osteoblasts, chondrocytes, and platelets (Pećina and Vukičević 2007; Sipe et al. 2004), later it was discovered that this family of proteins is not restricted to mineralized tissues but they play an essential role in various functions such as proliferation, differentiation, apoptosis, morphogenesis, patterning of various organs and organogenesis (Wozney 1998; Ebendal et al. 1998; Graff 1997). Interestingly, several of the BMPs do not induce de novo bone formation in classic subcutaneous implantation assay in animal models (Wolfman et al. 1997). For example, BMP1 is a metalloprotease capable of cleaving the C-terminus polypeptide of procollagen I, II, and III and several extracellular matrix proteins (Kessler et al. 1996). Also, BMP3 and BMP13 act as an inhibitor of bone mineralization (Daluiski et al. 2001). It was in the late 1980s that the first BMPs were cloned and characterized. Upon cloning human BMPs in 1988, the importance of BMPs especially BMP2 and BMP7 were demonstrated in orthopedic applications for bone repair enhancement. BMPs have similarities at the amino acid level among different species. They dimerize with the help of conserved cysteine residues called cysteine knots and by intramolecular disulfide bonds. The formation of cysteine–cysteine bridge stabilizes the dimer formation. However, some members of the family do not form this classical covalent dimerization due to the lack of required cysteine residue (Liao et al. 2003). BMPs are made as inactive precursor proteins with export signal peptides in the cell. The export signal peptide in the monomer protein is essential for the proper folding and export process associated with the secretory pathway (Miyazono et al. 1988). Upon dimerization, the pro-domain is cleaved by proteolytic enzymes (proprotein convertase family) at a consensus site, Arg-X-X-Arg, to produce active and mature homodimers or heterodimers (Constam and Robertson 1999). The BMPs are grouped into subfamilies based on the sequence homology, and ability to form either homodimers or heterodimers. The class I BMPs, which include BMP2 and BMP4 can heterodimerize with class II BMPs, which include BMPs 5–8. Heterodimers with class I and class II BMPs have higher specific activity than the homodimers. For example, BMP2/7 or BMP4/7 heterodimers are significantly more potent than homodimers in osteogenic induction (Kaito et al. 2018; Aono et al. 1995). Besides osteogenic induction, heterodimers are also more potent in the activation of BMP pathways leading to stem cell differentiation or fate determination in xenopus and zebrafish (Schmid et al. 2000; Nishimatsu and Thomsen 1998). The choice of a homodimer or heterodimer formation is regulated by the expression patterns. For example, BMP4 and BMP7 preferentially form heterodimers rather than either homodimer when ectopically coexpressed in *Xenopus* embryos (Neugebauer et al. 2015) and also the expression pattern of BMP4 and BMP7 overlap in many tissues of developing mouse (Danesh et al. 2009), suggesting that they may form heterodimers.

Table 4.1 Characteristics of different BMP family members. Adapted from Carreira et al. (Archives of Biochemistry and Biophysics 561 (2014) 64–73)

BMP	Function	Expression	Human Chromosome	Ref (PMID)	Receptors
BMP2 (BMP2A, BDA)	Induces bone and cartilage formation	Intestine; kidney; larynx; connective tissue; spleen; prostate; amniotic fluid; eye; heart; ganglia; cervix; placenta; lung; pancreas; mixed; brain; stomach; embryonic tissue; uncharacterized tissue; liver; bone; uterus; testis; mammary gland	20p12	3201241	ALK-2, 3, 6 BMPR-II; ActR-IIA, ActR-IIB
BMP3A (Osteogenin, BMP3)	Negatively regulates bone density; antagonizes the ability of certain osteogenic BMPs to induce osteoprogenitor differentiation and ossification	Bone marrow; eye; uncharacterized tissue; intestine; lung; pharynx; muscle; mixed; embryonic tissue; prostate; blood	4q21	3201241	AlK4, ActR-IIA, ActR-IIB
BMP4 (BMP2B, BMP2B, MCOPS6, OFC11, ZYME)	Induces cartilage and bone formation; acts in mesoderm induction, tooth development, limb formation, and fracture repair; acts in concert with PTHLH/PTHRP to stimulate ductal outgrowth during embryonic mammary development and to inhibit hair follicle induction. (by similarity)	Intestine; embryonic tissue; placenta; liver; stomach; brain; bone; mixed; prostate; vascular; eye; uncharacterized tissue; testis; mouth; mammary gland; lung; pancreas; ovary; heart; skin; connective tissue; lymph node; lymph; adipose tissue; salivary gland; kidney	14q22–q23	3201241	ALK-2,3,5,6 BMPRII, ActR-IIA
BMP5	Induces bone and cartilage formation	Placenta; heart; thymus; uncharacterized tissue; uterus; testis; connective tissue; ovary; mammary gland; embryonic tissue; lung; mixed; pancreas; liver; brain; muscle; trachea; prostate; eye; intestine	6p12.1	2263636	ALK-3 BMPRII; ActR-IIA, ActR-IIB
BMP6 (VGR, VGR-1)	Induces cartilage and bone formation; proposed role in early development	Vascular; eye; embryonic tissue; intestine; blood; uncharacterized	6p24–p23	2263636	

(continued)

Table 4.1 (continued)

BMP	Function	Expression	Human Chromosome	Ref (PMID)	Receptors
		tissue; placenta; brain; ovary; lung; mixed; prostate; uterus; testis; umbilical cord; larynx; cervix; kidney; mammary gland; connective tissue; mouth; parathyroid			ALK-2,3,6 BMPR-II; ActR-IIA, ActR-IIB
BMP7 (OP-1)	Induces cartilage and bone formation; may be the osteoinductive factor responsible for the phenomenon of epithelial osteogenesis; plays a role in calcium regulation and bone homeostasis	Brain; uncharacterized tissue; placenta; eye; testis; mixed; kidney; prostate; muscle; heart; embryonic tissue; intestine; mammary gland; stomach; ovary; mouth; thymus; trachea; connective tissue; uterus; skin; tonsil; lymph; lymph node; lung; ganglia; pancreas; bone	20q13	2263636	ALK 2,3,6 BMPR-II
BMP8A (OP-2)	Induces cartilage and bone formation; may be the osteoinductive factor responsible for the phenomenon of epithelial osteogenesis; plays a role in calcium regulation and bone homeostasis. (by similarity)	Skin; thyroid; mammary gland; brain; testis; uncharacterized tissue; mixed; intestine; thymus; larynx; placenta; connective tissue; bone; spleen; embryonic tissue; stomach; prostate	1p34.3	1460021	ALK 2, 3, 4,6,7; BMPRII
BMP8B (BMP8)	Induces cartilage and bone formation; may be the osteoinductive factor responsible for the phenomenon of epithelial osteogenesis; plays a role in calcium regulation and bone homeostasis (by similarity)	Skin; testis; liver; brain; mixed; ascites; uncharacterized tissue; intestine; pancreas; muscle; uterus; ovary; bone; bone marrow; spinal cord; eye; embryonic tissue; ganglia; prostate	1p35–p32	12477932	ALK3–6; BMPR-II; ActR-IIA, ActR-IIB

BMP9 (GDF2)	Potent circulating inhibitor of angiogenesis; could be involved in bone formation; signals through the type I activin receptor ACVRL1 but not other Alks	Liver	10q11.22	10849432	ALK-1,2; BMPR-II; ActR-IIA, ActR-IIB
BMP10	Required for maintaining the proliferative activity of embryonic cardiomyocytes by preventing premature activation of the negative cell cycle regulator CDKN1C/p57KIP and maintaining the required expression levels of cardiogenic factors such as MEF2C and NKX2–5; acts as a ligand for ACVRL1/ALK1, BMPR1A/ALK3 and BMPR1B/ALK6, leading to activation of SMAD1, SMAD5, and SMAD8 transcription factors; inhibits endothelial cell migration and growth	Ascites	2p13.3	10072785	ALK-1,3,6; ActR-IIA, ActR-IIB
BMP11 (GDF11)	Secreted signal that acts globally to specify positional identity along the anterior/posterior axis during development; plays critical roles in patterning both mesodermal and neural tissues and in establishing the skeletal pattern	Eye; brain; uncharacterized tissue; mixed; skin; mammary gland; mouth; intestine; lung; uterus; amniotic fluid; muscle; blood; testis; liver; pancreas; tonsil; kidney; placenta; prostate; connective tissue; parathyroid; nerve; bone; embryonic tissue; thymus; umbilical cord; heart; adrenal gland; lymph node	12q13.2	10075854	ALK-3,4,5,7; BMPR-II, ActR-IIA, ActR-IIB
BMP12 (GDF7)	May play an active role in the motor area of the primate neocortex (by similarity)	Kidney; brain; testis; uncharacterized tissue	2p24.1	8145850	ALK-3,6; BMPR-II, ActR-IIA

(continued)

Table 4.1 (continued)

BMP	Function	Expression	Human Chromosome	Ref (PMID)	Receptors
BMP13 (GDF6, CDPM2, MCOP4)	Required for normal formation of bones and joints in the limbs, skull, and axial skeleton. Plays a key role in establishing boundaries between skeletal elements during development (by similarity)	Embryonic tissue; mixed; testis; bone; brain; placenta	8q22.1	8145850	ALK-3,6; BMPR-II, ActR-IIA, ActR-IIB
BMP14 (GDF5, CDPM1, LAP4OS5, SYNS2)	Could be involved in bone and cartilage formation; chondrogenic signaling is mediated by the high-affinity receptor BMPR1B	Eye; embryonic tissue; lung; heart; bone; pituitary gland; brain; salivary gland; mixed; connective tissue; prostate; skin; uterus	20q11.2	8145850	ALK-3,6; BMPR-II, ActR-IIA
BMP15 (GDF9B, ODG2, POF4)	May be involved in follicular development; oocyte-specific growth/differentiation factor that stimulates folliculogenesis and granulosa cell (GC) growth. Essential for mesoderm formation and axial patterning during embryonic development by similarity	Ovary	Xp11.2	9849956	ALK-6
BMP16 (Nodal)	Essential for mesoderm formation and axial patterning during embryonic development by similarity	Embryonic tissue; testis	10q22.1	US patent no. 596503	–
BMP17 (Lefty1, leftyB)	Required for left-right axis determination as a regulator of LEFTY2 and NODAL	Embryonic tissue; joint; brain; mammary gland; lymph node; mixed; uncharacterized tissue; spleen; pancreas; testis; bladder; intestine; lung; liver; skin	1q42.1	US patent no. 6027917	–
BMP18 (Lefty2, lefty A)	Required for left-right (L-R) asymmetry determination of organ systems in mammals	Embryonic tissue; testis; mixed; brain	1q42.1	US patent no. 6027917	–

GDF1 (DORV, DTGA3	May mediate cell differentiation events during embryonic development	Brain	19p12	1704486	–
GDF3 (KFS3, MCOP7, MCOPCB6)	Negatively and positively control differentiation of embryonic stem cells; role in mesoderm and definitive endoderm formation during the pre-gastrulation stages of development	Uncharacterized tissue; testis; embryonic tissue; kidney; epididymis; mixed	12p13.1	8429021	ALK4,7
GDF8 (MSTN, myostatin)	Acts specifically as a negative regulator of skeletal muscle growth	Heart; brain;connective tissue; lung; eye; embryonic tissue; muscle; placenta	2q32.2	1997	ActRIIB; ALK3, 4
GDF9	Required for ovarian folliculogenesis; promotes primordial follicle development. Stimulates granulosa cell proliferation; promotes cell transition from G0/G1 to S and G2/M phases	Testis; brain; kidney; mammary gland; liver; adrenal gland; ovary	5q31.1	8429021	BMPRII; ALK5
GDF15	May be involved in follicular development; oocyte-specific growth/differentiation factor that stimulates folliculogenesis and granulosa cell (GC) growth	Placenta; skin; mixed; lung; intestine; kidney; prostate; liver; pancreas; muscle; uncharacterized tissue; stomach; connective tissue; uterus; trachea; eye; brain; ascites; embryonic tissue; amniotic fluid; blood; testis; mammary gland; ovary; lymph node; lymph; vascular; umbilical cord; heart; cervix; spleen; bladder; bone; parathyroid	19p13.11	9139826	GDNF family receptor α-like (GFRAL

Various BMPs, their function, tissue specific expression patterns, chromosome location and receptors is presented in Table 4.1.

The regulatory effects of BMPs depend on various parameters. The efficacy of the BMPs depends on the formation of dimers, posttranslational modifications (N- and O-glycosylation), the target cell type, differentiation stage of the target cell, concentrations of locally secreted BMPs, and the combinatorial effect of these factors with other growth factors in a complex multi-cell-type signaling system. List of ECM associated proteins that interact with BMPs and control downstream signaling pathway is presented in Table 4.2.

Table 4.2 Matrix-associated proteins and their interaction with BMP family members (Adapted from Development 136, 3715–3728 (2009) doi:https://doi.org/10.1242/dev.031534)

ECM component	Effect on signaling	Mechanism
Heparan sulfate proteoglycans (HSCPGs)		
Glypicans/Dally/Dlp/ LON-2, Syndecans	Inhibit/ Promote	Binds BMPs; Increases or decreases BMP movement
Chondroitin sulfate small leucine-rich proteins		
Biglycan	Inhibit/may promote?	Binds BMP4 and Chordin; enhances formation of BMP4- Chordin complex
Tsukushi	Inhibit/may promote?	Binds BMP4/7 and Chordin; forms BMP-Tsukushi-Chordin complex
T βRIII (Betaglycan)	Promote	Binds BMP2/4/7. Co-receptor?
Membrane-bound CR-containing proteins		
Crim1/Crm1	Inhibit/ promote	Binds BMPs, can block BMP processing and secretion
Procollagen-IIA	Inhibit	Strongly binds BMP2
CR-containing proteins with indirect membrane association		
Cv2 (BMPER)	Inhibit/ promote	Binds BMP2/4/7, HSPGs, Chordin/Sog, Vertebrate Tsg, type I Receptor
KCP/Kielin	Inhibit/ Promote	Binds BMP7, Activin A, TGFβ1
Other co-receptors and pseudoreceptors		
Dagon/RGMs	Inhibit?/ Promote	Binds type I and II receptors, BMP2/4
BAMBI	Inhibit	Binds type I receptor and blocks formation of the receptor complex
Bone or enamel extract-derived proteins		
GLA	Inhibit	Binds BMP2/4
SPP2	Inhibit	Unknown
Dermatopontin	Inhibit	Unknown
Ahsg	Inhibit	Binds BMP and blocks signalling
Amelogenin	Promote	Binds heparan sulphate and BMP2
Other matrix proteins		
Collagen IV	Inhibit/ Promote	Binds sog (Chordin) and Dpp (BMP)
Fibrillin	May Promote?	Binds BMP/Prodoman complexes

The action of specific BMP is exerted via hetero-tetrameric serine/threonine kinase receptors. There are two specific receptor subunits (type I/BMPR-I, and type II/BMPR-II) that bind to any specific BMPs. These receptors are transmembrane proteins with an N-terminal extracellular ligand-binding domain and intracellular region. The type I receptors (BMPR-I) include type IA (ALK3, BRK1) and type 1B (TSK7L/ALK2, BRKII, RPK1), while type II receptors (BMPR-II) include ActRI (Activin Receptor Type-I), ActRII, ActRIIB, and T-ALK. BMPs have the ability to bind to either type I or type II receptors independently; however, high-affinity binding of BMP ligand and efficient signaling is noted when both type I and type II receptors are involved. In the absence of BMP ligands, type I and type II receptors are at different localizations on the plasma membrane. However, in the presence of BMPs, type I and type II receptors get associated with BMPs as a bridge between the two receptors and get activated. The mechanism by which BMPs form the hetero-tetrameric complex varies. For example, BMP6 and BMP7 initiate interaction with BMPR-II and recruit BMPR-I. On the other hand, BMP2 and BMP4 bind to BMPR-I followed by BMPR-II recruitment (De Caestecker 2004). This layer of complexity further specifies the downstream signaling pathway selection (Nohe et al. 2002). Additionally, the type II receptor is constitutively phosphorylated and upon binding to the BMPs, the type I receptor is recruited followed by phosphorylation at glycine–serine-rich motif known as the GS domain. The phosphorylation of type I receptors and its kinase activity (activin receptor-like kinases, ALKs) further activates the cascade of events with the initiation of SMAD phosphorylation.

SMADs serve as signal transducers for various TGFβ super family ligands. The SMADs are sub-grouped into receptor SMADs (R-SMADs), common SMADs (Co-SMADs), and inhibitory SMADs (i-SMADs). There are eight SMAD proteins encoded in the mammalian genomes, of which SMAD1, SMAD2, SMAD3, SMAD5, and SMAD8 act as substrates for the TGFβ family of receptors and are commonly referred to as receptor-SMADs (R-SMADs). Specifically, SMADs 1, 5, and 8 serve as substrates for BMPRs, and SMADs 2 and 3 act as the substrate for the TGFβ, activin and Nodal receptors. Besides, there is also SMAD4, referred to as Co-SMAD which acts as a common partner for all R-SMADs. The binding of Co-SMAD to R-SMAD leads to the translocation of the complex to the nucleus to regulate gene expression of the target genes. In addition, SMAD6 and SMAD7 act as inhibitor SMADs that serve as decoys for competing with SMAD-receptor or SMAD-SMAD interactions.

Upon signal stimulation, SMAD complexes translocate to the nucleus, gets accumulated and retained for hours (ten Dijke and Hill 2004). In the nucleus, R-SMADs undergo dephosphorylation leading to disassociation from Co-SMADs and exported from the nucleus and undergoes recycling. The nuclear import of SMADs does not depend on the classical importin-dependent transport (Xu et al. 2000, 2002). Besides nucleocytoplasmic shuttling of SMADs, the duration of SMAD signaling pathway activation depends on the mode of receptor internalization. For example, clathrin-mediated endocytosis of receptors promotes SMAD signaling while lipid-raft mediated endocytosis of receptor leads to degradation via ubiquitin pathway (Di Guglielmo et al. 2003). Thus, the internalization of BMPR plays a critical role in the SMAD signaling pathway.

The commonality in the SMAD activation pathway for various TGFβ ligands raises an essential point on the specificity of the TGFβ superfamily ligands. It is noted that the expression pattern of accessory proteins involved in the SMAD pathway induces ligand-specific downstream effect. The availability of receptor subunits and the formation of hetero-tetrameric complex, in combination with accessory proteins, add a layer of specificity in the SMAD signaling pathway. For example, activin receptor type II combines with ALK4 to induce activin-related signaling pathway, while it combines with ALK3 or ALK6 to mediate BMP4 downstream signals (Derynck and Zhang 2003; Shi and Massagué 2003). In addition to these levels of BMP pathway regulation, co-secretion of antagonists of BMPs to the extracellular environment regulates the BMP pathway signals. The BMP antagonists are known to bind to BMP ligands and prevent them from binding to their receptors. Noggin is an antagonist to a variety of BMPs with varying affinity. Noggin is a glycosylated protein, which upon secretion forms homodimers. Structurally noggin has a cystenine-rich domain (cystenine knot) and a heparin-binding region that helps in the cellular surface attachment. The expression control of noggin by BMP2 during osteoblast differentiation indicates that it potentially acts as a security system to prevent overload of BMP2 signaling. Similarly in chondrocytes, the Indian hedgehog (IHH) controls the expression of noggin suggesting a tight control of noggin in the differentiation/proliferation of chondrocytes. Further, overexpression of noggin in mesenchymal cells inhibits osteoblast differentiation. Animal models lacking functional noggin are lethal with various developmental abnormalities (Balemans and Van Hul 2002). Several other BMP antagonists were characterized which include chordin, Gremlin, Sclerostin, Dan, USAG1, and others in mineralized tissues (Calvo et al. 2009).

4.6 Transforming Growth Factor Beta (TGF-β)

In the early 1980s, it is well known that many proteins; specifically secreted ones and hormones control cell growth. Some of these secreted factors cause malignant transformation of cells (Todaro et al. 1980). One such factor identified is called SGF (Sarcoma Growth Factor) discovered in cultures of transformed rat kidney fibroblasts (de Larco and Todaro 1978). Later it became evident that this factor is a mixture of at least two substances with different functions. They were called Transforming Growth Factor-α (TGF-α) and Transforming Growth Factor-β (TGF-β) (Roberts et al. 1981). TGF-β was further described by Roberts and Sporn as a secreted polypeptide capable of inducing fibroblast growth and collagen production (Roberts et al. 1985). Transforming growth factor-beta 1 or TGF-β1 is a secreted protein of the transforming growth factor superfamily that controls many cellular functions, such as growth, proliferation, differentiation, and apoptosis. There are three isoforms known in mammals as TGF-β1, TGF-β2, and TGF-β3; however,

TGF-β1 is the predominant form expressed. The chromosomal location of the gene associated with each isoform is distinct. In humans, the TGF-β1 gene is located at 19q13; the TGF-β2 gene at 1q41; and the TGF-β3 gene at 14q23-4.

TGF-β1, the most predominant expressed isoform, was cloned from human placenta mRNA (Derynck et al. 1985). In developing mice, TGF-β1 mRNA and/or protein have been localized in cartilage, endochondral and membrane bone and skin, indicating a role in the growth and differentiation of cells in these tissues (Dickinson et al. 1990).

TGF-β2 was first reported in human glioblastoma cells as a suppressor for the interleukin-2-dependent growth of T lymphocytes. Hence named as glioblastoma-derived T cell suppressor factor (G-TsF). Physiologically, TGF-β2 is expressed by neurons and astroglial cells in the embryonic nervous system (Flanders et al. 1991). The mature form ofTGF-β2, which consist of the C-terminal 112 amino acids, share 71% sequence similarity with TGF-β1 (Ten Dijke et al. 1988).

The third isoform, TGF-β3, was isolated from a cDNA library of human rhabdomyosarcoma cell lines sharing 80% of amino acid sequence homology with TGF-β1 and TGF-β2. Studies on mice demonstrated the essential function of Tgf-β3 in normal palate and lung morphogenesis and implicate this cytokine in epithelial–mesenchymal interaction (Proetzel et al. 1995; Kaartinen et al. 1995). Its mRNA is present in lung adenocarcinoma and kidney carcinoma cell lines; interestingly, umbilical cord expresses very high level of TGF-β3 (Ten Dijke et al. 1988).

The genes associated with TGF-β isoforms encode 390–412 amino acids-long precursor polypeptides containing three distinct regions, namely an N-terminal signal domain which targets the precursor protein to its cellular secretory pathways; a propeptide domain, which helps in the dimerization of the mature cytokine; and a C-terminal "TGF-beta-like" domain of about 100–114 amino acids that is conserved across the superfamily. Upon cleavage in the Golgi apparatus, the mature C-terminal protein fragment is secreted as a homodimer, which has two identical subunits linked by four internal disulfide bonds and a single cross-linking disulfide bond. Structurally, each monomer of the dimer pair is comprised of several β strands. TGF-β synthesized as a precursor with a signal peptide, and the precursor is processed in the Golgi by a furin-like peptidase that removes the N terminus of the immature protein. The TGF-β precursor forms an inactive complex composed of a TGF-β dimer in association with the latency-associated protein (LAP). The complex is secreted to the extracellular compartment and form a complex with latent-TGF-β-binding protein (LTBP) that mediates its deposition to the extracellular matrix (ECM). TGF-β becomes activated after binding of a TGF-β activator (TA) that induces LAP degradation or alters LAP's conformation. Active TGF-β binds to a hetero-tetrameric complex composed of TGF-β receptor II (TGF-βRII) and TGF-β receptor I (TGF-βRI) and initiates signaling pathways associated with the kinase activity of the receptors. The canonical TGF-b pathway is initiated by the binding of ligand to hetero-tetrameric complexes of type I and type II receptors activating their serine-threonine kinase. The activated receptor further phosphorylates R-SMADs (SMAD2 and 3). Followed by endocytosis, R-SMADs form complex with Co-SMADs, gets translocated to the nucleus, and regulates several target genes.

4.6.1 Extracellular Matrix Regulation of BMPs and TGFβ Signaling

Several ECM components are capable of binding directly to signaling molecules and thereby modulating their ability to successfully control signaling in various ways (Umulis et al. 2009). The interaction of BMPs with ECM molecules depends on their distinct biochemical properties. Both TGFβ and BMPs are regulated tightly at the transcription level. In spite of this, an additional layer of their downstream signaling exists in several systems. Intracellularly, BMP signaling is negatively regulated by the inhibitory SMADs (SMAD6 and SMAD7) or by the SMURF family of ubiquitin ligases. On the other hand, extracellularly, a variety of cell-secreted antagonists bind to distinct BMPs with varying affinities (Mulloy and Rider 2015). The activity, bioavailability, and diffusion of extracellular BMP signaling agonists and antagonists are further regulated by extracellular matrix (ECM) components.

The ECM and milieu are the noncellular compartments that surround cells. It comprises a variety of proteins and polysaccharides that are unique for a specific tissue. The unique composition of ECM molecules enables the binding of cell-type-specific binding of growth factors, receptors, and co-receptors through their biochemical properties (Schultz and Wysocki 2009). This enables the availability of these proteins to the cells at the time of requirement. In addition, the interactions of non-collagenous proteins with ECM components can play an instructive role by controlling the temporal, spatial, and kinetics of non-collagenous proteins (Ramirez et al. 2007). Besides, the ECM is a structural determinant with defined physical properties. The characteristics of ECM such as stiffness, elasticity, and porosity are directly related to the molecular composition of the ECM and modulates the diffusion of macromolecules (Kihara et al. 2013). Cells are capable of sensing these physical properties and reacting with distinct cellular responses via mechanosignaling pathways (Wang et al. 1993).

Many ECM components are proteoglycans (PGs), proteins that are covalently linked to glycosaminoglycan (GAG) chains such as heparin, heparan sulfate, chondroitin sulfate (CS), dermatan sulfate (DS), and keratan sulfate. BMPs bind to negatively charged heparins via the N-terminal region containing basic amino acids (Ohkawara et al. 2002; Ruppert et al. 1996). Further, the degree of sulfation has an impact on the strength of the electrostatic interactions between heparan sulfate proteoglycans and BMPs (Gandhi and Mancera 2012). However, the binding of BMPs to HSPGs does not entirely depend on sulfated GAGs (Kirkpatrick et al. 2006). There are several BMP antagonists that bind to ECM. Small leucine-rich proteoglycans (SLRP) are a family of ECM proteoglycans that consist of a protein core made up of leucine-rich repeats to which GAG chains of either CS or DS are attached [24]. The members of the SLRP family can regulate BMP signaling extracellularly and all of them show inhibitory functions on BMP signaling. For example, decorin was demonstrated to inhibit BMP7-mediated signaling in nephron progenitor cell differentiation, while it inhibits BMP2 activity in osteoblast differentiation (Kirkpatrick et al. 2006, Fetting et al. 2014). Thus, cell-specific inhibition

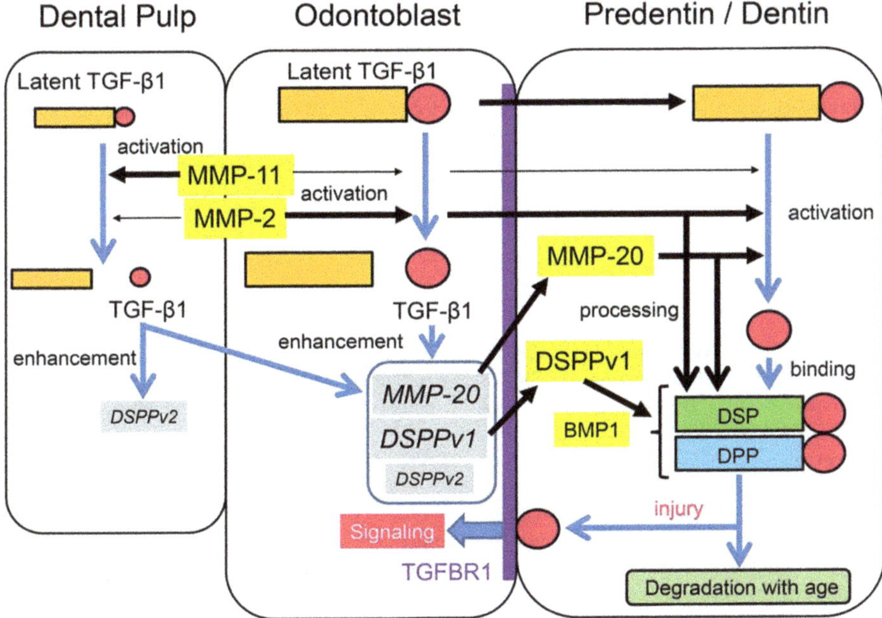

Fig. 4.3 Schematic diagram showing the processing and activation of TGFβ. Latent TGF-β1 synthesized in dental pulp is primarily activated by MMP11 and MMP2. In odontoblasts, latent TGF-β1 is mainly activated by MMP2. Activated TGF-β1 increased expression of MMP20 and DSPP, leading to odontoblast differentiation. Latent TGF-β1 in the dentin matrix is activated by MMP2, MMP20, and MMP11. In dentin, active TGF-β1 binds to DSPP and released during matrix degradation by BMP1, MMP2, and MMP20

of specific BMP pathways is achieved via the release of deposited antagonists from the ECM (Fig. 4.3).

4.7 Conclusions

Mineralized matrix formation is a cellular event and is orchestrated by the synthesis and secretion of several proteins. The identification of genes regulating matrix mineralization, followed by the analysis of their functions and their complex interactions with each other and other binding partners is therefore of fundamental interest for understanding the process of biomineralization. The remarkable assembly of dentin is attributed not only to the two predominant components namely, mineral and collagen but also to the numerous non-collagenous proteins including growth factors and cytokines. NCPs exhibit multifunctional roles as depicted in this review and alterations in their expression levels can have detrimental effects on both cell differentiation and matrix mineralization. The mechanistic role of NCPs is another interesting facet and is an important contribution to the microstructure of the mineralized tissues.

Acknowledgements We are pleased to acknowledge the support from the National Institutes of Health. A.G has been supported by Grants DE011657, DE 028531and the Brodie Endowment Fund; A.M has been supported by NIDCR F30 DE027601: E.G by NIDCR F30 DE028193.

References

Allam L, Ghrifi F, Mohammed H, El Hafidi N, El Jaoudi R, El Harti J, Lmimouni B, Belyamani L, Ibrahimi A (2020) Targeting the GRP78-Dependant SARS-CoV-2 cell entry by peptides and small molecules. Bioinformatics and Biology Insights 14:117793222096550. https://doi.org/10.1177/1177932220965505

Alvares K, Veis A (2012) Commentary on "biomineralization--an active or passive process?". Connect Tissue Res 53:437

Angker L, Nockolds C, Swain MV, Kilpatrick N (2004) Correlating the mechanical properties to the mineral content of carious dentine--a comparative study using an ultra-micro indentation system (UMIS) and SEM-BSE signals. Arch Oral Biol 49(5):369–378. https://doi.org/10.1016/j.archoralbio.2003.12.005

Aono A, Hazama M, Notoya K, Taketomi S, Yamasaki H, Tsukuda R, Sasaki S, Fujisawa Y (1995) Potent ectopic bone-inducing activity of bone morphogenetic protein-4/7 heterodimer. Biochem Biophys Res Commun 210:670–677

Asano K, Kinzy TG, Merrick WC, Hershey JW (1997) Conservation and diversity of eukaryotic translation initiation factor eIF3. J Biol Chem 272:1101–1109

Bahar E, Kim H, Yoon H (2016) ER stress-mediated signaling: action potential and Ca2+ as key players. Int J Mol Sci 17(9):1558. MDPI AG. https://doi.org/10.3390/ijms17091558

Balemans W, Van Hul W (2002) Extracellular regulation of BMP signaling in vertebrates: a cocktail of modulators. Dev Biol 250:231–250

Behlke J, Bommer UA, Lutsch G, Henske A, Bielka H (1986) Structure of initiation factor eIF-3 from rat liver. Hydrodynamic and electron microscopic investigations. Eur J Biochem 157:523–530

Beniash E, Deshpande AS, Fang PA, Lieb NS, Zhang X, Sfeir CS (2011) Possible role of DMP1 in dentin mineralization. J Struct Biol 174:100–106

Boskey AL (1989) Noncollagenous matrix proteins and their role in mineralization. Bone Miner 6:111–123

Boskey AL (1990) Bone mineral and matrix. Are they altered in osteoporosis? Orthop Clin North Am 21:19–29

Boskey AL (1992) Mineral-matrix interactions in bone and cartilage. Clin Orthop Relat Res 281:244–274

Boskey AL (1998) Biomineralization: conflicts, challenges, and opportunities. J Cell Biochem Suppl 30–31:83–91

Boskey AL (2003a) Biomineralization: an overview. Connect Tissue Res 44(Suppl 1):5–9

Boskey AL (2003b) Mineral analysis provides insights into the mechanism of biomineralization. Calcif Tissue Int 72:533–536

Boskey AL (2013) Bone composition: relationship to bone fragility and antiosteoporotic drug effects. Bonekey Rep 2:447

Boskey AL, Villarreal-Ramirez E (2016) Intrinsically disordered proteins and biomineralization. Matrix Biol 52–54:43–59

Calvo MB, Fernández VB, Villaamil VM, Gallego GA, Prado SD, Pulido EG (2009) Biology of BMP signalling and cancer. Clin Transl Oncol 11:126–137

Chen Y, George A (2018) TRIP-1 promotes the assembly of an ECM that contains extracellular vesicles and factors that modulate angiogenesis. Front Physiol 9:1092

Chen J, Shapiro HS, Sodek J (1992) Developmental expression of bone sialoprotein mRNA in rat mineralized connective tissues. J Bone Miner Res 7(8):987–997. https://doi.org/10.1002/jbmr.5650070816

Chen RH, Miettinen PJ, Maruoka EM, Choy L, Derynck R (1995) A WD-domain protein that is associated with and phosphorylated by the type II TGF-beta receptor. Nature 377:548–552

Conner C, Lager TW, Guldner IH, Wu MZ, Hishida Y, Hishida T, Ruiz S, Yamasaki AE, Gilson RC, Belmonte JCI, Gray PC, Kelber JA, Zhang S, Panopoulos AD (2020) Cell surface GRP78 promotes stemness in normal and neoplastic cells. Sci Rep 10(1):3474. https://doi.org/10.1038/s41598-020-60269-y

Constam DB, Robertson EJ (1999) Regulation of bone morphogenetic protein activity by pro domains and proprotein convertases. J Cell Biol 144:139–149

D'souza RN, Cavender A, Sunavala G, Alvarez J, Ohshima T, Kulkarni AB, MacDougall M (1997) Gene expression patterns of murine dentin matrix protein 1 (Dmp1) and dentin Sialophosphoprotein (DSPP) suggest distinct developmental functions in vivo. J Bone Miner Res 12(12):2040–2049. https://doi.org/10.1359/jbmr.1997.12.12.2040

Daluiski A, Engstrand T, Bahamonde ME, Gamer LW, Agius E, Stevenson SL, Cox K, Rosen V, Lyons KM (2001) Bone morphogenetic protein-3 is a negative regulator of bone density. Nat Genet 27:84–88

Danesh SM, Villasenor A, Chong D, Soukup C, Cleaver O (2009) BMP and BMP receptor expression during murine organogenesis. Gene Expr Patterns 9:255–265

David V, Quarles LD (2010) ASARM mineralization hypothesis: a bridge too far? J Bone Miner Res 25(4):692–694. https://doi.org/10.1002/jbmr.69

De Caestecker M (2004) The transforming growth factor-β superfamily of receptors. Cytokine Growth Factor Rev 15:1–11

de Larco JE, Todaro GJ (1978) Growth factors from murine sarcoma virus-transformed cells. Proc Natl Acad Sci 75(8):4001–4005

Derynck R, Zhang YE (2003) Smad-dependent and Smad-independent pathways in TGF-β family signalling. Nature 425:577–584

Derynck R, Jarrett JA, Chen EY, Eaton DH, Bell JR, Assoian RK, Roberts AB, Sporn MB, Goeddel DV (1985) Human transforming growth factor-β complementary DNA sequence and expression in normal and transformed cells. Nature 316:701–705

Di Guglielmo GM, Le Roy C, Goodfellow AF, Wrana JL (2003) Distinct endocytic pathways regulate TGF-β receptor signalling and turnover. Nat Cell Biol 5:410–421

Dickinson ME, Kobrin MS, Silan CM, Kingsley DM, Justice MJ, Miller DA, Ceci JD, Lock LF, Lee A, Buchberg AM (1990) Chromosomal localization of seven members of the murine TGF-β superfamily suggests close linkage to several morphogenetic mutant loci. Genomics 6:505–520

Dorvee JR, Veis A (2013) Water in the formation of biogenic minerals: peeling away the hydration layers. J Struct Biol 183:278–303

Duer M, Veis A (2013) Bone mineralization: water brings order. Nat Mater 12:1081–1082

Dutta D, Williamson CD, Cole NB, Donaldson JG (2012) Pitstop 2 is a potent inhibitor of Clathrin-independent endocytosis. PLoS One 7(9):e45799. https://doi.org/10.1371/journal.pone.0045799

Ebendal T, Bengtsson H, Söderström S (1998) Bone morphogenetic proteins and their receptors: potential functions in the brain. J Neurosci Res 51:139–146

Estroff LA, Cohen I (2011) Biomineralization: micelles in a crystal. Nat Mater 10:810–811

Feng Q (2011) Principles of calcium-based biomineralization. Prog Mol Subcell Biol 52:141–197

Fetting JL, Guay JA, Karolak MJ, Iozzo RV, Adams DC, Maridas DE, Brown AC, Oxburgh L (2014) FOXD1 promotes nephron progenitor differentiation by repressing decorin in the embryonic kidney. Development 141:17–27

Fisher LW, Fedarko NS (2003) Six genes expressed in bones and teeth encode the current members of the SIBLING family of proteins. Connect Tissue Res 44:33–40. https://doi.org/10.1080/03008200390152061

Fisher LW, Whitson SW, Avioli LV, Termine JD (1983) Matrix sialoprotein of developing bone. J Biol Chem 258(20):12723–12727

Fisher LW, Torchia DA, Fohr B, Young MF, Fedarko NS (2001) Flexible structures of SIBLING proteins, bone sialoprotein, and osteopontin. Biochem Biophys Res Commun 280(2):460–465. https://doi.org/10.1006/bbrc.2000.4146

Flanders KC, Ludecke G, Engels S, Cissel DS, Roberts AB, Kondaiah P, Lafyatis R, Sporn MB, Unsicker K (1991) Localization and actions of transforming growth factor-beta s in the embryonic nervous system. Development 113:183–191

Foster BL, Nagatomo KJ, Tso HW, Tran AB, Nociti FH Jr, Narisawa S, Yadav MC, Mckee MD, Millan JI, Somerman MJ (2013) Tooth root dentin mineralization defects in a mouse model of hypophosphatasia. J Bone Miner Res 28:271–282

Foster BL, Ao M, Salmon CR, Chavez MB, Kolli TN, Tran AB, Chu EY, Kantovitz KR, Yadav M, Narisawa S, Millán JL, Nociti FH, Somerman MJ (2018) Osteopontin regulates dentin and alveolar bone development and mineralization. Bone 107:196–207. https://doi.org/10.1016/j.bone.2017.12.004

Franzén A, Heinegård D (1985) Isolation and characterization of two sialoproteins present only in bone calcified matrix. Biochem J 232(3):715–724

Gandhi NS, Mancera RL (2012) Prediction of heparin binding sites in bone morphogenetic proteins (BMPs). Biochim Biophys Acta Proteins Proteom 1824:1374–1381

Ganss B, Kim RH, Sodek J (1999) Bone Sialoprotein. Crit Rev Oral Biol Med 10(1):79–98. https://doi.org/10.1177/10454411990100010401

George A, Veis A (1991) FTIRS in H2O demonstrates that collagen monomers undergo a conformational transition prior to thermal self-assembly in vitro. Biochemistry 30:2372–2377

George A, Veis A (2008) Phosphorylated proteins and control over apatite nucleation, crystal growth, and inhibition. Chem Rev 108:4670–4693

George A, Sabsay B, Simonian PA, Veis A (1993) Characterization of a novel dentin matrix acidic phosphoprotein. Implications for induction of biomineralization. J Biol Chem 268 (17):12624–12630

George A, Gui J, Jenkins NA, Gilbert DJ, Copeland NG, Veis A (1994) In situ localization and chromosomal mapping of the AG1 (Dmp1) gene. J Histochem Cytochem 42(12):1527–1531. https://doi.org/10.1177/42.12.7983353

George A, Silberstein R, Veis A (1995) In situ hybridization shows Dmp1 (AG1) to be a developmentally regulated dentin-specific protein produced by mature odontoblasts. Connect Tissue Res 33(1–3):67–72

George A, Srinivasan R, Thotakura SR, Liu K, Veis A (1999) Rat dentin matrix protein 3 is a compound protein of rat dentin Sialoprotein and Phosphophoryn. Connect Tissue Res 40 (1):49–57. https://doi.org/10.3109/03008209909005277

George A, Guirado E, Chen Y (2018) DMP1 binds specifically to type I collagen and regulates mineral nucleation and growth. In: Endo K, Kogure T, Nagasawa H (eds) Biomineralization. Springer, Singapore. https://doi.org/10.1007/978-981-13-1002-7_15

Gericke A, Qin C, Spevak L, Fujimoto Y, Butler WT, Sørensen ES, Boskey AL (2005) Importance of phosphorylation for Osteopontin regulation of biomineralization. Calcif Tissue Int 77 (1):45–54. https://doi.org/10.1007/s00223-004-1288-1

Giachelli CM (2005) Inducers and inhibitors of biomineralization: lessons from pathological calcification. Orthod Craniofac Res 8:229–231

Goldberg M, Lacerda-Pinheiro S, Priam F, Jegat N, Six N, Bonnefoix M, Septier D, Chaussain-Miller C, Veis A, Denbesten P, Poliard A (2008) Matricellular molecules and odontoblast progenitors as tools for dentin repair and regeneration. Clin Oral Investig 12:109–112

Goldberg M, Kulkarni AB, Young M, Boskey A (2011) Dentin: structure, composition and mineralization. Front Biosci (Elite Ed) 3:711–735

Gorski JP (2011) Biomineralization of bone: a fresh view of the roles of non-collagenous proteins. Front Biosci (Landmark Ed) 16:2598–2621

Graff JM (1997) Embryonic patterning: to BMP or not to BMP, that is the question. Cell 89:171–174

Gullard A, Gluhak-Heinrich J, Papagerakis S, Sohn P, Unterbrink A, Chen S, MacDougall M (2016) MEPE localization in the craniofacial complex and function in tooth dentin formation. J Histochem Cytochem 64(4):224–236. https://doi.org/10.1369/0022155416635569

Hagiwara M, Shirai Y, Nomura R, Sasaki M, Kobayashi K-I, Tadokoro T, Yamamoto Y (2009) Caveolin-1 activates Rab5 and enhances endocytosis through direct interaction. Biochem Biophys Res Commun 378(1):73–78. https://doi.org/10.1016/j.bbrc.2008.10.172

Halvorsen YD, Franklin D, Bond AL, Hitt DC, Auchter C, Boskey AL, Paschalis EP, Wilkison WO, Gimble JM (2001) Extracellular matrix mineralization and osteoblast gene expression by human adipose tissue-derived stromal cells. Tissue Eng 7:729–741

Hao J, Ramachandran A, George A (2009) Temporal and spatial localization of the dentin matrix proteins during dentin biomineralization. J Histochem Cytochem 57(3):227–237. https://doi.org/10.1369/jhc.2008.952119

He G, Dahl T, Veis A, George A (2003) Dentin matrix protein 1 initiates hydroxyapatite formation in vitro. Connect Tissue Res 44(Suppl 1):240–245

Helseth DL Jr, Veis A (1981) Collagen self-assembly in vitro. Differentiating specific telopeptide-dependent interactions using selective enzyme modification and the addition of free amino telopeptide. J Biol Chem 256:7118–7128

Hershey JW, Asano K, Naranda T, Vornlocher HP, Hanachi P, Merrick WC (1996) Conservation and diversity in the structure of translation initiation factor EIF3 from humans and yeast. Biochimie 78:903–907

Hetz C (2012) The unfolded protein response: controlling cell fate decisions under ER stress and beyond. Nat Rev Mol Cell Biol 13(2):89–102. https://doi.org/10.1038/nrm3270

Hirst KL, Simmons D, Feng J, Aplin H, Dixon MJ, Macdougall M (1997) Elucidation of the sequence and the genomic Organization of the Human Dentin Matrix Acidic Phosphoprotein 1 (DMP1) gene: exclusion of the locus from a causative role in the pathogenesis of Dentinogenesis Imperfecta type II. Genomics 42(1):38–45. https://doi.org/10.1006/geno.1997.4700

Hu H, Tian M, Ding C, Yu S (2019) The C/EBP homologous protein (CHOP) transcription factor functions in endoplasmic reticulum stress-induced apoptosis and microbial infection. Front Immunol 10(JAN):3083). Frontiers Media S.A. https://doi.org/10.3389/fimmu.2018.03083

Jacob A, Zhang Y, George A (2014) Transcriptional regulation of dentin matrix protein 1 (DMP1) in odontoblasts and osteoblasts. Connect tissue Res 55(Supp 1):107–112. https://doi.org/10.3109/03008207.2014.923850

Kaito T, Morimoto T, Mori Y, Kanayama S, Makino T, Takenaka S, Sakai Y, Otsuru S, Yoshioka Y, Yoshikawa H (2018) BMP-2/7 heterodimer strongly induces bone regeneration in the absence of increased soft tissue inflammation. Spine J 18:139–146

Kalmar L, Homola D, Varga G, Tompa P (2012) Structural disorder in proteins brings order to crystal growth in biomineralization. Bone 51:528–534

Kaartinen V, Voncken JW, Shuler C, Warburton D, Bu D, Heisterkamp N, Groffen J (1995) Abnormal lung development and cleft palate in mice lacking TGF–β3 indicates defects of epithelial–mesenchymal interaction. Nat Genet 11:415–421

Kawashima N, Okiji T (2016) Odontoblasts: specialized hard-tissue-forming cells in the dentin-pulp complex. Congenit Anom 56(4):144–153. https://doi.org/10.1111/cga.12169

Kessler E, Takahara K, Biniaminov L, Brusel M, Greenspan DS (1996) Bone morphogenetic protein-1: the type I procollagen C-proteinase. Science 271:360–362

Kihara T, Ito J, Miyake J (2013) Measurement of biomolecular diffusion in extracellular matrix condensed by fibroblasts using fluorescence correlation spectroscopy. PLoS One 8:e82382

Kim HM, Himeno T, Kawashita M, Kokubo T, Nakamura T (2004) The mechanism of biomineralization of bone-like apatite on synthetic hydroxyapatite: an in vitro assessment. J R Soc Interface 1:17–22

Kirkpatrick CA, Knox SM, Staatz WD, Fox B, Lercher DM, Selleck SB (2006) The function of a Drosophila glypican does not depend entirely on heparan sulfate modification. Dev Biol 300:570–582

Koehne T, Marshall RP, Jeschke A, Kahl-Nieke B, Schinke T, Amling M (2013) Osteopetrosis, osteopetrorickets and hypophosphatemic rickets differentially affect dentin and enamel mineralization. Bone 53:25–33

Liao WX, Moore RK, Otsuka F, Shimasaki S (2003) Effect of intracellular interactions on the processing and secretion of bone morphogenetic protein-15 (BMP-15) and growth and differentiation factor-9 implication of the aberrant ovarian phenotype of BMP-15 mutant sheep. J Biol Chem 278:3713–3719

Linde A, Goldberg M (1993) Dentinogenesis. Crit Rev Oral Biol Med 4(5):679–728

Lorenz-Depiereux B, Bastepe M, Benet-Pagès A, Amyere M, Wagenstaller J, Müller-Barth U, Badenhoop K, Kaiser SM, Rittmaster RS, Shlossberg AH, Olivares JL, Loris C, Ramos FJ, Glorieux F, Vikkula M, Jüppner H, Strom TM (2006) DMP1 mutations in autosomal recessive hypophosphatemia implicate a bone matrix protein in the regulation of phosphate homeostasis. Nat Genet 38(11):1248–1250. https://doi.org/10.1038/ng1868

Ma L, Pang AP, Luo Y, Lu X, Lin F (2020) Beneficial factors for biomineralization by ureolytic bacterium Sporosarcina pasteurii. Microb Cell Factories 19:12

MacDougall M, Simmons D, Luan X, Nydegger J, Feng J, Gu TT (1997) Dentin Phosphoprotein and dentin Sialoprotein are cleavage products expressed from a single transcript coded by a gene on human chromosome 4 dentin phosphoprotein dna sequence determination. J Biol Chem 272 (2):835–842. https://doi.org/10.1074/jbc.272.2.835

MacDougall M, Gu TT, Luan X, Simmons D, Chen J (1998) Identification of a novel isoform of mouse dentin matrix protein 1: spatial expression in mineralized tissues. J Bone Miner Res Off J Am Soc Bone Miner Res 13(3):422–431. https://doi.org/10.1359/jbmr.1998.13.3.422

Mark MP, Butler WT, Prince CW, Finkelman RD, Ruch J-V (1988) Developmental expression of 44-kDa bone phosphoprotein (osteopontin) and bone γ-carboxyglutamic acid (Gla)-containing protein (osteocalcin) in calcifying tissues of rat. Differentiation 37(2):123–136. https://doi.org/10.1111/j.1432-0436.1988.tb00804.x

Martin A, David V, Laurence JS, Schwarz PM, Lafer EM, Hedge A-M, Rowe PSN (2008) Degradation of MEPE, DMP1, and release of SIBLING ASARM-peptides (Minhibins): ASARM-peptide(s) are directly responsible for defective mineralization in HYP. Endocrinology 149(4):1757–1772. https://doi.org/10.1210/en.2007-1205

Mckee MD, Hoac B, Addison WN, Barros NM, Millán JL, Chaussain C (2013) Extracellular matrix mineralization in periodontal tissues: noncollagenous matrix proteins, enzymes, and relationship to hypophosphatasia and X-linked hypophosphatemia. Periodontol 2000(63):102–122

Melvyn W, Leblond CP (1973) Radioautographic visualization of the deposition of a phosphoprotein at the mineralization front in the dentin of the rat incisor. J Cell Biol 56(3):838–845

Merkel A, Chen Y, George A (2019) Endocytic trafficking of DMP1 and GRP78 complex facilitates Osteogenic differentiation of human periodontal ligament stem cells. Front Physiol 10:1175. https://doi.org/10.3389/fphys.2019.01175

Miyazono K, Hellman U, Wernstedt C, Heldin C (1988) Latent high molecular weight complex of transforming growth factor beta 1. Purification from human platelets and structural characterization. J Biol Chem 263:6407–6415

Mulloy B, Rider CC (2015) The bone morphogenetic proteins and their antagonists. Vitamins & Hormones. Elsevier

Nabi IR, Le PU (2003) Caveolae/raft-dependent endocytosis. J Cell Biol 161(4):673–677. https://doi.org/10.1083/jcb.200302028

Narayanan K, Ramachandran A, Hao J, He G, Park KW, Cho M, George A (2003) Dual functional roles of dentin matrix protein 1. Implications in biomineralization and gene transcription by activation of intracellular Ca2+ store. J Biol Chem 278(19):17500–17508. https://doi.org/10.1074/jbc.M212700200

Neer EJ, Smith TF (1996) G protein heterodimers: new structures propel new questions. Cell 84:175–178

Neer EJ, Schmidt CJ, Nambudripad R, Smith TF (1994) The ancient regulatory-protein family of WD-repeat proteins. Nature 371:297–300

Neugebauer JM, Kwon S, Kim H-S, Donley N, Tilak A, Sopory S, Christian JL (2015) The prodomain of BMP4 is necessary and sufficient to generate stable BMP4/7 heterodimers with enhanced bioactivity in vivo. Proc Nat Acad Sci 112:E2307–E2316

Ni M, Zhang Y, Lee AS (2011) Beyond the endoplasmic reticulum: atypical GRP78 in cell viability, signalling and therapeutic targeting. Biochem J 434(2):181–188. https://doi.org/10.1042/BJ20101569

Nishimatsu S-I, Thomsen GH (1998) Ventral mesoderm induction and patterning by bone morphogenetic protein heterodimers in Xenopus embryos. Mech Dev 74:75–88

Nogami H, Urist MR (1970) A morphogenetic matrix for differentiation of cartilage in tissue culture. Proc Soc Exp Biol Med 134:530–535

Nohe A, Hassel S, Ehrlich M, Neubauer F, Sebald W, Henis YI, Knaus P (2002) The mode of bone morphogenetic protein (BMP) receptor oligomerization determines different BMP-2 signaling pathways. J Biol Chem 277:5330–5338

Nudelman F, Pieterse K, George A, Bomans PH, Friedrich H, Brylka LJ, Hilbers PA, De With G, Sommerdijk NA (2010) The role of collagen in bone apatite formation in the presence of hydroxyapatite nucleation inhibitors. Nat Mater 9:1004–1009

Ohkawara B, Iemura S-I, Ten Dijke P, Ueno N (2002) Action range of BMP is defined by its N-terminal basic amino acid core. Curr Biol 12:205–209

Opsahl Vital S, Gaucher C, Bardet C, Rowe PS, George A, Linglart A, Chaussain C (2012) Tooth dentin defects reflect genetic disorders affecting bone mineralization. Bone 50:989–997

Paine ML, Snead ML (2005) Tooth developmental biology: disruptions to enamel-matrix assembly and its impact on biomineralization. Orthod Craniofac Res 8:239–251

Pećina M, Vukičević S (2007) Biological aspects of bone, cartilage and tendon regeneration. Int Orthop 31:719–720

Pfaffenbach KT, Lee AS (2011) The critical role of GRP78 in physiologic and pathologic stress. Curr Opin Cell Biol 23(2):150–156. https://doi.org/10.1016/j.ceb.2010.09.007

Pohjolainen V, Taskinen P, Soini Y, Rysa J, Ilves M, Juvonen T, Ruskoaho H, Leskinen H, Satta J (2008) Noncollagenous bone matrix proteins as a part of calcific aortic valve disease regulation. Hum Pathol 39:1695–1701

Pragnère S, Auregan JC, Bosser C, Linglart A, Bensidhoum M, Hoc T, Nouguier-Lehon C, Chaussain C (2021) Human dentin characteristics of patients with osteogenesis imperfecta: insights into collagen-based biomaterials. Acta Biomater 119:259–267. https://doi.org/10.1016/j.actbio.2020.10.033. Epub 2020 Oct 26

Proetzel G, Pawlowski SA, Wiles MV, Yin M, Boivin GP, Howles PN, Ding J, Ferguson MW, Doetschman T (1995) Transforming growth factor–β3 is required for secondary palate fusion. Nat Genet 11:409–414

Qin C, Brunn JC, Jones J, George A, Ramachandran A, Gorski JP, Butler WT (2001) A comparative study of sialic acid-rich proteins in rat bone and dentin. Eur J Oral Sci 109(2):133–141. https://doi.org/10.1034/j.1600-0722.2001.00001.x

Qin C, Baba O, Butler WT (2004) Post-translational modifications of sibling proteins and their roles in osteogenesis and dentinogenesis. Crit Rev Oral Biol Med Off Publ Am Assoc Oral Biol 15 (3):126–136

Ramachandran A, Ravindran S, George A (2012) Localization of transforming growth factor beta receptor II interacting protein-1 in bone and teeth: implications in matrix mineralization. J Histochem Cytochem 60:323–337

Ramachandran A, Ravindran S, Huang CC, George A (2016) TGF beta receptor II interacting protein-1, an intracellular protein has an extracellular role as a modulator of matrix mineralization. Sci Rep 6:37885

Ramachandran A, He K, Huang CC, Shahbazian-Yassar R, Shokuhfar T, George A (2018) TRIP-1 in the extracellular matrix promotes nucleation of calcium phosphate polymorphs. Connect Tissue Res 59:13–19

Ramirez F, Sakai L, Rifkin D, Dietz HC (2007) Extracellular microfibrils in development and disease. Cell Mol Life Sci 64:2437–2446

Ravindran S, Narayanan K, Eapen AS, Hao J, Ramachandran A, Blond S, George A (2008) Endoplasmic reticulum chaperone protein GRP-78 mediates endocytosis of dentin matrix protein 1. J Biol Chem 283(44):29658–29670. https://doi.org/10.1074/jbc.M800786200

Ravindran S, Gao Q, Ramachandran A, Sundivakkam P, Tiruppathi C, George A (2012) Expression and distribution of grp-78/bip in mineralizing tissues and mesenchymal cells. Histochem Cell Biol 138(1):113–125. https://doi.org/10.1007/s00418-012-0952-1

Roberts AB, Anzano MA, Lamb LC, Smith JM, Sporn MB (1981) New class of transforming growth factors potentiated by epidermal growth factor: isolation from non-neoplastic tissues. Proc Natl Acad Sci 78(9):5339–5343

Roberts AB, Anzano MA, Wakefield LM, Roche NS, Stern DF, Sporn MB (1985) Type beta transforming growth factor: a bifunctional regulator of cellular growth. Proc Natl Acad Sci 82 (1):119–123

Rowe PS, de Zoysa PA, Dong R, Wang HR, White KE, Econs MJ, Oudet CL (2000) MEPE, a new gene expressed in bone marrow and tumors causing osteomalacia. Genomics 67(1):54–68. https://doi.org/10.1006/geno.2000.6235

Ruppert R, Hoffmann E, Sebald W (1996) Human bone morphogenetic protein 2 contains a heparin-binding site which modifies its biological activity. Eur J Biochem 237:295–302

Salmon B, Bardet C, Khaddam M, Naji J, Coyac BR, Baroukh B, Letourneur F, Lesieur J, Decup F, Denmat DL, Nicoletti A, Poliard A, Rowe PS, Huet E, Vital SO, Linglart A, McKee MD, Chaussain C (2013) MEPE-derived ASARM peptide inhibits Odontogenic differentiation of dental pulp stem cells and impairs mineralization in tooth models of X-linked hypophosphatemia. PLoS One 8(2):e56749. https://doi.org/10.1371/journal.pone.0056749

Salmon B, Bardet C, Coyac BR, Baroukh B, Naji J, Rowe PS, Vital SO, Linglart A, Mckee MD, Chaussain C (2014) Abnormal osteopontin and matrix extracellular phosphoglycoprotein localization, and odontoblast differentiation, in X-linked hypophosphatemic teeth. Connect Tissue Res 55(Sup 1):79–82. https://doi.org/10.3109/03008207.2014.923864

Saunders M, Kong C, Shaw JA, Clode PL (2011) Matrix-mediated biomineralization in marine mollusks: a combined transmission electron microscopy and focused ion beam approach. Microsc Microanal 17:220–225

Schmid B, Furthauer M, Connors SA, Trout J, Thisse B, Thisse C, Mullins MC (2000) Equivalent genetic roles for bmp7/snailhouse and bmp2b/swirl in dorsoventral pattern formation. Development 127:957–967

Schultz GS, Wysocki A (2009) Interactions between extracellular matrix and growth factors in wound healing. Wound Repair Regen 17(2):153–162

Shi Y, Massagué J (2003) Mechanisms of TGF-β signaling from cell membrane to the nucleus. Cell 113:685–700

Sipe JB, Zhang J, Waits C, Skikne B, Garimella R, Anderson HC (2004) Localization of bone morphogenetic proteins (BMPs)-2, −4, and-6 within megakaryocytes and platelets. Bone 35:1316–1322

Stenmark H (2009) Rab GTPases as coordinators of vesicle traffic. Nat Rev Mol Cell Biol 10 (8):513–525. https://doi.org/10.1038/nrm2728

Sun Y, Lu Y, Chen L, Gao T, D'Souza R, Feng JQ, Qin C (2011) DMP1 processing is essential to dentin and jaw formation. J Dent Res 90(5):619–624. https://doi.org/10.1177/0022034510397839

Surdilovic D, Natarajan PM, Ille T, Shetty SR, Adtani P (2018) Non-collagen protein in the dentin tissue – the role in the process of Dentinogenesis. Biomed Pharmacol J 11(2):843–849

Takagi Y, Veis A, Sauk JJ (1983) Relation of mineralization defects in collagen matrices to noncollagenous protein components. Identification of a molecular defect in dentinogenesis imperfecta. Clin Orthop 176:282–290

Tavafoghi M, Cerruti M (2016) The role of amino acids in hydroxyapatite mineralization. J R Soc Interface 13:20160462

Ten Dijke P, Hill CS (2004) New insights into TGF-β–Smad signalling. Trends Biochem Sci 29:265–273

Ten Dijke P, Hansen P, Iwata KK, Pieler C, Foulkes JG (1988) Identification of another member of the transforming growth factor type beta gene family. Proc Natl Acad Sci 85:4715–4719

Tian E, Watanabe F, Martin B, Zangari M (2020) Innate Biomineralization. Int J Mol Sci 21:4820

Todaro GJ, Fryling C, De Larco JE (1980) Transforming growth factors produced by certain human tumor cells: polypeptides that interact with epidermal growth factor receptors. Proc Natl Acad Sci 77(9):5258–5262

Tsai Y-L, Lee AS (2018) Cell surface GRP78. In cell surface GRP78, a new paradigm in signal transduction biology. Elsevier, Amsterdam, pp 41–62. https://doi.org/10.1016/B978-0-12-812351-5.00003-9

Umulis D, O'connor MB, Blair SS (2009) The extracellular regulation of bone morphogenetic protein signaling. Development 136:3715–3728

Urist MR, Peltier LF (2002) Bone: formation by autoinduction. Clin Orthop Relat Res 395:4–10

Urist MR, Strates BS (1971) Bone morphogenetic protein. J Dent Res 50:1392–1406

Veis A, Dorvee JR (2013) Biomineralization mechanisms: a new paradigm for crystal nucleation in organic matrices. Calcif Tissue Int 93:307–315

Veis A, Spector AR, Zamoscianyk H (1972) The isolation of an Edta-soluble phosphoprotein from mineralizing bovine dentin. Biochim Biophys Acta BBA - Protein Struct 257(2):404–413. https://doi.org/10.1016/0005-2795(72)90293-0

Wald T, Spoutil F, Osickova A, Prochazkova M, Benada O, Kasparek P, Bumba L, Klein OD, Sedlacek R, Sebo P, Prochazka J, Osicka R (2017) Intrinsically disordered proteins drive enamel formation via an evolutionarily conserved self-assembly motif. Proc Natl Acad Sci U S A 114: E1641–e1650

Walter P, Ron D (2011) The unfolded protein response: from stress pathway to homeostatic regulation. Science 334(6059):1081–1086). American Association for the Advancement of Science. https://doi.org/10.1126/science.1209038

Wang N, Butler JP, Ingber DE (1993) Mechanotransduction across the cell surface and through the cytoskeleton. Science 260:1124–1127

Whyte MP, McAlister WH, Fallon MD, Pierpont ME, Bijanki VN, Duan S, Otaify GA, Sly WS, Mumm S (2017) Raine syndrome (OMIM #259775), caused by FAM20C mutation, is congenital Sclerosing Osteomalacia With cerebral calcification (OMIM 259660). J Bone Miner Res Off J Am Soc Bone Miner Res 32(4):757–769. https://doi.org/10.1002/jbmr.3034

Wolfman NM, Hattersley G, Cox K, Celeste AJ, Nelson R, Yamaji N, Dube JL, Diblasio-Smith E, Nove J, Song JJ (1997) Ectopic induction of tendon and ligament in rats by growth and differentiation factors 5, 6, and 7, members of the TGF-beta gene family. J Clin Invest 100:321–330

Wozney JM (1998) The bone morphogenetic protein family: multifunctional cellular regulators in the embryo and adult. Eur J Oral Sci 106:160–166

Wrana JL, Attisano L, Cárcamo J, Zentella A, Doody J, Laiho M, Wang XF, Massagué J (1992) TGF beta signals through a heteromeric protein kinase receptor complex. Cell 71(6):1003–1014

Xu L, Chen Y-G, Massagué J (2000) The nuclear import function of Smad2 is masked by SARA and unmasked by TGFb-dependent phosphorylation. Nat Cell Biol 2:559–562

Xu L, Kang Y, Çöl S, Massagué J (2002) Smad2 nucleocytoplasmic shuttling by nucleoporins CAN/Nup214 and Nup153 feeds TGFβ signaling complexes in the cytoplasm and nucleus. Mol Cell 10:271–282

Yao S, Xu Y, Shao C, Nudelman F, Sommerdijk N, Tang R (2019) A biomimetic model for mineralization of type-I collagen fibrils. Methods Mol Biol 1944:39–54

Young MF, Kerr JM, Ibaraki K, Heegaard AM, Robey PG (1992) Structure, expression, and regulation of the major noncollagenous matrix proteins of bone. Clin Orthop Relat Res 281:275–294

Zhang Y-R, Du W, Zhou X-D, Yu H-Y (2014) Review of research on the mechanical properties of the human tooth. Int J Oral Sci 6(2):61–69. https://doi.org/10.1038/ijos.2014.21

Zurick KM, Qin C, Bernards MT (2013) Mineralization induction effects of osteopontin, bone sialoprotein, and dentin phosphoprotein on a biomimetic collagen substrate. J Biomed Mater Res A 101:1571–1581

Chapter 5
Odontoblast Processes: New Insights into Its Role in Dentin Mineralization

Yan Jing, Chaoyuan Li, and Jian Q. Feng

Abstract The formation of dentin is a highly regulated and well-controlled process, in which odontoblasts form a massive dentin tubular system with one main stalk. These tubules secrete collagenous proteins followed by the deposition of inorganic calcium phosphate in the form of mineral crystals. Two non-collagenous matrix proteins (dentin matrix protein 1, DMP1; and dentin sialophosphoprotein, DSPP) play an important role in the control of mineralization. The current theory, named "mineralization front," is that minerals are deposited at the edge of the dental pulp and the dentin. In this chapter, we briefly introduced an overall background on dentin mineralization and key roles of DMP1 and DSPP. We then focus on two new findings in researches of dentin mineralization: a recent breakthrough in mineralization mechanism studies because of applications of mineral labeling in a cell lineage-tracing line (the *Gli1*-cre^{ERT2}; R26R^{tdTomato}) and the 2.3 Col1a1-GFP reporter line; and discoveries of a novel function of bone morphogenetic protein 1 (BMP1) and tolloid-like 1 (TLL1) proteinase in control of DMP1 and DSPP during dentinogenesis. These findings challenge the current dogma and will shed new light on our understanding of dentin structure and function as well as the mechanisms of mineralization.

Y. Jing (✉)
Department of Orthodontics, Texas A&M College of Dentistry, Dallas, TX, USA
e-mail: yjing@tamu.edu

C. Li
Department of Biomedical Sciences, Texas A&M College of Dentistry, Dallas, TX, USA

Department of Oral Implant, School of Stomatology, Tongji University, Shanghai Engineering Research Center of Tooth Restoration and Regeneration, Shanghai, China

J. Q. Feng
Department of Biomedical Sciences, Texas A&M College of Dentistry, Dallas, TX, USA

M. Goldberg, P. Den Besten (eds.), *Extracellular Matrix Biomineralization of Dental Tissue Structures*, Biology of Extracellular Matrix 10,
https://doi.org/10.1007/978-3-030-76283-4_5

5.1 Introduction

Dentin is the most voluminous mineralized tissue of the tooth. It is usually covered by enamel on the crown, cementum at the root, and surrounds the entire pulp. The formation of dentin is a highly regulated and well-controlled process, in which odontoblasts form a collagenous organic matrix followed by the deposition of an inorganic calcium phosphate in the form of mineral crystals (Linde and Goldberg 1993). For decades, the mechanism of dentin mineralization was believed to be much like that of osteogenesis. However, a deeper examination of the two processes reveals the pitfalls of this dogma. Unlike osteoblasts, the odontoblast cell body permanently remains in a non-mineralized dental pulp environment. Only one extended cytoplasmic process from its cell body inserts into the mineralized dentin matrix to form different sizes of dentinal tubules (Sasaki and Garant 1996; Kawashima and Okiji 2016). In addition, there are many mini branches sprouting from the processes and connecting them as a network within the dentinal tubules. Numerous studies have reported the structure of the odontoblast process in different species, but its biological functions are still unclear. This oversight is especially evident when prior researchers uncovered the mechanisms involved with mineralization, yet neglected the most voluminous part of the tooth.

The mechanisms of mineralization in osteogenesis and dentinogenesis have been a research focus due to their critical role in physiological and pathological processes. The current theory of the mineralization front is described as a newly formed, superposed mineral layer on either the bone edge and the osteoid layer (Frost et al. 1961a, b; Frost and Villaneuva 1960) or the dentin edge close to the predentin (Goldberg et al. 2011; Salomon et al. 1990). This theory was developed using techniques such as double tetracycline-labeling assays. The colors from the labeling represent the location of previous mineralization events (Wu and Frost 1969; Frost et al. 1969), which are usually found in the pulp-end dentin near the predentin. However, this theory cannot explain the nature of dentin, which displays evenly distributed mineral in the dentin matrix.

Recently, remarkable progress in imaging techniques (for instance the invention of the confocal microscope, a more powerful fluorescent microscope that has higher resolution and contrast) and cell lineage-tracing technologies have precisely revealed the secrets of odontoblasts and their processes. This occurrence helps in solving the above puzzles. In this chapter, we briefly described the basic knowledge of dentin formation. Next, we summarized recent breakthroughs in two areas: studies of the dentin mineralization mechanism and the roles of proteases. The "mineralization front" (the dominant theory in the mineralization field for over a half century) believes that mineralization takes place at the edge of dentin and predentin. Studies using the advanced confocal microscope and different combinations of the 2.3 Col1a1-GFP report line with Calcein labeling or the Gli1 cell lineage tracing with Alizarin Red labeling clearly demonstrated two key points: (1) the so-called mineralization front line is essentially composed of many individual mineral collars that are closely associated with dentin tubules, which are wide in diameter and rich in a high density of odontoblast processes at the edge of dentin and predentin; and (2) an identical mineral deposition takes place in the small dentin tubules. Finally, we reviewed two well-documented players in mineralization: DMP1 and DSPP, as well as bone morphogenetic protein 1 (BMP1) and tolloid-like 1 (TLL1) (two functionally overlapped proteinases) that cleave secretory proteins during mineralization.

5.2 The Entire Dentinal Tubule Is Full of Numerous Mini-branches with Significantly Higher Dentinal Tubules in the Edge of the Dentin Pulp

Studies of the dentine structure date back to the early history of light microscopy (Hanazawa 1917). A variety of techniques including light microscopy, immunofluorescence microscopy, microradiography, transmission and scanning electron microscopy, and histochemical techniques have been used to reveal the detailed structure of both demineralized and undemineralized dentin tissue. Although much attention has focused on the extent, structure, and function of the odontoblast process (Holland 1985; Mjor and Nordahl 1996), different views have been reported.

The branching and extent of the odontoblast processes in dentin matrices were discovered decades ago. Early electron transmission work showed that odontoblast processes were only located in the inner dentin layer, with no extension to the outer half region between the pulp and the dentin-enamel joint (DEJ)(Yoshiba et al. 2002). Khatibi et al. recently reported that mini-branches of the odontoblast processes only exist in the start and end terminals of the odontoblast process (Khatibi Shahidi et al. 2015). On the other hand, it is commonly believed that the dentin tubules in the inner side of dentin are larger than that in the outer side of dentin. However, the acid-etched and non-etched SEM studies in bovine dentin tissues showed a controversial finding: the diameter of dentin tubules is more than twofold larger in the outer region than in the inner region (i.e., adjacent to pulp) (Dutra-Correa et al. 2007). Mjor et al. examined the density and branching of dentinal tubules in human teeth via light and scanning electron microscopy and reported statistically significant differences in the density of tubules depending on location (Mjor and Nordahl 1996). Three types of branches (major, fine, and microbranches) were identified based on their size, direction, and location (Mjor and Nordahl 1996). Despite these advances, there is still a knowledge gap for the function of these branches in dentin formation and remodeling.

To uncover the complexity of the dentinal tubule structure in dentin and reveal its biological functions, we used multiple staining and visualization methods on both mouse and human samples. By staining the mouse incisor with FITC (all of the unmineralized tissue including pulp, odontoblasts, and their processes were labeled using the color green) and examining the sample under a confocal microscope, we found that FITC-labeled small and middle branches of dentin tubules were

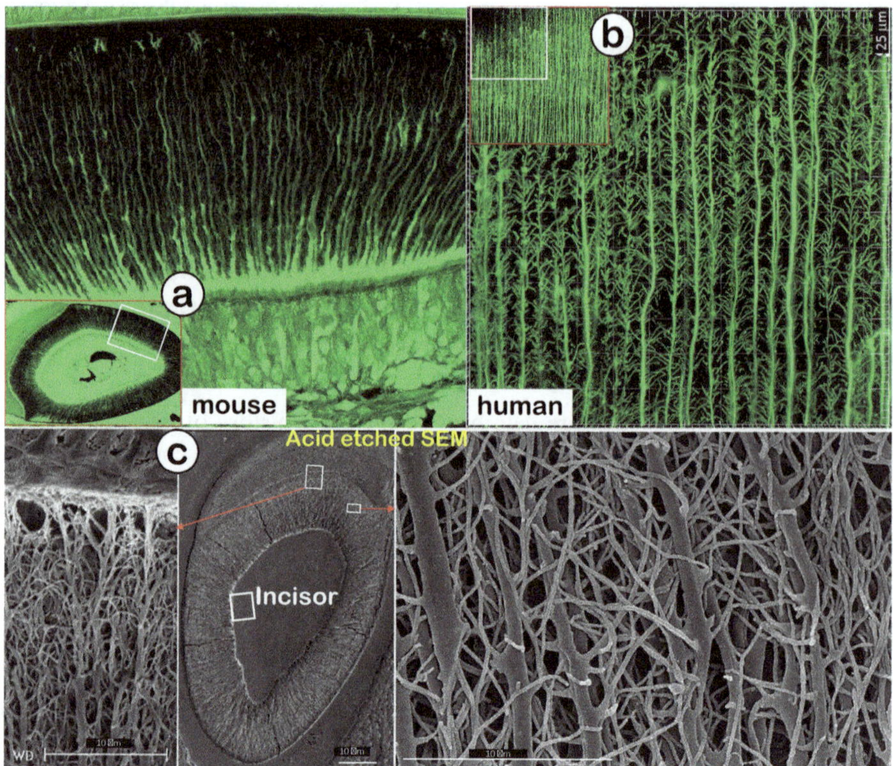

Fig. 5.1 Distributions of mini-branches along with the entire odontoblast processes. (**a**) The FITC confocal photograph image obtained from a 4-week-old mouse incisor, exhibiting numerous mini-dentin branches throughout the entire dentin layer with a large diameter and a high density at the edge close to the predentin layer; (**b**) The 1st molar dentin from a 43-year-old human displayed an identical distribution pattern of mini-branches along odontoblast processes as that in the mouse incisor; (**c**) The acid-etched SEM image obtained from a 4-week-old mouse incisor displayed numerous dentin branches from the area adjacent to the middle (left) to the area adjacent to enamel (right). Data are adapted from Li et al. Int J Biol Sci 14, 693–704, 2018

distributed throughout dentin matrices; the pattern stretched from predentin to the enamel-dentin junction (Li et al. 2018) (Fig. 5.1a). Similar patterns of dentinal tubules and their branches were also observed in adult human dentin (Li et al. 2018) (Fig. 5.1b). Subsequently, acid-etched SEM imaging confirmed that numerous mini-branches were sprouted from the whole odontoblast process (Li et al. 2018) (Fig. 5.1c). Together, these results from mice and humans demonstrate that dentin tubules are composed of numerous mini-branches throughout the entire dentin tubules. Importantly, the dentin tubule density at the pulp-end dentin is two-fold higher than that in the outer dentin area. This pattern will partially explain why mineral labels are strong in the "mineralization front," as demonstrated in the following section.

5.3 Mineralization Is Not Limited to the Mineralization Front but Occurs Throughout Dentin Matrices Surrounding Odontoblast Processes

The theory of the mineralization front is based on the double tetracycline-labeling assay that took place under a fluorescence light microscope decades ago. The incorporation of tetracycline into bone or dentin reflects the site of mineralization at the time of the tetracycline administration. This method is also widely used to quantify the mineral deposition rate (Wu and Frost 1969; Frost et al. 1969). In the study of dentin mineralization, additional methods using ^{33}P and ^{3}H-serine labeling phosphoproteins showed incorporation of these molecules into the predentin and dentin junction, which was also interpreted as the mineralization front by these authors (Weinstock and Leblond 1973; Inage and Toda 1988). In addition, Lundgren et al. demonstrated significant high Ca^{2+} ion activity in the mineralization front using calcium-specific micro-electrode methodology (Lundgren and Linde 1992). Nevertheless, as mentioned in the introduction, the theory of the mineralization front cannot explain why dentin mineral is evenly distributed in dentin matrices and how the mineral migrates a long distance to the outer dentin matrices from the pulp-end.

During the initial dentinogenesis, the mature odontoblasts (OD) deposit a collagenous dentin matrix at the dentin-enamel junction (DEJ). The forming intertubular dentin matrix pushes the main OD cell body away from the DEJ, leaving behind a thin connecting cell process attached to the DEJ as the OD moves in the direction of the pulp chamber. This process eventually retracts partially in the pulpal direction, leaving a tubule surrounded by collagen. Thus, the collagenous dentin is penetrated by a dense network of tubules, connected to the odontoblast layer, and partially filled by the OD processes. The perimeter of the tubule space then becomes filled with a material called peritubular dentin. The peritubular dentin collar around the OD processes is higher in mineral content than the surrounding intertubular dentin (Dorvee et al. 2014). On the other hand, it has been reported that calcium is accumulated in both the distal cell body and the odontoblast process (Boyde 1977). Calcium deposits inside organelles also have been demonstrated in the cell body as well as the odontoblast process (Reith 1976; Appleton and Morris 1979). Later, data obtained by using an autoradiographic technique aimed at avoiding Ca^{2+} ion redistribution in the tissue further supported a transport route through the odontoblasts and along their processes with the appearance of calcium in the dentin mineral phase (Lundgren and Linde 1992). These authors found that disturbing odontoblast microtubules involved in intracellular transport processes greatly impaired the Ca^{2+} ion transport into dentin mineral. Furthermore, a sulfated component (chondroitin sulfate) was recently identified within the hypermineralized bovine peritubular dentin (Dorvee et al. 2016). These authors speculate that the presence of sulfur co-localized with calcium may sequester calcium ions, allowing for the entire complex to later become mineralized. Collectively, more data indicate a likely

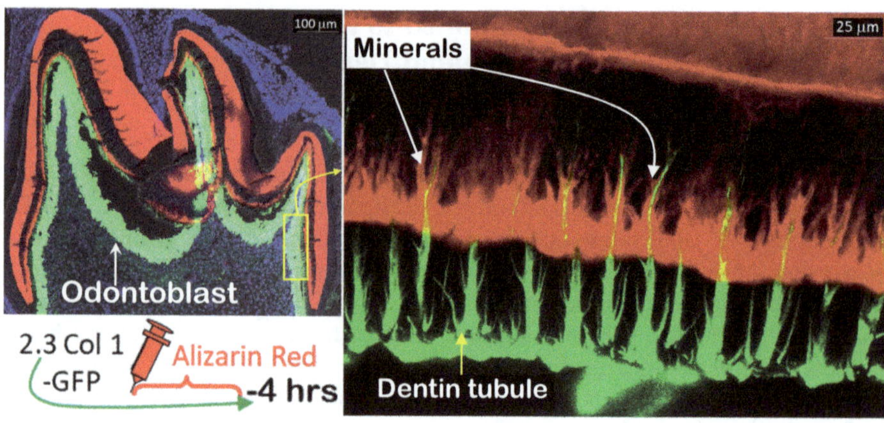

Fig. 5.2 The alizarin red labeling is directly distributed outside of the 2.3 Col1a1-GFP marked odontoblast processes. The GFP signal represents the dentin tubules, which were surrounded by alizarin red dye, supporting the notion that mineralization takes place along with entire odontoblast processes. Data are adapted from Li et al. Int J Biol Sci 14, 693–704, 2018

mechanism by which mineralization occurs through odontoblast processes instead of being limited to the adjacent line between predentin and pulp.

To further strengthen this new hypothesis, we extended the mineralization assay time frame by covering both short-term (minutes and hours) and long-term (days) labeling periods: (1) from 5 min to 24 h and (2) from 2 to 7 days. Likewise, to gain a precise definition of the relationship between dentin tubules and mineral deposition sites, we injected Alizarin Red into the 2.3 Col1a1-GFP mouse (a reporter line to label both odontoblast and bone cells (Kalajzic et al. 2002)), in which the dentin tubules were labeled with GFP (Fig. 5.2). We also improved the imaging quality by using a modern confocal microscope capable of collecting high-quality multispectral images across the visible range. Furthermore, we used advanced camera technology that captured a full-frame confocal image at high speed. For best imaging quality, we used both frozen and MMA methods to handle non-decalcified tooth section slides. As a result, the captured images displayed not only the expected two strong double-labeled lines (the green line on the distal side and the red line adjacent to the dentin-predentin edge) but also many thin lines along with the dentin tubules during a long labeling period (Fig. 5.3a). In the short-term labeling period, the images precisely revealed a close link between the labeled mineral and odontoblast processes by both sagittal and cross views (Fig. 5.3b). Particularly, the cross view provided a clear mineral deposition pattern of the labeled collars in the intertubular area on both the pulp-end dentin and outer dentin. In contrast, because of the limited imaging capabilities of old microscopes, the dimensions of thick and wide dentin tubules and the labeled mineralized collars (clearly separated by a confocal microscope) can only be displayed in thick “mineralization fronts.” Furthermore, those small and thin dentin tubules and surrounding mineral depositions cannot be revealed by the old imaging technique.

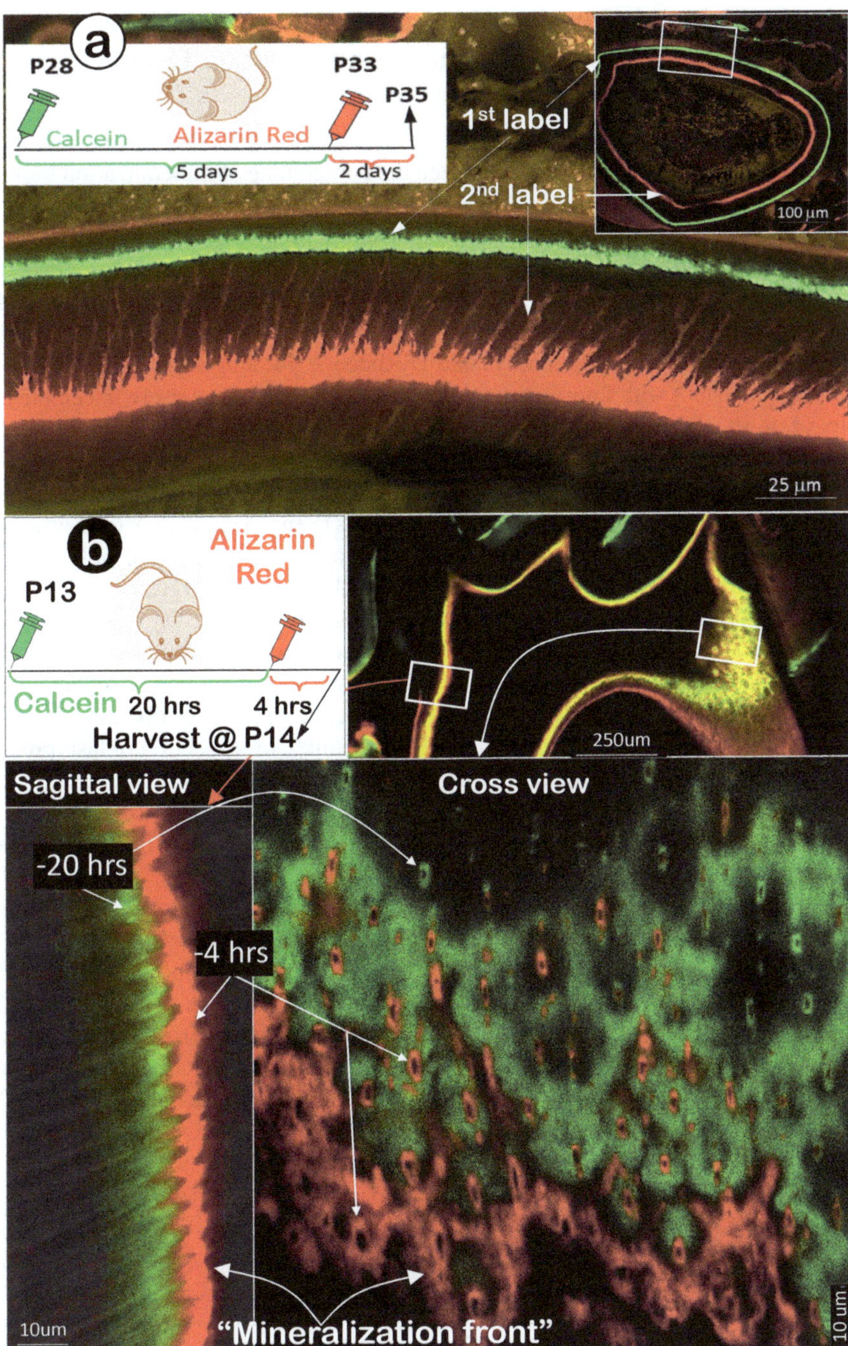

Fig. 5.3 Mineralization occurs along odontoblast processes. (**a**) A schematic illustration of long time injections of Calcein and Alizarin red at –7 days and –2 days, respectively (top left panel); the low magnification view of the cross-section of incisor (top right panel) and the enlarged confocal view of numerous labeled odontoblast processes with the green line on the top and red line at the

5.4 Cell Lineage Tracing Combined with Mineral Dye Injections Demonstrate that Mineralization Occurs Along with the Entire Dentin Tubules

The “mineralization front theory” has been the cornerstone of mineralization biology for more than a half century. Replacing this dogma with a new theory requires more stringent evidence. Thus, we took an additional approach to address this issue by combining the mineral labeling with a cell lineage-tracing technique. The cell lineage-tracing technique is the gold standard for investigating the fates of particular cells in vivo within their native environment (Kretzschmar and Watt 2012; Romagnani et al. 2015; Humphreys and DiRocco 2014; Jing et al. 2016). We used a Cre-loxP system with fluorescent markers as reporters that precisely traced the cell fate of a specific cell line and all progeny in vivo (Romagnani et al. 2015; Humphreys and DiRocco 2014; Jing et al. 2016). Specifically, we selected *Gli1* (a transcriptional factor expressed by the mesenchymal stem cells in the dental pulp which subsequently differentiate into odontoblasts (Feng et al. 2017)). *Gli1* was used to activate td tomato (the brightest fluoresce reporter) in the dental pulp cells of compound mice containing *Gli1-cre*^{ERT2}; R26R^{tdTomato} mice (Henry et al. 2009; Akiyama et al. 2005) after the injections of tamoxifen. Then, we injected calcein green into these tracing line mice for 4 h, with red colors representing odontoblasts in pulp and their processes (dentin tubules) in dentin and green colors reflecting the mineralization sites, respectively (Fig. 5.4). The confocal images obtained from the tracing line documented that (1) “the main trunk” of the odontoblast process gradually reduces its diameter when it penetrates in outer dentin matrix; (2) there are many branches along with the main process that directly contribute to dentin mineralization; (3) all odontoblast processes are wrapped with a newly deposited mineral layer labeled by a calcein stain (Fig. 5.4).

Taken together, to unlock the true dentin mineralization mechanism, we modified current research approaches by (a) extending the mineral deposition windows of time to include hours instead of limiting the current mineralization assay on days and weeks only. Next, we (b) switched a regular fluorescent microscope (with low-resolution capabilities) to a more powerful confocal microscope to separate minerals labeled along with dentin with large and small diameters. Lastly, (c) we used reporter mice, including the *Gli1-cre*^{ERT2}; R26R^{tdTomato} tracing line; and the 2.3 Col1a1-GFP to mark odontoblast processes combined with mineral dye injections,

---

Fig. 5.3 (continued) bottom; (**b**) The confocal images of the 1st molar labeled with Calcein at –20 h and Alizarin Red at –4 h revealed a clear pattern of the mineral deposited along odontoblast processes by the enlarged sagittal view (lower left panel), and the cross view (lower right panel). Of note, the only difference of the labeled mineral collars between the surface and the inside dentin is the diameter and density of odontoblast processes. As a result of the merged mineral collars on the surface, the mineralization front is observed. Data are adapted from Li et al. Int J Biol Sci 14, 693–704, 2018

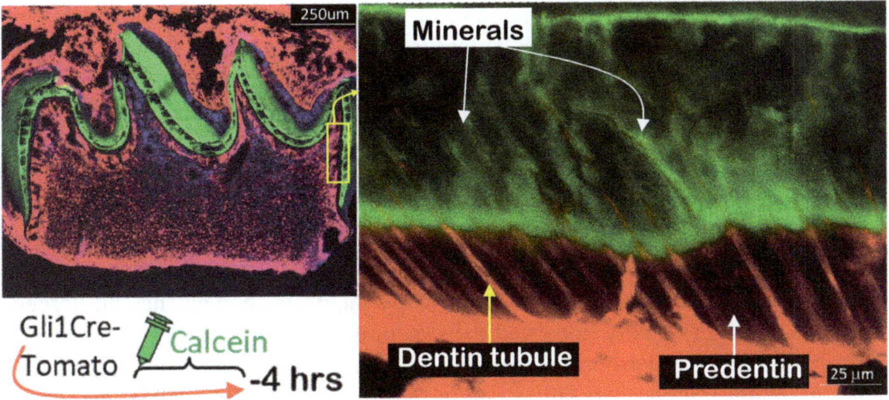

Fig. 5.4 The labeled mineral layer is directly distributed outside of the Gli1-tomato marked odontoblast processes. The red dentin tubules reflect the color of the *Gli1-cre*^{ERT2} activated tomato expression in odontoblast processes and the 4-h calcein labeled mineral matrix distributes to throughout dentin areas with a close link to all odontoblast processes. Data are adapted from Li et al. Int J Biol Sci 14, 693–704, 2018

respectively. For the first time, we demonstrated that mineralization takes place along with the entire dentin tubules in both long and short windows of time points. Our data suggest that the following two factors are likely responsible for the mineralization front phenomenon: the large diameter of the odontoblast processes and a high process density along the predentin-pulp edge leading to the merge of the labeled minerals. On the other hand, in the outer dentin, the labeled mineral collars are small in diameter and far apart in distance. In other words, the "mineralization front" only reflects the high activity of mineralization at the edge of the predentin-pulp edge.

5.5 Vital Roles of Molecules in Odontoblast Differentiation and Dentin Mineralization

5.5.1 Dentin Sialophosphoprotein (DSPP) Is Critical for Proper Dentin Mineralization

Dentin is the mineralized tissue that comprises the body of a tooth. On a weight basis, dentin is about 70% mineral, 20% organic matrix, and 10% water. Type I collagen and dentin sialophosphoprotein (DSPP)-derived proteins are the most abundant proteins in dentin (Yamakoshi 2009). Dentin sialophosphoprotein (DSPP) is required for the normal mineralization of teeth. However, mutations in the dentin sialophosphoprotein (DSPP) gene causes dentinogenesis imperfecta. After synthesis, DSPP is proteolytically processed into NH2- and COOH-terminal fragments. The NH2-terminal fragment of DSPP is highly glycosylated but not phosphorylated, whereas the COOH-terminal fragment (named "dentin phosphoprotein" or "DPP") is highly phosphorylated but not glycosylated. These two fragments are believed to perform distinct roles in dentin formation. To specifically define the function of DPP in dentinogenesis, Zhang et al. created "*Dspp–/–*; DPP Tg mice," in which the transgenic DPP (driven by a Type I collagen promoter) was targeted into the *Dspp*-null background (Zhang et al. 2018). At postnatal 6 months, the targeted expression of DPP greatly reduced predentin-thickness, doubled the dentin thickness in the *Dspp*-null mice, and restored the dentin material density by 29.5%. In fact, scanning electron microscopy analyses showed that the targeted expression of DPP greatly improved the structure of dentinal tubules in the rescued *Dspp*-null mice. Furthermore, double fluorochrome labeling analyses demonstrated that the dentin mineral deposition rate in the *Dspp–/–*; DPP Tg mice was significantly improved compared to that of non-rescued *Dspp*-null mice. Thus, these investigators concluded that DPP may promote dentin formation and coordinate dentinogenesis.

5.5.2 Dentin Matrix Protein 1 Is Critical for Postnatal Dentin Formation

Dentin matrix protein 1 (DMP1) is a non-collagenous phosphoprotein that belongs to the small integrin-binding ligand in the N-linked glycoproteins (SIBLING) family (Fisher and Goldney 2003; Fisher et al. 2001; Ravindran and George 2014). *Dmp1* was originally isolated from a rat incisor and was highly expressed in osteocytes and odontoblasts (George et al. 1993; Feng et al. 2003, 2006). Mutations of DMP1 in humans (Feng et al. 2006; Lorenz-Depiereux et al. 2006) or deletion of *Dmp1* in mice and rabbits (Ye et al. 2005; Liu et al. 2019) causes a low level of phosphorus, leading to abnormalities in bone maturation (Lu and Feng 2011; Carpenter et al. 2017; Feng et al. 2013) and defects in odontogenesis and mineralization (Ye et al. 2004). We previously reported a double-labeling assay on *Dmp1*-null mice that showed a diffused labeling pattern and great reduction in the mineralization rate. There was also a sharp decrease in mineral labeling for both the mineralization front and inside dentin matrices (Li et al. 2018; Lu et al. 2007). These labeled minerals appeared along with dentin tubules, in which they were sparse and thin in the *Dmp1* null dentin tubules compared to the age-matched controls. This information further strengthens the current hypothesis that dentin tubules are essential for dentin mineralization, which is not simply limited to the mineralization front.

Unlike the mineralization defect in Vitamin D receptor null mice, the mineralization defect in *Dmp1*-null mice was not able to be rescued by a high calcium and phosphate diet, suggesting that the effects of DMP1 on mineralization are locally mediated (Lu et al. 2007). Reexpression of *Dmp1* in early and late odontoblasts under the control of the Col1a1 promoter rescued the defects in mineralization as well as the defects in the dentinal tubules and third molar development. In contrast, reexpression in mature odontoblasts using the *Dspp* promoter produced only a partial rescue of the mineralization defects (Li et al. 2018). On the contrary, the expression of DMP1 was not altered in the *Dspp* KO mice (Lu et al. 2007; Gibson et al. 2013). These data suggest that DMP1 is a key regulator of odontoblast differentiation, formation of the dentin tubular system, and mineralization. Further, its expression is required in both early and late odontoblasts for normal odontogenesis to proceed; DSPP is a downstream effector molecule that mediates the roles of DMP1 in dentinogenesis.

5.5.3 BMP1/TLL1 Plays Crucial Roles in Maintaining Extracellular Matrix Homeostasis Essential to Root Formation and Dentin Mineralization

Bone morphogenetic protein 1 (BMP1) and tolloid-like 1 (TLL1) belong to the BMP1/tolloid-like proteinase family, which cleaves secretory proteins. It has been reported that the odontoblasts in the tooth and cells of the periodontal ligament express both *Bmp1* and *Tll1*. Zhang et al. created 2.3 kb Col1a1-Cre; *Bmp1* flox/flox; *Tll1* flox/flox compound mice, in which both *Bmp1* and *Tll1* were inactivated in Type I collagen-expressing cells such as odontoblasts (Zhang et al. 2017). Their results showed that the molars of the double KO mice had wider predentin, thinner dentin, and larger pulp chambers than those of the normal controls. The dentinal tubules of the molars in the double KO mice also appeared disorganized, with decreased expression of DSPP. These findings indicate that the proteinases encoded by *Bmp1* and *Tll1* genes play essential roles in the development and maintenance of mouse dentin. Wang et al. deleted the BMP1 and TLL1 genes with a ubiquitously expressed Cre induced by a tamoxifen treatment beginning at 3 days of age (harvested at 3 weeks of age) or beginning at 4 weeks of age (harvested at 8 weeks of age) and found similar results. The effects of this work included short and thin root dentin, defects in dentin mineralization, and delayed tooth eruption in the dKO group. Likewise, molecular mechanism studies revealed the accumulation of collagens in dentin and a sharp reduction in non-collagenous proteins such as DMP1 and DSPP (Wang et al. 2017). These findings indicate that BMP1 and TLL1 (two extracellular proteinases) maintain homeostasis for the levels of collagen and non-collagenous proteins, which is required for normal mineralization. Removing these proteinases destroys this homeostatic balance, leading to abnormal collagen accumulations, but reductions in those non-collagenous proteins (DMP1 and DSPP) are important to mineralization.

Taken together, the DMP1, DSPP, and BMP1/TLL1 signaling pathway play vital roles in odontoblast differentiation. The defects of dentin mineralization in the whole layer after mutating these genes further strengthens the current hypothesis that dentin tubules are essential for dentin mineralization, which is not simply limited in the mineralization front.

5.6 Conclusion

The theory of the mineralization front at the edge of the dental pulp and the dentin has dominated the mineralization research field for many decades. A recent breakthrough by applications of Calcein (green color) in the *Gli1*-cre^{ERT2}; R26R^{tdTomato} mouse (to label dentin tubules) and Alizarin Red in the 2.3 Col1a1-GFP mouse (also to label dentin tubules) greatly challenges this dogma. The so-called mineralization front is composed of the mineral matrix surrounding an individual dentin tubule, which cannot be separated by the old fluorescent light microscope but is clearly visualized by the high resolution of the confocal microscope. In addition, an identical mineral deposition pattern was also demonstrated in mini-branches of dentin tubules (Fig. 5.5a). This new finding raises a new theory: mineralization takes place throughout the entire dentin matrices surrounding both large and small dentin tubules. Furthermore, two new players in dentin mineralization were identified. They are BMP1 and tolloid-like proteinases, which cleave secretory proteins such as DMP1 and DSPP. The loss of these two proteinases will lead to a drastic decrease in dentin mineralization (Fig. 5.5b). Together, these findings will shed new

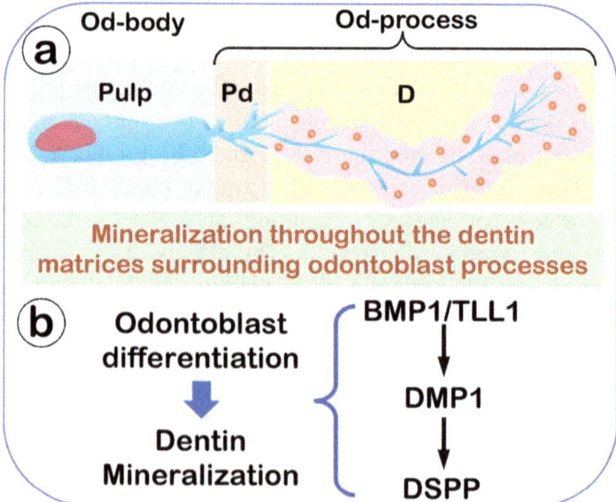

Fig. 5.5 The schematic diagram outlines a new theory of dentin mineralization: (**a**) the odontoblast process that directly contributes to mineralization (represented by pink color around Od-process), which is not only limited in the mineralization front at the edge of dentin and predentin, leading to an even distribution of mineralization in the whole dentin; (**b**) deletions of the above key molecules impair not only odontoblast differentiation, but also the whole length of dentin mineralization. Pd, predentin; D, dentin; Od, odontoblast

light on our understanding of dentin structure and function as well as the mechanisms of mineralization.

Acknowledgment This study is partially supported by NIH to JF (DE028291, DE025014, and DE025659), and NIH to Yan Jing (DE028593 and DE029541). The authors have no conflict of interest

References

Akiyama H et al (2005) Osteo-chondroprogenitor cells are derived from Sox9 expressing precursors. Proc Natl Acad Sci U S A 102:14665–14670. https://doi.org/10.1073/pnas.0504750102

Appleton J, Morris DC (1979) An ultrastructural investigation of the role of the odontoblast in matrix calcification using the potassium pyroantimonate osmium method for calcium localization. Arch Oral Biol 24:467–475. https://doi.org/10.1016/0003-9969(79)90010-4

Boyde A (1977) Qualitative electron probe analysis of secretory ameloblasts and odontoblasts in the rat incisor. Histochemistry 50:347–354. https://doi.org/10.1007/BF00507128

Carpenter TO et al (2017) Rickets. Nat Rev Dis Primers 3:17101. https://doi.org/10.1038/nrdp.2017.101

Dorvee JR, Deymier-Black A, Gerkowicz L, Veis A (2014) Peritubular dentin, a highly mineralized, non-collagenous, component of dentin: isolation and capture by laser microdissection. Connect Tissue Res 55(Suppl 1):9–14. https://doi.org/10.3109/03008207.2014.923876

Dorvee JR, Gerkowicz L, Bahmanyar S, Deymier-Black A, Veis A (2016) Chondroitin sulfate is involved in the hypercalcification of the organic matrix of bovine peritubular dentin. Arch Oral Biol 62:93–100. https://doi.org/10.1016/j.archoralbio.2015.11.008

Dutra-Correa M, Anauate-Netto C, Arana-Chavez VE (2007) Density and diameter of dentinal tubules in etched and non-etched bovine dentine examined by scanning electron microscopy. Arch Oral Biol 52:850–855. https://doi.org/10.1016/j.archoralbio.2007.03.003

Feng JQ et al (2003) The Dentin matrix protein 1 (Dmp1) is specifically expressed in mineralized, but not soft, tissues during development. J Dent Res 82:776–780. https://doi.org/10.1177/154405910308201003

Feng JQ et al (2006) Loss of DMP1 causes rickets and osteomalacia and identifies a role for osteocytes in mineral metabolism. Nat Genet 38:1310–1315. https://doi.org/10.1038/ng1905

Feng JQ, Clinkenbeard EL, Yuan B, White KE, Drezner MK (2013) Osteocyte regulation of phosphate homeostasis and bone mineralization underlies the pathophysiology of the heritable disorders of rickets and osteomalacia. Bone 54:213–221. https://doi.org/10.1016/j.bone.2013.01.046

Feng J et al (2017) BMP signaling orchestrates a transcriptional network to control the fate of mesenchymal stem cells in mice. Development 144:2560–2569. https://doi.org/10.1242/dev.150136

Fisher LJ, Goldney RD (2003) Differences in community mental health literacy in older and younger Australians. Int J Geriatr Psychiatry 18:33–40. https://doi.org/10.1002/gps.769

Fisher LW, Torchia DA, Fohr B, Young MF, Fedarko NS (2001) Flexible structures of SIBLING proteins, bone sialoprotein, and osteopontin. Biochem Biophys Res Commun 280:460–465. https://doi.org/10.1006/bbrc.2000.4146

Frost HM, Villaneuva AR (1960) Tetracycline staining of newly forming bone and mineralizing cartilage in vivo. Stain Technol 35:135–138. https://doi.org/10.3109/10520296009114729

Frost HM, Roth H, Villanueva AR, Stanisavljevic S (1961a) Experimental multiband tetracycline measurement of lamellar osteoblastic activity. Henry Ford Hosp Med Bull 9:312–329

Frost HM, Villanueva AR, Roth H, Stanisavljevic S (1961b) Tetracycline bone labeling. J New Drugs 1:206–216. https://doi.org/10.1177/009127006100100503

Frost HM, Vilanueva AR, Jett S, Eyring E (1969) Tetracycline-based analysis of bone remodelling in osteopetrosis. Clin Orthop Relat Res 65:203–217

George A, Sabsay B, Simonian PA, Veis A (1993) Characterization of a novel dentin matrix acidic phosphoprotein. Implications for induction of biomineralization. J Biol Chem 268:12624–12630

Gibson MP et al (2013) The rescue of dentin matrix protein 1 (DMP1)-deficient tooth defects by the transgenic expression of dentin sialophosphoprotein (DSPP) indicates that DSPP is a downstream effector molecule of DMP1 in dentinogenesis. J Biol Chem 288:7204–7214. https://doi.org/10.1074/jbc.M112.445775

Goldberg M, Kulkarni AB, Young M, Boskey A (2011) Dentin: structure, composition and mineralization. Front Biosci (Elite Ed) 3:711–735. https://doi.org/10.2741/e281

Hanazawa K (1917) A study of the minute structures of dentine, especially of the relationship between dentinal tubules and fibrils. Dent. Cosmos 59:374–405

Henry SP et al (2009) Generation of aggrecan-CreERT2 knockin mice for inducible Cre activity in adult cartilage. Genesis 47:805–814. https://doi.org/10.1002/dvg.20564

Holland GR (1985) The odontoblast process: form and function. J Dent Res 64(Spec No):499–514. https://doi.org/10.1177/002203458506400403

Humphreys BD, DiRocco DP (2014) Lineage-tracing methods and the kidney. Kidney Int 86:481–488. https://doi.org/10.1038/ki.2013.368

Inage T, Toda Y (1988) Phosphoprotein synthesis and secretion by odontoblasts in rat incisors as revealed by electron microscopic radioautography. Am J Anat 182:369–380. https://doi.org/10.1002/aja.1001820408

Jing Y, Hinton RJ, Chan KS, Feng JQ (2016) Co-localization of cell lineage markers and the tomato signal. J Vis Exp. https://doi.org/10.3791/54982

Kalajzic Z et al (2002) Directing the expression of a green fluorescent protein transgene in differentiated osteoblasts: comparison between rat type I collagen and rat osteocalcin promoters. Bone 31:654–660. https://doi.org/10.1016/s8756-3282(02)00912-2

Kawashima N, Okiji T (2016) Odontoblasts: Specialized hard-tissue-forming cells in the dentin-pulp complex. Congenit Anom (Kyoto) 56:144–153. https://doi.org/10.1111/cga.12169

Khatibi Shahidi M et al (2015) Three-dimensional imaging reveals new compartments and structural adaptations in odontoblasts. J Dent Res 94:945–954. https://doi.org/10.1177/0022034515580796

Kretzschmar K, Watt FM (2012) Lineage tracing. Cell 148:33–45. https://doi.org/10.1016/j.cell.2012.01.002

Li C et al (2018) Dentinal mineralization is not limited in the mineralization front but occurs along with the entire odontoblast process. Int J Biol Sci 14:693–704. https://doi.org/10.7150/ijbs.25712

Linde A, Goldberg M (1993) Dentinogenesis. Crit Rev Oral Biol Med 4:679–728. https://doi.org/10.1177/10454411930040050301

Liu T et al (2019) DMP1 ablation in the rabbit results in mineralization defects and abnormalities in haversian canal/osteon microarchitecture. J Bone Miner Res 34:1115–1128. https://doi.org/10.1002/jbmr.3683

Lorenz-Depiereux B et al (2006) DMP1 mutations in autosomal recessive hypophosphatemia implicate a bone matrix protein in the regulation of phosphate homeostasis. Nat Genet 38:1248–1250. https://doi.org/10.1038/ng1868

Lu Y, Feng JQ (2011) FGF23 in skeletal modeling and remodeling. Curr Osteoporos Rep 9:103–108. https://doi.org/10.1007/s11914-011-0053-4

Lu Y et al (2007) Rescue of odontogenesis in Dmp1-deficient mice by targeted re-expression of DMP1 reveals roles for DMP1 in early odontogenesis and dentin apposition in vivo. Dev Biol 303:191–201. https://doi.org/10.1016/j.ydbio.2006.11.001

Lundgren T, Linde A (1992) Calcium ion transport kinetics during dentinogenesis: effects of disrupting odontoblast cellular transport systems. Bone Miner 19:31–44. https://doi.org/10.1016/0169-6009(92)90842-2

Mjor IA, Nordahl I (1996) The density and branching of dentinal tubules in human teeth. Arch Oral Biol 41:401–412. https://doi.org/10.1016/0003-9969(96)00008-8
Ravindran S, George A (2014) Multifunctional ECM proteins in bone and teeth. Exp Cell Res 325:148–154. https://doi.org/10.1016/j.yexcr.2014.01.018
Reith EJ (1976) The binding of calcium within the Golgi saccules of the rat odontoblast. Am J Anat 147:267–270. https://doi.org/10.1002/aja.1001470302
Romagnani P, Rinkevich Y, Dekel B (2015) The use of lineage tracing to study kidney injury and regeneration. Nat Rev Nephrol 11:420–431. https://doi.org/10.1038/nrneph.2015.67
Salomon JP, Lecolle S, Roche M, Septier D, Goldberg M (1990) A radioautographic comparison of in vivo 3H-proline and 3H-serine incorporation in the pulpal dentine of rat molars: variations according to the different zones. J Biol Buccale 18:307–312
Sasaki T, Garant PR (1996) Structure and organization of odontoblasts. Anat Rec 245:235–249. https://doi.org/10.1002/(SICI)1097-0185(199606)245:2<235::AID-AR10>3.0.CO;2-Q
Wang J et al (2017) Essential roles of bone morphogenetic protein-1 and mammalian tolloid-like 1 in postnatal root dentin formation. J Endod 43:109–115. https://doi.org/10.1016/j.joen.2016.09.007
Weinstock M, Leblond CP (1973) Radioautographic visualization of the deposition of a phosphoprotein at the mineralization front in the dentin of the rat incisor. J Cell Biol 56:838–845. https://doi.org/10.1083/jcb.56.3.838
Wu K, Frost HM (1969) Bone formation in osteoporosis. Appositional rate measured by tetracycline labeling. Arch Pathol 88:508–510
Yamakoshi Y (2009) Dentinogenesis and dentin sialophosphoprotein (DSPP). J Oral Biosci 51:134. https://doi.org/10.2330/joralbiosci.51.134
Ye L et al (2004) Deletion of dentin matrix protein-1 leads to a partial failure of maturation of predentin into dentin, hypomineralization, and expanded cavities of pulp and root canal during postnatal tooth development. J Biol Chem 279:19141–19148. https://doi.org/10.1074/jbc.M400490200
Ye L et al (2005) Dmp1-deficient mice display severe defects in cartilage formation responsible for a chondrodysplasia-like phenotype. J Biol Chem 280:6197–6203. https://doi.org/10.1074/jbc.M412911200
Yoshiba K, Yoshiba N, Ejiri S, Iwaku M, Ozawa H (2002) Odontoblast processes in human dentin revealed by fluorescence labeling and transmission electron microscopy. Histochem Cell Biol 118:205–212. https://doi.org/10.1007/s00418-002-0442-y
Zhang H, Jani P, Liang T, Lu Y, Qin C (2017) Inactivation of bone morphogenetic protein 1 (Bmp1) and tolloid-like 1 (Tll1) in cells expressing type I collagen leads to dental and periodontal defects in mice. J Mol Histol 48:83–98. https://doi.org/10.1007/s10735-016-9708-x
Zhang H et al (2018) Transgenic expression of dentin phosphoprotein (DPP) partially rescued the dentin defects of DSPP-null mice. PLoS One 13:e0195854. https://doi.org/10.1371/journal.pone.0195854

Chapter 6
Small Leucine-Rich Proteoglycans (SLRPs) and Biomineralization

Yoshiyuki Mochida, Patricia Miguez, and Mitsuo Yamauchi

Abstract Small leucine-rich proteoglycans (SLRPs) represent the largest family of proteoglycans consisting of 18–19 members. They play diverse structural and biological roles by interacting with extracellular matrix components, growth factors, and cell surface receptors. One of the well-studied features of SLRPs is their structural control of collagen fibrillogenesis. Since fibrillar collagens are the structural basis for mineralization in bone, dentin, and cementum, such interactions may inhibit and/or facilitate mineralization. The ability to modulate specific growth factors can be used for effective mineralized tissue engineering. In this chapter, we will provide an overview of the basic structural and genomic features of SLRPs, their roles in collagen mineralization, cell signaling, and potential utility for mineralized tissue regeneration.

6.1 Introduction

Proteoglycans are a class of molecules consisting of a protein core with one or more covalently bound glycosaminoglycan (GAG) chains. GAG is a linear polymer of repeating disaccharides and, based on the combination of the two saccharide units, there are six types of GAGs, i.e., chondroitin sulfate (CS), dermatan sulfate (DS),

---

Y. Mochida
Department of Molecular and Cell Biology, Henry M. Goldman School of Dental Medicine, Boston University, Boston, MA, USA
e-mail: mochida@bu.edu

P. Miguez
Division of Comprehensive Oral Health, Adams School of Dentistry, University of North Carolina, Chapel Hill, NC, USA
e-mail: miguezp@unc.edu

M. Yamauchi (✉)
Division of Oral and Craniofacial Health Sciences, Adams School of Dentistry, University of North Carolina, Chapel Hill, NC, USA
e-mail: mitsuo_yamauchi@unc.edu

M. Goldberg, P. Den Besten (eds.), *Extracellular Matrix Biomineralization of Dental Tissue Structures*, Biology of Extracellular Matrix 10,
https://doi.org/10.1007/978-3-030-76283-4_6

keratan sulfate (KS), heparan sulfate (HS), heparin (Hep), and hyaluronic acid (HA). Except for HA, GAGs are sulfated. Due to their highly negative charged nature, they interact with a number of positively charged molecules (see a review by Vallet et al. 2020). Proteoglycans are highly heterogeneous molecules and localized inside, on the surface, and outside of the cells (Karamanos et al. 2018) playing both structural and biological roles.

"SLRPs", "*S*mall *L*eucine-*R*ich *P*roteoglycans" (Iozzo and Murdoch 1996), were originally defined as proteoglycans with a relatively small protein core (36–42 kDa) harboring tandem leucine-rich repeats (LRRs), a hallmark motif LxxLxLxxNxL (L: leucine which can be substituted by isoleucine, valine, and other hydrophobic amino acids; x: any amino acids), with several posttranslational modifications including substitution with GAG side chains of various types.

Through the protein core, SLRPs are capable of binding a number of proteins including fibrillar collagens, growth factors including transforming growth factor-β (TGF-β), platelet-derived growth factor (PDGF), vascular endothelial growth factor A (VEGFA), connective tissue growth factor, and cell surface receptors including epidermal growth factor receptor (EGFR), c-Met, insulin-like growth factor receptor I (IGF-IR), vascular endothelial growth factor receptor 2 (VEGFR2), Toll-like receptors (TLRs). As such they are involved in a plethora of biological and pathological processes such as development, inflammation, autophagy, angiogenesis, and tumorigenesis (Iozzo and Schaefer 2015).

Thus, it is not surprising that SLRP members play important roles in the formation and maintenance of skeletal tissues including mineralized tissues like bones and teeth (enamel, dentin, and cementum)—see reviews by Nikitovic et al. (2012), Zappia et al. (2020), and Kram et al. (2020). In this chapter, we will provide an overview of the basic structures and functions of SLRPs, then focus on the specific SLRPs and their roles in mineralized tissues, especially, collagen-based mineralized dental tissues such as dentin and cementum, and bone, and their therapeutic potentials.

6.2 SLRP Family

There has been a number of SLRP genes identified by biochemical and molecular approaches. As such, the SLRP represents the largest family of proteoglycans (Karamanos et al. 2018) composed of 18–19 members (Iozzo and Schaefer 2015) (see below). These members are secreted extracellular matrix (ECM) molecules except nyctalopin that is located at the outer leaflet of the plasma membrane. Based on the protein, genomic structure, and chromosomal organization (see below), they are classified into five major classes. With the increase in the number of SLRP members, the definition and classification of SLRPs became less clear. For instance, the size of the protein core of some members such as chondroadherin-like is relatively large (~83 kDa) when compared to those of typical SLRP members (36–42 kDa). Some of the members such as asporin and chondroadherin lack GAG chains; thus, by definition, they are not "proteoglycans". Nonetheless, in the following section, we will discuss SLRP members which (1) have "relatively" small protein core as compared to large proteoglycans like aggrecan and versican (>200 kDa), (2) do not have any transmembrane domains, and (3) have multiple LRR domains. Some other structural features of the protein core are also briefly discussed.

6.2.1 Protein Structures/Domains

1. BMP-1/mammalian tolloid proteinase-processing motif
   It has been reported that some SLRP family members are substrates for BMP-1, mammalian tolloid (mTLD), and mTLD-like proteinases (Scott et al. 2000; Von Marschall and Fisher 2010; Ge et al. 2004). It is well known that several ECM proteins including procollagen $\alpha 1(I)$, $\alpha 2(I)$, $\alpha 1(II)$, $\alpha 1(III)$, chordin, prolysyl oxidase are also processed by this protease family (Scott et al. 2000). BMP-1 and mTLD have been demonstrated to play important biological roles in the formation of ECM, for example, cleavage of the C-propeptides of procollagens I, II, and III (Kessler et al. 1996; Li et al. 1996) allowing collagen fibrillogenesis, and cleavage of the prodomain of prolysyl oxidase leading to its enzymatic activation to initiate cross-linking of collagen and elastin (Uzel et al. 2001). Among SLRP members, biglycan was the first that was shown to be secreted as a pro-form and cleaved by BMP-1 at the M(M/L)N-DEE site (Scott et al. 2000). Decorin also has a similar sequence of M(L/I)E-DE(A/G) that was susceptible to cleavage by BMP-1/mTLD members (Von Marschall and Fisher 2010). The immunoelectron microscopic study revealed that distribution of pro-decorin and pro-biglycan are more readily detected in predentin as compared to dentin (Septier et al. 2001) indicating that active enzymatic processing occurs prior to matrix mineralization. Osteoglycin, one of class III SLRP members, has also been reported that BMP-1/mTLD proteinases process its pro-form. Moreover, like decorin, the mature osteoglycin was shown to delay the rate of type I collagen fibrillogenesis in vitro as compared to the pro-form, indicating the cleavage of the prodomain enhances its biological activity (Ge et al. 2004). The significance of the pro versus mature forms of SLRPs in the process of mineralization is still not clear.
2. Cysteine residues cluster motif
   There are four cysteine residues commonly found at the N-terminal region, which is called a "capping motif". This motif is often used to classify the SLRP members mainly based on the different spacing between cysteine residues. In general, class I has CX_3CXCX_6C, class II CX_3CXCX_9C, class III CX_2CXCX_6C, class IV CX_3CXCX_8C or CX_3CXCX_6C or $CX_3CXCX_{17}C$, and class V CX_3CXCX_7C (Fig. 6.1). Interestingly, chondroadherin-like, a newly identified SLRP member, has this capping motif not only at the N-terminal but also at the mid-region of the protein. The four cysteine residues are known to form a disulfide bond between the first LRR domain (LRR1) and β-hairpin structure.

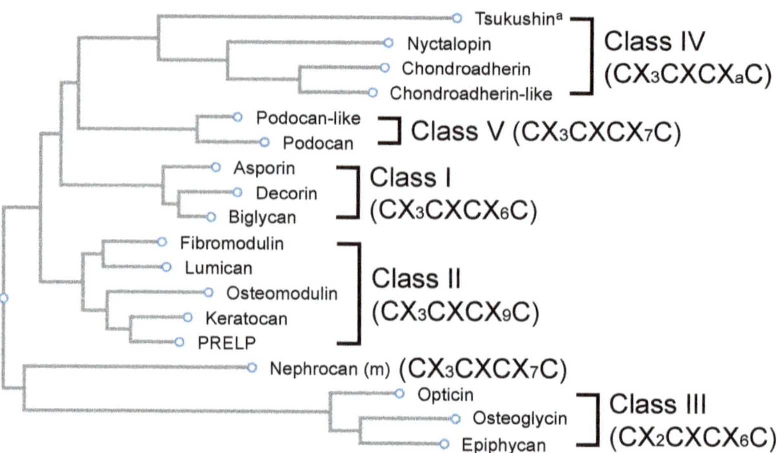

Fig. 6.1 Phylogenetic analysis of all SLRP members. The SLRP protein sequences were obtained from the public database as the NCBI reference sequences. Phylogenetic analysis was performed using PhyML software program based on the multiple sequence alignment of these protein sequences by the Clustal-W program (Guindon and Gascuel 2003). The NCBI reference sequences used are: mouse Decorin (NP_001177380.1), mouse Biglycan (NP_031568.2), mouse Asporin (NP_001165952.1), mouse Lumican (NP_032550.2), mouse Fibromodulin (NP_067330.1), mouse Osteomodulin (NP_001347637.1), mouse PRELP (NP_473418.3), mouse Keratocan (NP_032464.1), mouse Osteoglycin (NP_032786.1), mouse Opticin (NP_473417.2), mouse Epiphycan (NP_001366385.1), mouse Nyctalopin (NP_775591.1), mouse Chondroadherin (NP_031715.1), mouse Chondroadherin-like (NP_001157792.1), mouse Podocan (NP_001272885.1), mouse Podocan-like (NP_001013402.2), mouse Tsukushin (NP_001019790.1), mouse Nephrocan (NP_079960.1). [a]Tsukushin is a mouse ortholog of human Tsukushi. (m) indicates that Nephrocan is identified in mouse but not human. Note that the N-terminal Cysteine residues cluster varies in class IV members, i.e. CX_3CXCX_8C found in Chondroadherin and Chondroadherin-like, CX_3CXCX_6C in Nyctalopin and $CX_3CXCX_{17}C$ in Tsukushin

At the C-terminal region, SLRP members have the capping motif where two cysteine residues form a disulfide bond (Mcewan et al. 2006). One of the two C-terminal cysteines usually located at one LRR domain and another at the subsequent LRR. This has been well presented by the crystal structural analysis of decorin (Scott et al. 2004). There is one cysteine residue in LRR11, which is longer than other LRR motifs in decorin, bonded with another cysteine residue in LRR12. This longer LRR has been called "ear repeat" (Mcewan et al. 2006). It is considered that both N-terminal and C-terminal cysteine motifs of SLRPs would help stabilize the central LRR domain (Iozzo and Schaefer 2015).

There are several SLRP members expressed in the cornea, a fibrillar collagen-rich tissue. The mice lacking lumican, a collagen-binding SLRP, exhibits abnormal size and shape of collagen fibrils leading to cornea opacity (Chakravarti et al. 2000). Congenital stromal corneal dystrophy (CSCD) is a rare autosomal dominant eye disease characterized by corneal stromal opacification and visual loss with increasing corneal thickness and abnormal fibrils in the stroma. This disease is caused by the heterozygous mutation in the human decorin gene and the mutations cause frameshift leading to a truncation of the decorin protein (Bredrup et al. 2005; Rodahl et al. 2006; Jing et al. 2014). The truncated decorin lacks the C-terminal 33 amino acids, which disrupted the "ear repeat" domain. It is thus likely that loss of stabilization of the central LRR domain by truncation of the ear repeat domain could lead to abnormal collagen assembly.

3. Glycosaminoglycan (GAG) attachment and other unique domains

   As described above, many of the SLRP members are proteoglycans with covalently bound GAG(s) but some are known as "part-time" proteoglycans, i.e., partially or completely lacking GAG components. It is known that the CS/DS-type GAG attachment is initiated by transferring a UDP-xylose residue to specific serine moieties. This rate-limiting step is catalyzed by two enzymes, namely xylosyltransferase (XYLT) I and II. The consensus sequence for xylosylation has been reported as aaaa-GSG-aa/G-a, where a is aspartic acid (D) or glutamic acid (E) (Roch et al. 2010). For instance, human decorin contains a sequence of DEASGIGP and human biglycan, which has two GAG attachment sites, contains a sequence of DEEASGADTSGVLD (underlined amino acids are met with consensus). KS type GAG attachment is a more complicated process and there have been three types of KS structures identified. Keratan sulfate in the cornea (KS-I) is attached to asparagine via a complex-type N-linked oligosaccharide, while in cartilage, KS-II is O-linked via GalNAc to serine or threonine residues via the mucin core-2 structure. KS-III identified in the brain is attached mannose O-linked to serine (Funderburgh 2002). The GAG attachment is also required for efficient secretion of the proteoglycan (Seo et al. 2005), collagen interaction (Ruhland et al. 2007; Raspanti et al. 2008), and interaction with Lyme disease spirochete Borrelia burgdorferi (Benoit et al. 2011).

   Within the protein core, some SLRP members have unique acidic amino acid structures outside of LRRs. Fibromodulin, osteoadherin/osteomodulin, and lumican, for instance, are known to have tyrosine sulfate residues, i.e., fibromodulin has up to nine sites at the N-terminal region prior to LRR1, osteoadherin has up to six sites at the N-terminal and two at the C-terminal regions, and lumican has up to two sites at the N-terminal region (Antonsson et al. 1991; Onnerfjord et al. 2004). Tyrosine sulfation is considered to enhance protein interaction. Another example is asporin, a class I SLRP member, whose name reflects the unique poly aspartate-stretch at the N terminus. It has been reported, in osteoarthritis, the human asporin allele having 14 aspartic acid repeats, designated D14, was overrepresented relative to the common allele having 13 aspartic acid repeats (D13), and its frequency increased with disease severity (Kizawa et al. 2005).

4. LRR motifs

   The core proteins of all SLRP members contain a central region composed of LRRs flanked by disulfide-bonded terminal domains as described above.

   There are two types of LRR motifs that are based on the general consensus sequence (Matsushima et al. 2000), i.e., type T (zzxxaxxxxFxxaxxLxxLxLxxNxL) and type S (xxaPzxLPxxLxxLxLxxNxI),

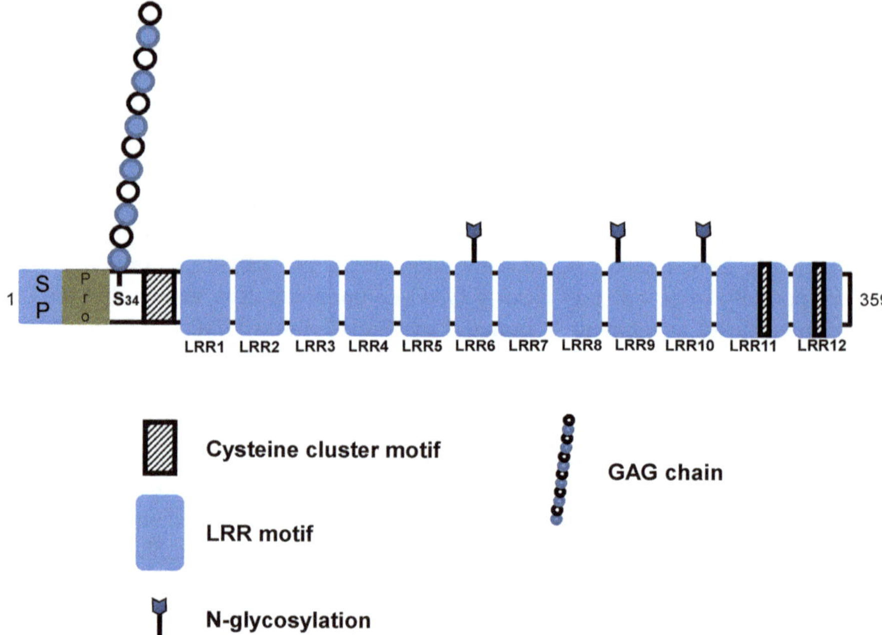

Fig. 6.2 Decorin domain structure. Schematic representation of protein domain structures in human decorin. Human decorin protein structure is illustrated (359 amino acids based on NCBI Reference Sequence; NP_001911.1). Each domain is symbolized such as Signal peptide domain (SP), prodomain (Pro), glycosaminoglycan (GAG) attachment site, cysteine cluster motifs at N and C terminal parts, Leucine-Rich Repeat (LRR) motif and N-glycosylation sites. S^{34} indicates serine residue with a GAG attachment site

where “z” is frequently a gap, “x” indicates variable residues, “a” is valine, leucine, or isoleucine and I is isoleucine or leucine. Based on the criteria, class I and II SLRPs have 12 LRR motifs composed of four tandem STT “super-motifs,” i.e., (STT)4. Class III SLRPs, including epiphycan and osteoglycin, have seven LRRs composed of (ST)T(ST)2. Chondroadherin and nyctalopin, class IV members, contain only type T LRRs.

The LRR domains of many SLRP members bind collagens and modulate collagen fibrillogenesis in vitro. Indeed, decorin is named because of its ability to bind, thus, “decorate” collagen fibrils (Krusius and Ruoslahti 1986). Figure 6.2 shows a schematic representation of a human decorin protein structure containing 12 LRRs. SLRP-collagen binding has been extensively studied for the last two decades. It has been demonstrated that a small region of LRR6-LRR7 in decorin is responsible for binding type I collagen and the synthetic peptide of SYIRIADTNIT derived from decorin inhibits the binding of full-length decorin to type I collagen in vitro (Kalamajski et al. 2007). In another report, a region of LRR1 in fibromodulin (RLDGNEIKR) was shown to bind type I collagen (Kalamajski and Oldberg 2007). Apparently, lumican and fibromodulin compete with each other for the collagen binding indicating that both bind the same or close region of collagen (Svensson et al. 2000). It has also been demonstrated that a region of LRR7 in lumican binds type I collagen (Kalamajski and Oldberg 2009). These findings suggest that the binding of multiple SLRPs may regulate collagen fibrillogenesis in a concerted manner that may contribute to the formation of specific ECM architecture.

6.3 SLRPs in Mineralized Tissues and Their Potential Functions

Biglycan and decorin, two of the most well-studied SLRP members, were first identified in bone (Fisher et al. 1983; Fisher et al. 1987) and named PG1 and 2, respectively. Extraction and identification of proteins in mineralized tissues such as bones and teeth are challenging because the proteins are embedded and protected by the minerals (see below). Thus, in order to isolate and characterize SLRPs in mineralized tissues, an extra step of "demineralization" is often required.

During the last several decades, several GAG components and SLRP members in dentin, cementum, and bone were characterized by biochemical, histochemical, immunohistochemical, immunogold-labeling, and spectroscopic (FTIR/Raman) approaches.

6.3.1 Identification and Localization of SLRPs in Pre- and Mineralizing-Fractions and Available Imaging Techniques

Several studies have employed multiple approaches to identify and localize SLRP members in predentin-dentin, precementum-cementum as well as in osteoid-mineralized bone matrices (Ababneh et al. 1999; Cheng et al. 1996, 1999; Goldberg et al. 2003; Hall and Embery 1997; Hoshi et al. 2001). Some of the studies on SLRPs in these tissues have been done biochemically by differential fractionation of mineralized matrices into mineral-unbound and -bound fractions by using chaotropic agents, e.g., guanidine-HCl, followed by demineralization with EDTA. Then, SLRPs could be isolated and identified from each fraction by a series of chromatography, Western blot, and N-terminal amino acid sequence analyses (Cheng et al. 1996, 1999). Studies in the late 1990s by Embery and Yamauchi's groups (Ababneh et al. 1999; Cheng et al. 1996) are examples of two early and thorough evaluations of SLRPs in cementum. Below, we describe some of the approaches to localize GAGs, proteoglycans, and SLRPs in these mineralized tissues:

1. Histochemical approach: Proteoglycans and their GAGs can be detected in tissues by histological staining using cationic dyes such as Safranin O, Cuprolinic blue, Alcian blue, and Toluidine blue (Hyllested et al. 2002). These dyes can bind

anionic components such as GAGs and proteoglycans. One of the early methods was the staining with Cuprolinic blue (Sarathchandra et al. 2002). The native structure of proteoglycans appears as a stellate rod-like structure (Goldberg and Takagi 1993). Alcian blue staining which has a copper atom in its chromophore can be used for electron and light microscopy (Scott 1972; Tice and Barrnett 1965). Toluidine blue reacts with the negatively charged groups to produce a different intensity of colors depending on the extent of carbohydrates or other negatively charged molecules. In dental tissues, Goldberg and Takagi provided an extensive review on GAGs/proteoglycans in predentin and dentin characterized by using multiple cationic dyes, hyaluronidase-gold labeling, lectin-labeling, specific antibodies, and autoradiography. Based on the staining patterns, they proposed that there are two distinct groups of proteoglycans, one mineral inhibitor abundant in predentin and another mineralization promotor localized in dentin (Goldberg and Takagi 1993). Goldberg and Septier showed another method to stain GAGs in predentin and dentin by using various concentrations of magnesium chloride and showed how to best identify the presence of GAGs, although the type of GAGs could not be determined by this method (Goldberg and Septier 1992). Alcian blue, colloidal iron for acidic proteoglycans, and periodic acid-Schiff (PAS) for glycoproteins have also been used in combination to characterize bone proteoglycans in cortical versus trabecular; highly mineralized versus osteoid. Tissues rich in proteoglycans interfibrillarly (such as medullary bone area) is highly stained by these dyes (Bonucci and Gherardi 1975). GAGs can also be characterized by radiolabeling sulfate groups and quantified (Maccarana et al. 2017), by fluorophore-assisted carbohydrate electrophoresis (FACE) (Karousou et al. 2014), or by liquid chromatography combined with mass spectroscopy (Huang et al. 2013; Kubaski et al. 2017).

2. Immunohistochemistry (IHC): When antibodies against specific proteoglycan molecules with no clear cross-reactivity to other molecules are used, IHC is a powerful approach to localize and semi-quantify the molecules within tissues. One of the early reports is by Embery's group who utilized antibodies for versican (a large CS proteoglycan), biglycan, decorin, and lumican to examine their distribution in cementum. Samples were from human teeth with various age groups and healthy versus periodontitis groups. There was no difference among different age groups or disease conditions. But it was found through IHC that acellular cementum did not have those proteoglycans but cellular cementum did. Biglycan was concentrated around cementocytes and incremental lines (Ababneh et al. 1999). The distribution of fibromodulin was described by IHC in rat dental and periodontal tissues during cementogenesis and root development corroborating its role in mineralization (Matias et al. 2003).

   Hall and co-workers reported that lumican is highly concentrated in predentin compared to dentin matrix in human teeth (Hall and Embery 1997). This is similar to the results reported by Cheng et al. showing that fibromodulin and lumican are almost exclusively localized in precementum and lacunae housing cementocytes, thus, pre-/non-mineralized matrices (Cheng et al. 1996). The latter group further reported that decorin and biglycan were also preferentially localized in pre-/non-mineralized matrix in cementum (Cheng et al. 1999). The tissue distribution of decorin and fibromodulin in bone, periodontal ligament, and cementum were examined using biglycan KO and wild-type mice revealing a compensatory function for decorin in periodontal tissues in the absence of biglycan (Chiu et al. 2012).

3. Electron microscopy: In the last 20 years, electron microscopy has been widely used to localize proteoglycans in dental tissues by labeling SLRPs with colloidal gold particles. Using antibodies against pro- and mature-forms of decorin and biglycan, Septier and co-workers semi-quantified the immunoreactivities of these SLRP forms in rat incisors and demonstrated that the distribution pattern of the pro- and mature-forms of these SLRPs are distinct from each other and their concentrations change from predentin to dentin transition (Septier et al. 2001). If double labeling of an SLRP and collagen using different sizes of gold particles is used, a spatial correlation between the specific SLRP and collagen fibrils can be assessed. Using this technique, Orsini and co-workers examined the distribution of decorin and biglycan in human dentin and found that both SLRPs are associated with collagen fibrils in predentin and their immunoreactivities are weak in mineralized intertubular dentin (Orsini et al. 2007). Decorin appears to be abundant around odontoblast processes (Orsini et al. 2007).
4. Fourier transform infrared imaging spectroscopy (FT-IRIS) and Raman spectroscopy: FT-IRIS imaging is sensitive to detect matrix composition including collagens and proteoglycans, and can be used alone or in combination with Near-Infrared spectroscopy which is highly sensitive to water content. Proteoglycans are measured in areas of absorbance at 985–1140 cm^{-1} (while collagen CH2 side chain is at 1326–1356 cm^{-1}). These absorbance bands have been validated against histological staining in native, diseased, and engineered cartilage (Boskey and Pleshko Camacho 2007). Proteoglycan content and distribution can be evaluated across tissue and even correlated with mineral distribution and collagen maturation. With this technique, relative collagen content increased from cartilage toward bone while proteoglycan content is higher in deeper areas of the cartilage tissue (Khanarian et al. 2014). Raman Spectroscopy is also capable of assessing proteoglycan content in soft and mineralized tissues and may have an advantage over FT-IRIS because its spectral band for proteoglycan (~1375 cm^{-1}) does not overlap with the phosphate band that is close to that of proteoglycan in the case of FT-IRIS (Gamsjaeger et al. 2014). Although both techniques can provide relative proteoglycan content, they cannot identify the specific proteoglycan species. However, Raman spectral signature comes from the glycosaminoglycan chains which are mostly of CS nature (~90% of total GAG content in bone and almost 100% in mineralized dentin) in mineralized tissues such as bone and dentin and are associated mostly with decorin and biglycan (Paschalis et al. 2017; Waddington et al. 2003).

6.3.2 Collagen-Based Mineralization and SLRPs

Bone, dentin, and cementum are primarily composed of two phases, inorganic minerals (mostly, carbonated hydroxyapatite) and organic fibrillar type I collagen. In these tissues, collagen fibrils are formed prior to mineralization creating a pre-mineralizing matrix, i.e., osteoid, predentin, and precementum, then mineralization occurs in and around the collagen fibrils (Mahamid et al. 2010). Though the exact mechanism of mineralization and the location of minerals are still controversial (Boskey 1998; Landis et al. 1996; Schwarcz 2015; Georgiadis et al. 2016; Xu et al. 2020), a number of early and recent in vitro and in vivo observations indicate that the majority of minerals are first deposited in the contiguous hole zones in the collagen fibrils, i.e., intermolecular channels (Katz and Li 1973; Weiner and Traub 1986, Landis et al. 1993, Nudelman et al. 2010, Wang et al. 2012, Zhou et al. 2016, Xu et al. 2020). Thus, minerals in the fibrillar collagens are deposited and organized in a highly specific manner resulting in a very stiff and durable biomaterial (Glimcher 1984). Unquestionably, collagen fibrils control the spatial aspect of mineralization in these tissues by defining the space for mineral deposition and growth. Apparently, as minerals grow in the fibrils, the fibrils are structurally distorted as observed in periodontal ligament-cementum interface (Quan and Sone 2015), turkey leg tendon (Fratzl and Daxer 1993; Yamauchi and Katz 1993), and bone (Burger et al. 2008) resulting in increasing intrafibrillar space to accommodate the final mass of minerals. There has been controversial if non-collagenous proteins are required to induce collagen mineralization, but some of the recent in vitro studies show that collagen alone is capable of initiating nucleation (Nudelman et al. 2010; Xu et al. 2020).

Though collagen fibrils alone are capable of directing crystal nucleation and growth (Silver and Landis 2011; Wang et al. 2012; Xu et al. 2020), the presence of non-collagenous proteins that bind collagen fibrils including SLRPs has a significant impact on the mineral ions sequestration and release, and the hydration state around collagen fibrils. The calcium-binding phosphorylated acidic proteins such as SIBLINGs (small integrin-binding ligand N-linked glycoproteins) including dentin sialophosphoprotein, dentin matrix protein 1, and bone sialoprotein were shown to bind both calcium ion and fibrillar type I collagen (Traub et al. 1992; George et al. 1996; He and George 2004; George and Veis 2008). By binding near the gap regions/hole channels of collagen fibrils, they may efficiently facilitate the initiation of collagen mineralization. The fact that various mutations in the human dentin sialophosphoprotein gene lead to type II dentinogenesis imperfecta supports this notion (Kim and Simmer 2007).

The proteoglycans in the dentin matrix represent less than 3% by volume (Bertassoni 2017), however, they have been found to play significant roles in tissue mineralization. In dentin and cementum, the most well-characterized and the major SLRPs are in class I (decorin and biglycan), and class II (fibromodulin, lumican, and osteoadherin/osteomodulin) members. In dentin and cementum, mineralization occurs in and around type I collagen fibrils, thus, collagen-binding SLRPs such as decorin, fibromodulin, and lumican likely play an important role in this process (Chen and Birk 2013). Though other members including biglycan and osteoadherin (aka osteomodulin) have been reported to bind collagen fibrils (Schonherr et al. 1995; Svensson et al. 1995; Tashima et al. 2015, 2018), their binding is relatively weak. Indeed, gain- and loss-of-function experiments indicated that the amounts of biglycan do not significantly affect collagen fibrillogenesis like decorin does (see below) (Parisuthiman et al. 2005; Mochida et al. 2009).

6.3.3 GAGs in Mineralization

In the predentin, biglycan and decorin are mostly modified with DS, however, in the predentin-dentin interface, CS becomes the major form, and in mineralized dentin. CS is the only GAG chain identified (Waddington et al. 2003), while KS distribution in the predentin forms a gradient with a maximum concentration toward the mineralization front (Goldberg et al. 2003). Moreover, the position of the sulfate within the GAG also varies within these tissues, and the length of GAG chains is longer in dentin than those in predentin and in the dentin/predentin interface (Waddington et al. 2003). For the early studies on the distribution and potential functions of various GAG species in dentin mineralization, see Embery's review (Embery et al. 2001). Apparently, GAG components contribute to the mechanical properties of the dental tissues (De Mattos Pimenta Vidal et al. 2017) because of their physicochemical nature, capabilities of occupying a large space and retaining water, and forming bridges between collagen fibrils. GAGs are also thought to play a role in mineralization; in dentin, CS may facilitate the mineralization process by sequestering calcium ions.

6.3.4 SLRP Functions in Biomineralization

In our early studies, we identified several SLRP members in tooth cementum and dentin. In the former, lumican and fibromodulin were identified by immunohistochemical, Western blot, and amino acid sequence analyses. The former was the major species and the latter a minor, but both SLRPs were almost exclusively located in pre-mineralizing, i.e., precementum, and non-mineralized, i.e., pericementocytes, matrices (Cheng et al. 1996). In the same tissue, we also identified chondroitin-4-sulfate SLRPs, decorin, and biglycan. The former is closely associated with collagen fibers in the periodontal ligament and cementum, the latter with cementoblasts and precementum (Cheng et al. 1999). By employing a similar approach, we also identified and immunolocalized lumican and fibromodulin in predentin and dentin, and demonstrated that both are also mainly localized in pre- (predentin) and non-mineralized (peri-odontoblast processes) matrices (Yamauchi et al. 2000). Thus, in both mineralized dental tissues, these SLRP members appear to locate mainly in the "pre-mineralized" matrices, i.e., precementum and predentin, and "non-mineralized" matrices, i.e., lacunae housing cementocytes and intratubular dentin. These preferential localizations of SLRPs (and proteoglycans in general) in dental tissues have been reported by a number of groups (Takagi et al. 1990; Hall et al. 1997; Goldberg et al. 2005; Orsini et al. 2007; Nikdin et al. 2012; Randilini et al. 2020).

It has been reported that decorin is removed prior to bone mineralization (Hoshi et al. 1999). Similarly, fibromodulin that is abundant in articular cartilage surface decreases toward the cartilage–bone interface (Hedlund et al. 1994). Another class II SLRP member, ostoadherin, aka osteomodulin, is also almost exclusively localized in predentin but more concentrated toward the mineralization front (Nikdin et al. 2012).

All of these studies clearly demonstrate that these collagen-binding SLRPs are preferentially localized in pre-/non-mineralizing areas of mineralized tissues but at or prior to mineralization, they may be removed or partially degraded (Septier et al. 2001). Based on these observations and a gradient distribution pattern, some groups (Hall et al. 1997; Nikdin et al. 2012) speculate that some of these SLRP members play a critical role in facilitating matrix mineralization. However, on the contrary, some groups reported that these SLRPs rather inhibit premature collagen mineralization (Hoshi et al. 1999; Cheng et al. 1996; Yamauchi et al. 2000). Such inhibition can be done by binding and compressing the collagen fibrils through water-retaining GAG chains and/or by blocking the mineralization sites. Then after their removal/degradation, collagen fibrils get decompressed, thus, increases the intrafibrillar space to get mineralized (Cheng et al. 1996, 1999; Yamauchi et al. 2000) (Fig. 6.3). The decorin and biglycan gene knockout mouse models showed some phenotype in dentin, but it seems to be transient and self-repair (Goldberg et al. 2005). Biochemical studies in the past have indicated that there are two groups of proteoglycans: one in predentin matrix and the other transported to the mineralization front rapidly and incorporated into mineralized dentin matrix (Lormee et al. 1996). Apparently, fibromodulin and lumican belong to the former group and the regulation of collagen mineralization could be one of the functions of these proteoglycans.

The above hypotheses on the role of SLRPs in inhibition and/or facilitation of dentin and cementum (bone as well) mineralization, are largely based on observational and correlative studies. To prove these functions, one viable approach is a gain- and loss-of-function study.

6.3.5 *Effect of SLRPs on Biomineralization Through Gain- and Loss-of-Function Approaches* In Vitro

To understand the molecular function of SLRPs in the biomineralization of dental tissues, several in vitro studies have been conducted. However, such studies are very limited largely due to the unavailability of well-established odontogenic cell lines such as odontoblasts, ameloblasts, and cementoblasts. Thus, one approach would be

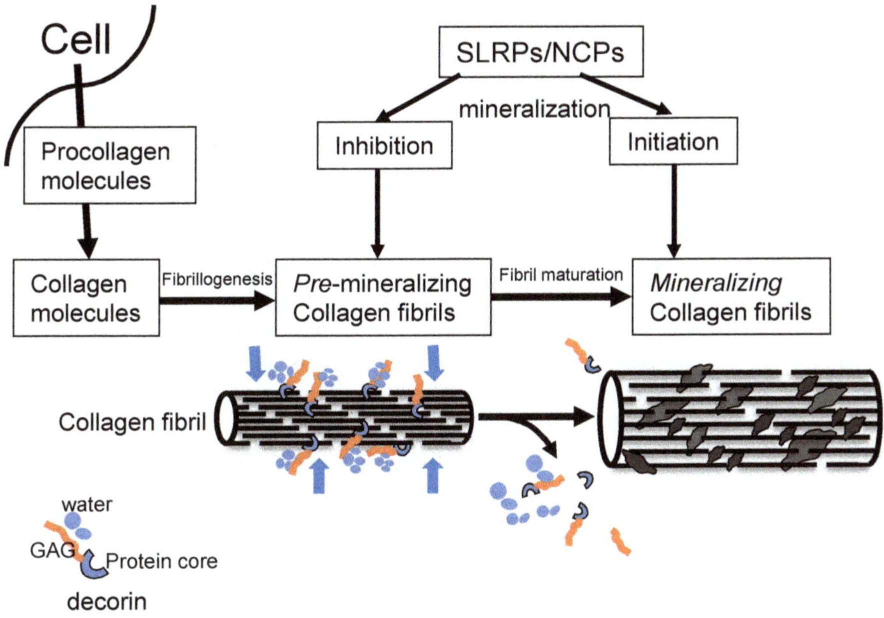

Fig. 6.3 Potential roles of collagen-binding SLRPs in collagen mineralization. In the pre-mineralizing state, they bind collagen fibril at or near the gap regions through the core proteins. Their GAG components attract and hold water molecules, thus, apply osmotic pressure to the fibril. These may inhibit premature collagen mineralization. Decorin is used as an example for this process in the illustration but other SLRPs such as fibromodulin and lumican may play this role as well. These SLRPs could be removed or degraded at/prior to mineralization, which decompresses the fibril and exposes the gap region of the fibril for mineralization. Some of the SLRPs/non-collagenous proteins such as SIBLINGs may, however, facilitate mineralization by sequestering and releasing calcium ions to collagen mineralization sites. Relative scale is not taken into consideration in this illustration

the use of a well-established osteoblastic cell line, e.g., MC3T3-E1 cells. These were originally established from newborn mouse calvaria (Sudo et al. 1983) and several subclones were further established (Wang et al. 1999). Subclone 26, which has a high differentiation/mineralization potential, has been widely used for functional studies (Mochida et al. 2003, 2009; Parisuthiman et al. 2005). Using this cell line, a single-cell-derived clones expressing higher or lower levels of decorin or biglycan were generated and characterized (Mochida et al. 2003; Parisuthiman et al. 2005). The levels of decorin did not affect cell proliferation or osteogenic marker expression. However, the decorin overexpression led to severely delayed mineralization, while its suppression resulted in accelerated mineralization. The timing of mineralization was inversely correlated to the expression levels of decorin in these clones (Mochida et al. 2003). The major phenotype in these clones was collagen fibrillogenesis. When decorin was overexpressed, collagen fibrils were markedly thinner, while decorin was suppressed, they were larger and irregular in shape. In the latter, the crystallinity of minerals was poor (Mochida et al. 2009). These results suggest that decorin inhibits mineralization by modulating collagen assembly. When the same approach was employed for biglycan, it was found that biglycan rather accelerates mineralization by positively modulating bone morphogenetic protein (BMP)2/4 function, thus, osteoblast differentiation. The effect of biglycan levels on collagen fibrillogenesis was minimal (Parisuthiman et al. 2005; Mochida et al. 2006). These results were confirmed by in vivo cell transplantation assays (Mochida et al. 2009; Parisuthiman et al. 2005). These studies indicate that these structurally similar class I SLRP members have distinct roles in matrix mineralization, i.e., decorin inhibits this process by controlling collagen fibrillogenesis while biglycan accelerates by positively modulating osteogenic BMP function.

To investigate the function of osteoadherin (aka osteomodulin), a class II SLRP member, in mineralization, a similar approach was also employed by using MC3T3-E1 cells (Rehn et al. 2008). Overexpression of osteoadherin reduced cell proliferation, increased alkaline phosphatase activity, and led to accelerated in vitro mineralization. The expression levels of osteocalcin (*Ocn*) and osteoglycin (*Ogn*), a class III SLRP, were upregulated, while bone sialoprotein (*Bsp*) level was unchanged in these overexpression clones. Repression of osteoadherin resulted in increased cell proliferation and reduced alkaline phosphatase activity; however, the extent of in vitro mineralization was unchanged. The expression of the level of *Ocn* and *Bsp* were unaffected while that of *Ogn* was downregulated. The results in this study appeared inconclusive possibly because these stably transfected clones were not established as a single-cell-derived, isolated clones, but maintained as a mixed population.

These studies described above may pose a limitation as only one cell type from calvaria/intramembranous ossification was investigated. However, the MC3T3-E1 cell is probably the only well-established non-transformed osteoblastic cell line that undergoes a normal process of differentiation and collagen mineralization with minimal inter-clonal variations. Thus, this is an excellent model to examine the roles of matrix molecules such as SLRPs in collagen fibrillogenesis and mineralization.

6.3.6 Gene KO Models: SLRPs Affect Bone/Tooth Structures/Mineralization

There have been 18 members of SLRPs identified in mice (Fig. 6.1), and almost all of the genes are targeted by conventional and/or conditional knockout (KO) technology based on the Mouse Genome Informatics (http://www.informatics.jax.org/). Among these KO mice, thus far, all of the published reports were based on the conventional KO approaches. In the following, we focus on the studies characterizing dental/bone phenotypes in several KO mouse models. Major dental/bone/collagen fibril phenotypes in these studies are summarized in Table 6.1.

Table 6.1 SLRP KO mouse phenotype. Dental tissue phenotype (enamel and dentin) and collagen fibril phenotype (dentin, bone and non-mineralized tissues) are summarized

Gene(s) in KO mouse	Enamel phenotype	References
Biglycan	Increased thickness of enamel matrix possibly due to increased amelogenin expression	Goldberg et al. (2005), Chiu et al. (2012)
Decorin	Enamel matrix is almost absent/delayed, but no structural abnormality	Goldberg et al. (2005)
Fibromodulin	Decreased thickness of enamel matrix, porous enamel	Goldberg et al. (2006, 2009)
Gene(s) in KO mouse	**Dentin phenotype**	**References**
Biglycan	Porous and hypomineralized	Goldberg et al. (2005)
Decorin	Porous and hypomineralized	Goldberg et al. (2005)
Fibromodulin	Hypomineralized, larger fibril diameter	Goldberg et al. (2006)
Gene(s) in KO mouse	**Collagen fibril phenotype in dentin**	**References**
Biglycan	No statistical difference between KO and WT	Goldberg et al. (2005)
Decorin	Increased collagen fibrils in distal predentin	Goldberg et al. (2005)
Gene(s) in KO mouse	**Collagen fibril phenotype in bone**	**References**
Biglycan	Larger fibril diameter, abnormal collagen fibrils	Corsi et al. (2002)
Decorin	Smaller fibril diameter	Corsi et al. (2002)
Biglycan/Decorin	Non-circular collagen fibril profile	Corsi et al. (2002)
Gene(s) in KO mouse	**Collagen fibril phenotype in non-mineralized tissues**	**References**
Biglycan	Larger fibril diameter and fused collagen fibrils in skin	Corsi et al. (2002)
Decorin	Larger fibril diameter and fused collagen fibrils in skin	Corsi et al. (2002)
Biglycan/ Decorin	More marked variability in diameter and shape, wider interfibrillar spaces in skin	Corsi et al. (2002)
Fibromodulin	Smaller fibril diameter, abnormal collagen fibrils in tendon	Svensson et al. (1999)
Lumican	Larger fibril diameter, abnormal collagen fibrils in tendon	Jepsen et al. (2002)
Fibromodulin/ Lumican	Fused collagen fibrils, smaller and larger fibril diameter observed in tendon	Jepsen et al. (2002)

For the soft-tissue phenotypes in the SLRPs KO mice, see the following reviews (Reed and Iozzo 2002; Kao and Liu 2002; Chakravarti 2002).

Although biglycan is structurally quite similar to decorin (56% identity), the dental phenotypes of biglycan KO are very different from those of decorin KO mice though this is not surprising considering their distinct roles (see above). The thickness of the enamel matrix in biglycan KO molars at post-natal day 1 was significantly increased as compared to that in wild-type (WT) control molars with no visible rod formation (Goldberg et al. 2005). This observation was further supported by micro CT analysis that the mineralized enamel volume of 8-week-old biglycan KO in molar crowns was increased as compared to that of WT

(Chiu et al. 2012), possibly by regulating amelogenin expression (Goldberg et al. 2005). On the other hand, the enamel matrix in decorin KO molars at post-natal day 1 was almost absent, although no structural difference was detected by the SEM study between decorin KO and WT mice at 6 weeks (Goldberg et al. 2005).

The TEM studies showed that dentin appeared to be porous and poorly mineralized in both biglycan and decorin KO molars. Metadentin in both biglycan KO and decorin KO molars at post-natal day 1 was enlarged as compared to that in WT molars. As compared to WT, the diameter of collagen fibrils in the proximal predentin of biglycan KO molars was smaller, however, significantly larger in the central and distal predentin. The density and spatial organization of collagen fibrils were not altered. On the other hand, the diameter of collagen fibrils in the decorin KO predentin was unchanged compared to that of WT, while decorin KO showed the increased density of collagen fibrils in the distal predentin (Goldberg et al. 2005). These observations suggest that biglycan and decorin play differential roles in dentin and enamel matrix formation as well as biomineralization, which is consistent with our previous studies (Mochida et al. 2009; Parisuthiman et al. 2005).

Among class II SLRPs, fibromodulin KO mice have also been characterized for their dental phenotype. The thickness of enamel in fibromodulin KO neonates at postnatal day 1 was significantly decreased while that of dentin and predentin was unchanged as compared to WT. The diameter of collagen fibrils in molar predentin in fibromodulin KO was significantly larger than that of WT. Ultrastructural analysis in molar dentin revealed that WT dentin showed homogeneously dense structure, while fibromodulin KO exhibited a heterogeneous appearance with hypomineralized electron-lucent areas (Goldberg et al. 2006). Three-week-old fibromodulin KO mice showed an increased size of pulp volume as compared to WT; however, the pulp volume appeared to be decreased and dentin volume was increased in 10-week-old fibromodulin KO mice (Goldberg et al. 2011). These results suggest that fibromodulin may play diverging roles in a mineralized tissue-specific manner or developmental stage-specific manner.

It has been more than 20 years since biglycan KO mice were generated and reported (Xu et al. 1998), demonstrating for the first time that a particular SLRP member exhibited a skeletal phenotype. As mouse biglycan gene is located on the X chromosome, characterization of mouse phenotype has been initially performed in males. Although apparently normal at birth, these mice displayed an age-dependent phenotype characterized by reduced growth rate and decreased bone mass. This may be the first report in which deficiency of a non-collagenous protein leads to a skeletal phenotype that is marked by low bone mass that becomes more obvious with age. Later, the bone phenotype in biglycan female KO mice was characterized upon ovariectomy (Nielsen et al. 2003). The results showed that biglycan deficiency had minimal effects on bone metabolism in females and biglycan deficiency protected trabecular bone loss caused by estrogen depletion, suggesting that there is a gender difference in response to its deficiency.

It has also been reported that using calvarial cells derived from biglycan KO neonatal mice, loss of biglycan caused reduced BMP4 binding to the osteoblast cell surface. This reduction of BMP4 binding resulted in lower sensitivity of BMP4 stimulation and reduced expression of *Cbfa1*, likely causing poor osteoblast differentiation (Chen et al. 2004). This notion was further supported by our study that responsiveness to BMP4 was enhanced in biglycan overexpressed MC3T3-E1 clones and impaired in biglycan suppressed clones (see above) (Parisuthiman et al. 2005). Later we found that biglycan directly binds BMP2 and its receptor, BMPR1B/ALK6, and enhances and sustains BMP2 signaling suggesting that biglycan may function as a bridge molecule for BMP2 signaling (Mochida et al. 2006). This study showed that biglycan also binds BMP4 and 6, but likely not via direct interaction.

Although decorin KO mice exhibited only mild ultrastructural changes in bone, when both decorin and biglycan were disrupted, bone phenotypes were synergized exhibiting severe osteopenia (Corsi et al. 2002), indicating a redundant function between decorin and biglycan in the biomineralization process in vivo.

More recently, a class IV SLRP member, tsukushin KO mice have been generated, and bone phenotype was reported (Yano et al. 2017). There have been many reports that tsukushin inhibits several signaling molecules including BMP, TGF-β, Wnt, and fibroblast growth factor (FGF) (Ohta et al. 2004, 2011; Morris et al. 2007; Niimori et al. 2012), suggesting the importance of tsukushin in skeletal development. Tsukushin KO exhibited reduced long bone phenotype as compared to WT showing significant decreases in trabecular number and thickness at both 3 and 20 weeks. The thickness of proliferating zone and hypertrophic zone, but not resting zone, in the tsukushin KO growth plate was markedly decreased accompanied by the reduced expression of *Sox9* and *Runx2* and increased expression of *type X collagen* and *Mmp13*. The results suggest that tsukushin plays an important role mainly in the endochondral ossification process.

Taken together, the dental/bone phenotypes of these SLRP KO mice are, overall, consistent with the results by in vitro cell culture studies, suggesting that biglycan in concert with decorin appears to be a major modulator in the assembly of the ECM in mineralized tissues and cell signaling.

6.4 SLRPs as Therapeutics and Tissue Engineering

As described above, several SLRP members interact with collagens including fibrillar and non-fibrillar collagens (Hocking et al. 1996; Pogany et al. 1994; Schonherr et al. 1995). SLRPs can also bind to growth factors such as TGF-β family, VEGF, PDGF, among others (Baghy et al. 2013; Berendsen et al. 2014; Tiedemann et al. 2005). The binding of decorin and biglycan to TGF-β1, 2, and 3, and TGF-β family members such as BMPs was the first description of SLRP binding to growth factors and their ability to regulate cell signaling (Hildebrand et al. 1994; Yamaguchi et al. 1990). Asporin also binds TGF-β1 and influences the expression of osteogenic markers, which highlights the importance of class I SLRPs on osteogenesis and mineralization (Hildebrand et al. 1994). Their involvement in other signaling pathways independent of TGF-β has been reported as well (Ameye and Young 2002;

Hocking et al. 1998; Kresse and Schonherr 2001; Yoon and Halper 2005) including signaling through IGF-1R and EGFR (Aggelidakis et al. 2018; Mohan et al. 2019). Most studies of SLRPs in this field have focused on decorin and biglycan as they interact with various receptors, secreted molecules, and ECM components (Babelova et al. 2009; Desnoyers et al. 2001; Elefteriou et al. 2001; Gendelman et al. 2003; Krumdieck et al. 1992; Mochida et al. 2006; Nadesalingam et al. 2003; Pogany et al. 1994). Through these interactions, they modulate a variety of signaling pathways including Wnt, Smads, and mitogen-activated protein kinase (MAPKs) (Babelova et al. 2009; Schaefer et al. 2005; Wang et al. 2010; Yan et al. 2009). Biglycan, for instance, directly binds BMP2 and its receptors, which enhances BMP2-induced osteoblast differentiation (Mochida et al. 2006; Parisuthiman et al. 2005). This positive effect of biglycan on BMP2 function is derived from the core protein, not the GAG component (Miguez et al. 2011). Further, the addition of recombinant biglycan core protein to a low dose BMP2 significantly accelerated BMP2-induced bone formation in a rodent model of craniofacial bone regeneration (Miguez et al. 2014). In an attempt to identify the effector domain within the biglycan core protein, we recently tested several synthetic peptides that correspond to various domains of biglycan and evaluated their effects on BMP2-induced osteoblast differentiation and mineralization in vitro. The results indicated that LRR2-3 of biglycan core protein significantly enhanced BMP2 osteogenic function (Jongwattanapisan et al. 2018). Since high-dose BMP2 may cause significant adverse side effects (see reviews by Carreira et al. 2014; James et al. 2016), use of the synthetic effector with low-dose BMP2 may provide a means to enhance BMP2-induced bone formation and to minimize the risk of side effects (Jongwattanapisan et al. 2018).

Because of the ability of SLRPs to interact with collagen and other molecules such as growth factors, several research groups have been exploring the potential use of these SLRPs not only for tissue engineering, as described above but also for other therapeutic applications. Decorin and biglycan, for instance, have been explored as antithrombotic molecules by forming a DS proteoglycan-heparin cofactor II complex that inactivates thrombin (Delorme et al. 1998; Whinna et al. 1993).

Due to the ability to delay collagen fibrillogenesis and act as a TGF-β1/2 antagonist, decorin is an attractive therapeutic candidate for anti-scarring treatment (Yamaguchi et al. 1990). Decorin has been tested for corneal wound healing in the form of eye drops or subconjunctival injection prior to and after glaucoma surgery on rabbits showing significantly less scarring (Chouhan et al. 2019; Grisanti et al. 2005; Hill et al. 2018). Other antifibrosis applications have been demonstrated in kidneys, muscle, and lungs through its negative TGF-β regulation (Fukushima et al. 2001; Isaka et al. 1996; Kolb et al. 2001). In mice with epidermolysis bullosa, decorin was used as a TGF-β inhibitor and fibrotic traits were significantly reduced in injured skin (Cianfarani et al. 2019). Decorin has been also investigated in bone-muscle injury models with impaired vascularization by its co-delivery with BMP2 in a collagen hydrogel (Ruehle et al. 2019). The rationale behind the use of decorin was because of its ability to bind TGF-β growth factor and support angiogenesis in vitro.

Biglycan has been studied as a therapeutic in musculoskeletal disorders such as Duchenne muscular dystrophy as it can modulate collagen fibrillogenesis and interact with other proteins to improve muscle function (Young and Fallon 2012). In severe muscular dystrophies, increased levels of biglycan mRNA and protein are detected in muscle with increased fibrosis (Zanotti et al. 2005). Biglycan has been found to bind to alpha- and gamma-sarcoglycan in muscle and regulate their expression which is important for muscle development (Rafii et al. 2006; Zanotti et al. 2005). The non-glycanated form of biglycan has been shown to have a specific function in Wnt signaling in muscle preventing muscle wasting (Amenta et al. 2011; Young and Fallon 2012). The non-glycanated biglycan delivered to dystrophic mice was able to improve muscle health and function (Fallon and Mcnally 2018; Ito et al. 2017).

Biglycan has also been studied as a regulator of angiogenesis during bone fracture repair. In biglycan KO mice, the callus formation is smaller than wild-type showing less cartilage and woven bone, less fibrillar collagens, and diminished vascular endothelial growth factor (Berendsen et al. 2014). The reduced callus size was likely due to Wnt availability affected by the absence of biglycan (Berendsen et al. 2011). Later, it has been reported that, based on the analyses by micro-computed tomography and angiography, biglycan KO mice exhibited fewer and reduced vessel size and volumes as biglycan inhibits endostatin, an anti-angiogenic protein (Myren et al. 2016).

The role of GAG chains in biglycan function is still poorly understood. As described above, biglycan's positive modulation of BMP2 function is likely derived from the specific LRR domains of its core protein, not GAG components (Miguez et al. 2011, 2014; Jongwattanapisan et al. 2018). The presence of GAGs reduced the ability of biglycan to promote osteogenesis suggesting that GAGs may negatively control the interaction between biglycan and BMP2/its receptors (Jongwattanapisan et al. 2018). A summary of biglycan in proteoglycan versus core versus peptide forms and their interactions with receptors associated with osteogenesis are depicted in Fig. 6.4. Biglycan proteoglycan form (with GAGs) has been implicated in inflammatory events interacting with TLR 2 and 4, and P2X receptors versus non-glycanated forms and peptides favoring interaction with BMP receptors (Fig. 6.4) (Babelova et al. 2009; Hsieh et al. 2014). The effect of bi- versus mono-glycanated forms of biglycan has not been investigated yet.

Fibromodulin has been studied for its role in angiogenesis and wound healing. Recombinant fibromodulin protein-enhanced human endothelial cell function in vitro (Jian et al. 2013) and increased capillary generation in in vivo chick embryo chorioallantoic membrane model (Zheng et al. 2014). In the rat Achilles tendon injury model, fibromodulin gene transfer improved biochemical and histological parameters of tendon injury (Delalande et al. 2015). Injection of fibromodulin protein reduced scar size and increased tensile strength of repaired tissue compared to a steroid treatment by local application in a pig wound model (Jiang et al. 2018). The osteogenic function of fibromodulin was also evaluated by implanting fibroblasts primed with fibromodulin in critical-sized mouse calvarial defects. The results showed that these cells lead to increased bone volume/total volume at 8-weeks

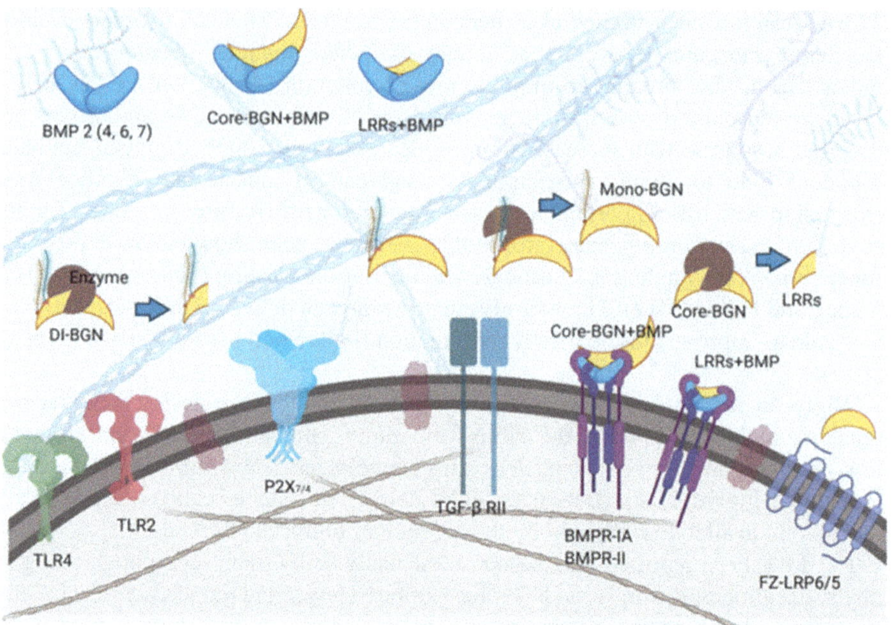

Fig. 6.4 The proteoglycan form of biglycan (yellow half-moon shape) can be cleaved by proteolytic enzymes (brown Pacman shape) leading to glycanated and non-glycanated peptides (chopped yellow half-moons) (Hou et al. 2007; Johnstone et al. 1993; Roughley et al. 1993). Glycanated form of biglycan (N-terminal region with glycosaminoglycans) may interact preferentially with receptors associated with inflammatory and immune response [i.e., Toll-like receptors (TLR) 2 and 4 and P2X purinoceptor 7/4 (P2X)] (Babelova et al. 2009). Biglycan core or peptides of biglycan composed of mainly leucine-rich repeat (LRR) interact with transforming growth factor (TGF)-β family receptors [bone morphogenetic protein receptor (BMPR) I and II, TGF-βR] and Wnt signaling trait [low-density lipoprotein receptor (LRP) 6/5 and frizzled receptor (FZ)] and bind to BMP molecules (blue boomerang) (Desnoyers et al. 2001; Hildebrand et al. 1994; Mochida et al. 2006; Moreno et al. 2005). Other forms of biglycan such as mono-glycanated (mono-biglycan) and biglycan modified with various types of glycosaminoglycans have not been explored in terms of preferred binding partners. Created with Biorender.com

post-surgery as compared to control groups. Possibly, fibromodulin is also a viable option for mineralized tissue reconstruction (Li et al. 2016).

Lumican, another SLRP member, also controls collagen fibrillogenesis and interacts with growth factors via its core protein (Rada et al. 1993). Lumican can modulate wound healing and innate immunity by interacting with receptors of immune cells such as macrophages (Shao et al. 2012). Its deficiency, increased expression of receptor activator of nuclear factor kappa-B ligand (RANKL), and elevated number of osteoclasts around teeth (Wang et al. 2014). As a therapeutic, the administration of lumican has a positive effect on epithelial proliferation and migration (Yeh et al. 2005). By applying recombinant lumican on wounded mouse skin, wound healing was enhanced (Liu et al. 2013) possibly by promoting contraction of fibroblasts through the $\alpha 2\beta 1$ integrin. Adenoviral expression of lumican in hypertrophic scarring animal as well as cell models showed the ability of lumican to reduce fibroblast proliferation via binding to $\alpha 2\beta 1$ integrin and subsequent reduction of focal adhesion kinase phosphorylation (Zhao et al. 2016). Gesteira and co-workers designed a peptide mimicking the activity of 13 C-terminal amino acids of lumican and showed that it effectively forms a complex with type I receptor for TGF-β1 and promoted corneal wound healing in mice (Gesteira et al. 2017; Yamanaka et al. 2013).

Much remains to be unveiled for the roles of SLRPs in soft and mineralized tissue maintenance and repair. Nonetheless, the above studies show promise in utilizing these molecules as therapeutics in disease treatments and reconstructive therapies. Further, as these SLRPs are rich in GAGs which can be degraded by proteases in the wounds, they can release GAG-attached peptide fragments which may modulate tissue healing (Peplow 2005). The function of GAGs and their fragments with and without peptides in tissue engineering or therapeutics could also be an important subject for further investigations.

A Perspective As described in this chapter, SLRPs may function as a modulator of mineralized tissue formation by controlling collagen fibrillogenesis, interacting with specific growth factors and their receptors. To further elucidate the specific roles of SLRP members in the process of mineralization, the generation of ameloblast/odontoblast/cementoblast/osteoblast-specific conditional knockout mice will be very helpful. Due to their multifaceted functions, their therapeutic applications are promising. Further understanding of the function of core versus proteoglycan forms of SLRPs may expand their applications. In addition, the utility of SLRPs, especially their short effector domains, for bone regeneration may advance therapies for bone defects.

Acknowledgments Fundings—Boston University Henry Goldman School of Dental Medicine (YM), NIH/NIDCR R03DE028035 and NIH CTSA – NCTraCs 550KR211924 (PM) and the University of North Carolina at Chapel Hill Adams School of Dentistry (MY).

References

Ababneh KT, Hall RC, Embery G (1999) The proteoglycans of human cementum: immunohistochemical localization in healthy, periodontally involved and ageing teeth. J Periodontal Res 34:87–96

Aggelidakis J, Berdiaki A, Nikitovic D, Papoutsidakis A, Papachristou DJ, Tsatsakis AM, Tzanakakis GN (2018) Biglycan regulates MG63 osteosarcoma cell growth through a LPR6/beta-catenin/IGFR-IR signaling axis. Front Oncol 8:470

Amenta AR, Yilmaz A, Bogdanovich S, Mckechnie BA, Abedi M, Khurana TS, Fallon JR (2011) Biglycan recruits utrophin to the sarcolemma and counters dystrophic pathology in mdx mice. Proc Natl Acad Sci U S A 108:762–767

Ameye L, Young MF (2002) Mice deficient in small leucine-rich proteoglycans: novel in vivo models for osteoporosis, osteoarthritis, Ehlers-Danlos syndrome, muscular dystrophy, and corneal diseases. Glycobiology 12:107R–116R

Antonsson P, Heinegard D, Oldberg A (1991) Posttranslational modifications of fibromodulin. J Biol Chem 266:16859–16861

Babelova A, Moreth K, Tsalastra-Greul W, Zeng-Brouwers J, Eickelberg O, Young MF, Bruckner P, Pfeilschifter J, Schaefer RM, Grone HJ, Schaefer L (2009) Biglycan, a danger signal that activates the NLRP3 inflammasome via toll-like and P2X receptors. J Biol Chem 284:24035–24048

Baghy K, Horvath Z, Regos E, Kiss K, Schaff Z, Iozzo RV, Kovalszky I (2013) Decorin interferes with platelet-derived growth factor receptor signaling in experimental hepatocarcinogenesis. FEBS J 280:2150–2164

Benoit VM, Fischer JR, Lin YP, Parveen N, Leong JM (2011) Allelic variation of the Lyme disease spirochete adhesin DbpA influences spirochetal binding to decorin, dermatan sulfate, and mammalian cells. Infect Immun 79:3501–3509

Berendsen AD, Fisher LW, Kilts TM, Owens RT, Robey PG, Gutkind JS, Young MF (2011) Modulation of canonical Wnt signaling by the extracellular matrix component biglycan. Proc Natl Acad Sci U S A 108:17022–17027

Berendsen AD, Pinnow EL, Maeda A, Brown AC, Mccartney-Francis N, Kram V, Owens RT, Robey PG, Holmbeck K, De Castro LF, Kilts TM, Young MF (2014) Biglycan modulates angiogenesis and bone formation during fracture healing. Matrix Biol 35:223–231

Bertassoni LE (2017) Dentin on the nanoscale: hierarchical organization, mechanical behavior and bioinspired engineering. Dent Mater 33:637–649

Bonucci E, Gherardi G (1975) Histochemical and electron microscopy investigations on medullary bone. Cell Tissue Res 163:81–97

Boskey AL (1998) Biomineralization: conflicts, challenges, and opportunities. J Cell Biochem 72 (Suppl 30–31):83–91

Boskey A, Pleshko Camacho N (2007) FT-IR imaging of native and tissue-engineered bone and cartilage. Biomaterials 28:2465–2478

Bredrup C, Knappskog PM, Majewski J, Rodahl E, Boman H (2005) Congenital stromal dystrophy of the cornea caused by a mutation in the decorin gene. Invest Ophthalmol Vis Sci 46:420–426

Burger C, Zhou HW, Wang H, Sics I, Hsiao BS, Chu B, Graham L, Glimcher MJ (2008) Lateral packing of mineral crystals in bone collagen fibrils. Biophys J 95:1985–1992

Carreira AC, Alves GG, Zambuzzi WF, Sogayar MC, Granjeiro JM (2014) Bone morphogenetic proteins: structure, biological function and therapeutic applications. Arch Biochem Biophys 561:64–73

Chakravarti S (2002) Functions of lumican and fibromodulin: lessons from knockout mice. Glycoconj J 19:287–293

Chakravarti S, Petroll WM, Hassell JR, Jester JV, Lass JH, Paul J, Birk DE (2000) Corneal opacity in lumican-null mice: defects in collagen fibril structure and packing in the posterior stroma. Invest Ophthalmol Vis Sci 41:3365–3373

Chen S, Birk DE (2013) The regulatory roles of small leucine-rich proteoglycans in extracellular matrix assembly. FEBS J 280:2120–2137

Chen XD, Fisher LW, Robey PG, Young MF (2004) The small leucine-rich proteoglycan biglycan modulates BMP-4-induced osteoblast differentiation. FASEB J 18:948–958

Cheng H, Caterson B, Neame PJ, Lester GE, Yamauchi M (1996) Differential distribution of lumican and fibromodulin in tooth cementum. Connect Tissue Res 34:87–96

Cheng H, Caterson B, Yamauchi M (1999) Identification and immunolocalization of chondroitin sulfate proteoglycans in tooth cementum. Connect Tissue Res 40:37–47

Chiu R, Li W, Herber RP, Marshall SJ, Young M, Ho SP (2012) Effects of biglycan on physico-chemical properties of ligament-mineralized tissue attachment sites. Arch Oral Biol 57:177–187

Chouhan G, Moakes RJA, Esmaeili M, Hill LJ, Decogan F, Hardwicke J, Rauz S, Logan A, Grover LM (2019) A self-healing hydrogel eye drop for the sustained delivery of decorin to prevent corneal scarring. Biomaterials 210:41–50

Cianfarani F, De Domenico E, Nystrom A, Mastroeni S, Abeni D, Baldini E, Ulisse S, Uva P, Bruckner-Tuderman L, Zambruno G, Castiglia D, Odorisio T (2019) Decorin counteracts disease progression in mice with recessive dystrophic epidermolysis bullosa. Matrix Biol 81:3–16

Corsi A, Xu T, Chen XD, Boyde A, Liang J, Mankani M, Sommer B, Iozzo RV, Eichstetter I, Robey PG, Bianco P, Young MF (2002) Phenotypic effects of biglycan deficiency are linked to collagen fibril abnormalities, are synergized by decorin deficiency, and mimic Ehlers-Danlos-like changes in bone and other connective tissues. J Bone Miner Res 17:1180–1189

Delalande A, Gosselin MP, Suwalski A, Guilmain W, Leduc C, Berchel M, Jaffres PA, Baril P, Midoux P, Pichon C (2015) Enhanced Achilles tendon healing by fibromodulin gene transfer. Nanomedicine 11:1735–1744

Delorme MA, Xu L, Berry L, Mitchell L, Andrew M (1998) Anticoagulant dermatan sulfate proteoglycan (decorin) in the term human placenta. Thromb Res 90:147–153

De Mattos Pimenta Vidal C, Leme-Kraus AA, Rahman M, Farina AP, Bedran-Russo AK (2017) Role of proteoglycans on the biochemical and biomechanical properties of dentin organic matrix. Arch Oral Biol 82:203–208

Desnoyers L, Arnott D, Pennica D (2001) WISP-1 binds to decorin and biglycan. J Biol Chem 276:47599–47607

Elefteriou F, Exposito JY, Garrone R, Lethias C (2001) Binding of tenascin-X to decorin. FEBS Lett 495:44–47

Embery G, Hall R, Waddington R, Septier D, Goldberg M (2001) Proteoglycans in dentinogenesis. Crit Rev Oral Biol Med 12:331–349

Fallon JR, Mcnally EM (2018) Non-glycanated biglycan and LTBP4: leveraging the extracellular matrix for Duchenne muscular dystrophy therapeutics. Matrix Biol 68–69:616–627

Fisher LW, Termine JD, Dejter SW Jr, Whitson SW, Yanagishita M, Kimura JH, Hascall VC, Kleinman HK, Hassell JR, Nilsson B (1983) Proteoglycans of developing bone. J Biol Chem 258:6588–6594

Fisher LW, Hawkins GR, Tuross N, Termine JD (1987) Purification and partial characterization of small proteoglycans I and II, bone sialoproteins I and II, and osteonectin from the mineral compartment of developing human bone. J Biol Chem 262:9702–9708

Fratzl P, Daxer A (1993) Structural transformation of collagen fibrils in corneal stroma during drying. An x-ray scattering study. Biophys J 64:1210–1214

Fukushima K, Badlani N, Usas A, Riano F, Fu F, Huard J (2001) The use of an antifibrosis agent to improve muscle recovery after laceration. Am J Sports Med 29:394–402

Funderburgh JL (2002) Keratan sulfate biosynthesis. IUBMB Life 54:187–194

Gamsjaeger S, Mendelsohn R, Boskey AL, Gourion-Arsiquaud S, Klaushofer K, Paschalis EP (2014) Vibrational spectroscopic imaging for the evaluation of matrix and mineral chemistry. Curr Osteoporos Rep 12:454–464

Ge G, Seo NS, Liang X, Hopkins DR, Hook M, Greenspan DS (2004) Bone morphogenetic protein-1/tolloid-related metalloproteinases process osteoglycin and enhance its ability to regulate collagen fibrillogenesis. J Biol Chem 279:41626–41633

Gendelman R, Burton-Wurster NI, Macleod JN, Lust G (2003) The cartilage-specific fibronectin isoform has a high affinity binding site for the small proteoglycan decorin. J Biol Chem 278:11175–11181

George A, Veis A (2008) Phosphorylated proteins and control over apatite nucleation, crystal growth, and inhibition. Chem Rev 108:4670–4693

George A, Bannon L, Sabsay B, Dillon JW, Malone J, Veis A, Jenkins NA, Gilbert DJ, Copeland NG (1996) The carboxyl-terminal domain of phosphophoryn contains unique extended triplet amino acid repeat sequences forming ordered carboxyl-phosphate interaction ridges that may be essential in the biomineralization process. J Biol Chem 271:32869–32873

Georgiadis M, Muller R, Schneider P (2016) Techniques to assess bone ultrastructure organization: orientation and arrangement of mineralized collagen fibrils. J R Soc Interface 13

Gesteira TF, Coulson-Thomas VJ, Yuan Y, Zhang J, Nader HB, Kao WW (2017) Lumican peptides: rational design targeting ALK5/TGFBRI. Sci Rep 7:42057

Glimcher MJ (1984) Recent studies of the mineral phase in bone and its possible linkage to the organic matrix by protein-bound phosphate bonds. Philos Trans R Soc Lond B Biol Sci 304:479–508

Goldberg M, Septier DS (1992) Differential staining of glycosaminoglycans in the predentine and dentine of rat incisor using cuprolinic blue at various magnesium chloride concentrations. Histochem J 24:648–654

Goldberg M, Takagi M (1993) Dentine proteoglycans: composition, ultrastructure and functions. Histochem J 25:781–806

Goldberg M, Rapoport O, Septier D, Palmier K, Hall R, Embery G, Young M, Ameye L (2003) Proteoglycans in predentin: the last 15 micrometers before mineralization. Connect Tissue Res 44(Suppl 1):184–188

Goldberg M, Septier D, Rapoport O, Iozzo RV, Young MF, Ameye LG (2005) Targeted disruption of two small leucine-rich proteoglycans, biglycan and decorin, excerpts divergent effects on enamel and dentin formation. Calcif Tissue Int 77:297–310

Goldberg M, Septier D, Oldberg A, Young MF, Ameye LG (2006) Fibromodulin-deficient mice display impaired collagen fibrillogenesis in predentin as well as altered dentin mineralization and enamel formation. J Histochem Cytochem 54:525–537

Goldberg M, Ono M, Septier D, Bonnefoix M, Kilts TM, Bi Y, Embree M, Ameye L, Young MF (2009) Fibromodulin-deficient mice reveal dual functions for fibromodulin in regulating dental tissue and alveolar bone formation. Cells Tissues Organs 189:198–202

Goldberg M, Marchadier A, Vidal C, Harichane Y, Kamoun-Goldrat A, Kellermann O, Kilts T, Young M (2011) Differential effects of fibromodulin deficiency on mouse mandibular bones and teeth: a micro-CT time course study. Cells Tissues Organs 194:205–210

Grisanti S, Szurman P, Warga M, Kaczmarek R, Ziemssen F, Tatar O, Bartz-Schmidt KU (2005) Decorin modulates wound healing in experimental glaucoma filtration surgery: a pilot study. Invest Ophthalmol Vis Sci 46:191–196

Guindon S, Gascuel O (2003) A simple, fast, and accurate algorithm to estimate large phylogenies by maximum likelihood. Syst Biol 52:696–704

Hall RC, Embery G (1997) The use of immunohistochemistry in understanding the structure and function of the extracellular matrix of dental tissues. Adv Dent Res 11:478–486

Hall RC, Embery G, Lloyd D (1997) Immunochemical localization of the small leucine-rich proteoglycan lumican in human predentine and dentine. Arch Oral Biol 42:783–786

He G, George A (2004) Dentin matrix protein 1 immobilized on type I collagen fibrils facilitates apatite deposition in vitro. J Biol Chem 279:11649–11656

Hedlund H, Mengarelli-Widholm S, Heinegard D, Reinholt FP, Svensson O (1994) Fibromodulin distribution and association with collagen. Matrix Biol 14:227–232

Hildebrand A, Romaris M, Rasmussen LM, Heinegard D, Twardzik DR, Border WA, Ruoslahti E (1994) Interaction of the small interstitial proteoglycans biglycan, decorin and fibromodulin with transforming growth factor beta. Biochem J 302(Pt 2):527–534

Hill LJ, Moakes RJA, Vareechon C, Butt G, Ng A, Brock K, Chouhan G, Vincent RC, Abbondante S, Williams RL, Barnes NM, Pearlman E, Wallace GR, Rauz S, Logan A, Grover LM (2018) Sustained release of decorin to the surface of the eye enables scarless corneal regeneration. NPJ Regen Med 3:23

Hocking AM, Strugnell RA, Ramamurthy P, Mcquillan DJ (1996) Eukaryotic expression of recombinant biglycan. Post-translational processing and the importance of secondary structure for biological activity. J Biol Chem 271:19571–19577

Hocking AM, Shinomura T, Mcquillan DJ (1998) Leucine-rich repeat glycoproteins of the extracellular matrix. Matrix Biol 17:1–19

Hoshi K, Kemmotsu S, Takeuchi Y, Amizuka N, Ozawa H (1999) The primary calcification in bones follows removal of decorin and fusion of collagen fibrils. J Bone Miner Res 14:273–280

Hoshi K, Ejiri S, Ozawa H (2001) Localizational alterations of calcium, phosphorus, and calcification-related organics such as proteoglycans and alkaline phosphatase during bone calcification. J Bone Miner Res 16:289–298

Hou S, Maccarana M, Min TH, Strate I, Pera EM (2007) The secreted serine protease xHtrA1 stimulates long-range FGF signaling in the early Xenopus embryo. Dev Cell 13:226–241
Hsieh LT, Nastase MV, Zeng-Brouwers J, Iozzo RV, Schaefer L (2014) Soluble biglycan as a biomarker of inflammatory renal diseases. Int J Biochem Cell Biol 54:223–235
Huang R, Liu J, Sharp JS (2013) An approach for separation and complete structural sequencing of heparin/heparan sulfate-like oligosaccharides. Anal Chem 85:5787–5795
Hyllested JL, Veje K, Ostergaard K (2002) Histochemical studies of the extracellular matrix of human articular cartilage—a review. Osteoarthritis Cartilage 10:333–343
Iozzo RV, Murdoch AD (1996) Proteoglycans of the extracellular environment: clues from the gene and protein side offer novel perspectives in molecular diversity and function. FASEB J 10:598–614
Iozzo RV, Schaefer L (2015) Proteoglycan form and function: a comprehensive nomenclature of proteoglycans. Matrix Biol 42:11–55
Isaka Y, Brees DK, Ikegaya K, Kaneda Y, Imai E, Noble NA, Border WA (1996) Gene therapy by skeletal muscle expression of decorin prevents fibrotic disease in rat kidney. Nat Med 2:418–423
Ito M, Ehara Y, Li J, Inada K, Ohno K (2017) Protein-anchoring therapy of biglycan for Mdx mouse model of duchenne muscular dystrophy. Hum Gene Ther 28:428–436
James AW, Lachaud G, Shen J, Asatrian G, Nguyen V, Zhang X, Ting K, Soo C (2016) A review of the clinical side effects of bone morphogenetic protein-2. Tissue Eng Part B Rev 22:284–297
Jepsen KJ, Wu F, Peragallo JH, Paul J, Roberts L, Ezura Y, Oldberg A, Birk DE, Chakravarti S (2002) A syndrome of joint laxity and impaired tendon integrity in lumican- and fibromodulin-deficient mice. J Biol Chem 277:35532–35540
Jian J, Zheng Z, Zhang K, Rackohn TM, Hsu C, Levin A, Enjamuri DR, Zhang X, Ting K, Soo C (2013) Fibromodulin promoted in vitro and in vivo angiogenesis. Biochem Biophys Res Commun 436:530–535
Jiang W, Ting K, Lee S, Zara JN, Song R, Li C, Chen E, Zhang X, Zhao Z, Soo C, Zheng Z (2018) Fibromodulin reduces scar size and increases scar tensile strength in normal and excessive-mechanical-loading porcine cutaneous wounds. J Cell Mol Med 22:2510–2513
Jing Y, Kumar PR, Zhu L, Edward DP, Tao S, Wang L, Chuck R, Zhang C (2014) Novel decorin mutation in a Chinese family with congenital stromal corneal dystrophy. Cornea 33:288–293
Johnstone B, Markopoulos M, Neame P, Caterson B (1993) Identification and characterization of glycanated and non-glycanated forms of biglycan and decorin in the human intervertebral disc. Biochem J 292(Pt 3):661–666
Jongwattanapisan P, Terajima M, Miguez PA, Querido W, Nagaoka H, Sumida N, Gurysh EG, Ainslie KM, Pleshko N, Perera L, Yamauchi M (2018) Identification of the effector domain of biglycan that facilitates BMP-2 osteogenic function. Sci Rep 8:7022
Kalamajski S, Oldberg A (2007) Fibromodulin binds collagen type I via Glu-353 and Lys-355 in leucine-rich repeat 11. J Biol Chem 282:26740–26745
Kalamajski S, Oldberg A (2009) Homologous sequence in lumican and fibromodulin leucine-rich repeat 5-7 competes for collagen binding. J Biol Chem 284:534–539
Kalamajski S, Aspberg A, Oldberg A (2007) The decorin sequence SYIRIADTNIT binds collagen type I. J Biol Chem 282:16062–16067
Kao WW, Liu CY (2002) Roles of lumican and keratocan on corneal transparency. Glycoconj J 19:275–285
Karamanos NK, Piperigkou Z, Theocharis AD, Watanabe H, Franchi M, Baud S, Brezillon S, Gotte M, Passi A, Vigetti D, Ricard-Blum S, Sanderson RD, Neill T, Iozzo RV (2018) Proteoglycan chemical diversity drives multifunctional cell regulation and therapeutics. Chem Rev 118:9152–9232
Karousou E, Asimakopoulou A, Monti L, Zafeiropoulou V, Afratis N, Gartaganis P, Rossi A, Passi A, Karamanos NK (2014) FACE analysis as a fast and reliable methodology to monitor the sulfation and total amount of chondroitin sulfate in biological samples of clinical importance. Molecules 19:7959–7980

Katz EP, Li ST (1973) The intermolecular space of reconstituted collagen fibrils. J Mol Biol 73:351–369

Kessler E, Takahara K, Biniaminov L, Brusel M, Greenspan DS (1996) Bone morphogenetic protein-1: the type I procollagen C-proteinase. Science 271:360–362

Khanarian NT, Boushell MK, Spalazzi JP, Pleshko N, Boskey AL, Lu HH (2014) FTIR-I compositional mapping of the cartilage-to-bone interface as a function of tissue region and age. J Bone Miner Res 29:2643–2652

Kim JW, Simmer JP (2007) Hereditary dentin defects. J Dent Res 86:392–399

Kizawa H, Kou I, Iida A, Sudo A, Miyamoto Y, Fukuda A, Mabuchi A, Kotani A, Kawakami A, Yamamoto S, Uchida A, Nakamura K, Notoya K, Nakamura Y, Ikegawa S (2005) An aspartic acid repeat polymorphism in asporin inhibits chondrogenesis and increases susceptibility to osteoarthritis. Nat Genet 37:138–144

Kolb M, Margetts PJ, Sime PJ, Gauldie J (2001) Proteoglycans decorin and biglycan differentially modulate TGF-beta-mediated fibrotic responses in the lung. Am J Physiol Lung Cell Mol Physiol 280:L1327–L1334

Kram V, Shainer R, Jani P, Meester JAN, Loeys B, Young MF (2020) Biglycan in the Skeleton. J Histochem Cytochem 68:747–762

Kresse H, Schonherr E (2001) Proteoglycans of the extracellular matrix and growth control. J Cell Physiol 189:266–274

Krumdieck R, Hook M, Rosenberg LC, Volanakis JE (1992) The proteoglycan decorin binds C1q and inhibits the activity of the C1 complex. J Immunol 149:3695–3701

Krusius T, Ruoslahti E (1986) Primary structure of an extracellular matrix proteoglycan core protein deduced from cloned cDNA. Proc Natl Acad Sci U S A 83:7683–7687

Kubaski F, Osago H, Mason RW, Yamaguchi S, Kobayashi H, Tsuchiya M, Orii T, Tomatsu S (2017) Glycosaminoglycans detection methods: applications of mass spectrometry. Mol Genet Metab 120:67–77

Landis WJ, Song MJ, Leith A, Mcewen L, Mcewen BF (1993) Mineral and organic matrix interaction in normally calcifying tendon visualized in three dimensions by high-voltage electron microscopic tomography and graphic image reconstruction. J Struct Biol 110:39–54

Landis WJ, Hodgens KJ, Arena J, Song MJ, Mcewen BF (1996) Structural relations between collagen and mineral in bone as determined by high voltage electron microscopic tomography. Microsc Res Tech 33:192–202

Li SW, Sieron AL, Fertala A, Hojima Y, Arnold WV, Prockop DJ (1996) The C-proteinase that processes procollagens to fibrillar collagens is identical to the protein previously identified as bone morphogenic protein-1. Proc Natl Acad Sci U S A 93:5127–5130

Li CS, Yang P, Ting K, Aghaloo T, Lee S, Zhang Y, Khalilinejad K, Murphy MC, Pan HC, Zhang X, Wu B, Zhou YH, Zhao Z, Zheng Z, Soo C (2016) Fibromodulin reprogrammed cells: a novel cell source for bone regeneration. Biomaterials 83:194–206

Liu XJ, Kong FZ, Wang YH, Zheng JH, Wan WD, Deng CL, Mao GY, Li J, Yang XM, Zhang YL, Zhang XL, Yang SL, Zhang ZG (2013) Lumican accelerates wound healing by enhancing alpha2beta1 integrin-mediated fibroblast contractility. PLoS One 8:e67124

Lormee P, Septier D, Lecolle S, Baudoin C, Goldberg M (1996) Dual incorporation of (35S)sulfate into dentin proteoglycans acting as mineralization promotors in rat molars and predentin proteoglycans. Calcif Tissue Int 58:368–375

Maccarana M, Svensson RB, Knutsson A, Giannopoulos A, Pelkonen M, Weis M, Eyre D, Warman M, Kalamajski S (2017) Asporin-deficient mice have tougher skin and altered skin glycosaminoglycan content and structure. PLoS One 12:e0184028

Mahamid J, Aichmayer B, Shimoni E, Ziblat R, Li C, Siegel S, Paris O, Fratzl P, Weiner S, Addadi L (2010) Mapping amorphous calcium phosphate transformation into crystalline mineral from the cell to the bone in zebrafish fin rays. Proc Natl Acad Sci U S A 107:6316–6321

Matias MA, Li H, Young WG, Bartold PM (2003) Immunohistochemical localization of fibromodulin in the periodontium during cementogenesis and root formation in the rat molar. J Periodontal Res 38:502–507

Matsushima N, Ohyanagi T, Tanaka T, Kretsinger RH (2000) Super-motifs and evolution of tandem leucine-rich repeats within the small proteoglycans—biglycan, decorin, lumican, fibromodulin, PRELP, keratocan, osteoadherin, epiphycan, and osteoglycin. Proteins 38:210–225

Mcewan PA, Scott PG, Bishop PN, Bella J (2006) Structural correlations in the family of small leucine-rich repeat proteins and proteoglycans. J Struct Biol 155:294–305

Miguez PA, Terajima M, Nagaoka H, Mochida Y, Yamauchi M (2011) Role of glycosaminoglycans of biglycan in BMP-2 signaling. Biochem Biophys Res Commun 405:262–266

Miguez PA, Terajima M, Nagaoka H, Ferreira JA, Braswell K, Ko CC, Yamauchi M (2014) Recombinant biglycan promotes bone morphogenetic protein-induced osteogenesis. J Dent Res 93:406–411

Mochida Y, Duarte WR, Tanzawa H, Paschalis EP, Yamauchi M (2003) Decorin modulates matrix mineralization in vitro. Biochem Biophys Res Commun 305:6–9

Mochida Y, Parisuthiman D, Yamauchi M (2006) Biglycan is a positive modulator of BMP-2 induced osteoblast differentiation. Adv Exp Med Biol 585:101–113

Mochida Y, Parisuthiman D, Pornprasertsuk-Damrongsri S, Atsawasuwan P, Sricholpech M, Boskey AL, Yamauchi M (2009) Decorin modulates collagen matrix assembly and mineralization. Matrix Biol 28:44–52

Mohan RR, Tripathi R, Sharma A, Sinha PR, Giuliano EA, Hesemann NP, Chaurasia SS (2019) Decorin antagonizes corneal fibroblast migration via caveolae-mediated endocytosis of epidermal growth factor receptor. Exp Eye Res 180:200–207

Moreno M, Munoz R, Aroca F, Labarca M, Brandan E, Larrain J (2005) Biglycan is a new extracellular component of the Chordin-BMP4 signaling pathway. EMBO J 24:1397–1405

Morris SA, Almeida AD, Tanaka H, Ohta K, Ohnuma S (2007) Tsukushi modulates Xnr2, FGF and BMP signaling: regulation of Xenopus germ layer formation. PLoS One 2:e1004

Myren M, Kirby DJ, Noonan ML, Maeda A, Owens RT, Ricard-Blum S, Kram V, Kilts TM, Young MF (2016) Biglycan potentially regulates angiogenesis during fracture repair by altering expression and function of endostatin. Matrix Biol 52–54:141–150

Nadesalingam J, Bernal AL, Dodds AW, Willis AC, Mahoney DJ, Day AJ, Reid KB, Palaniyar N (2003) Identification and characterization of a novel interaction between pulmonary surfactant protein D and decorin. J Biol Chem 278:25678–25687

Nielsen KL, Allen MR, Bloomfield SA, Andersen TL, Chen XD, Poulsen HS, Young MF, Heegaard AM (2003) Biglycan deficiency interferes with ovariectomy-induced bone loss. J Bone Miner Res 18:2152–2158

Niimori D, Kawano R, Felemban A, Niimori-Kita K, Tanaka H, Ihn H, Ohta K (2012) Tsukushi controls the hair cycle by regulating TGF-beta1 signaling. Dev Biol 372:81–87

Nikdin H, Olsson ML, Hultenby K, Sugars RV (2012) Osteoadherin accumulates in the predentin towards the mineralization front in the developing tooth. PLoS One 7:e31525

Nikitovic D, Aggelidakis J, Young MF, Iozzo RV, Karamanos NK, Tzanakakis GN (2012) The biology of small leucine-rich proteoglycans in bone pathophysiology. J Biol Chem 287:33926–33933

Nudelman F, Pieterse K, George A, Bomans PH, Friedrich H, Brylka LJ, Hilbers PA, De With G, Sommerdijk NA (2010) The role of collagen in bone apatite formation in the presence of hydroxyapatite nucleation inhibitors. Nat Mater 9:1004–1009

Ohta K, Lupo G, Kuriyama S, Keynes R, Holt CE, Harris WA, Tanaka H, Ohnuma SI (2004) Tsukushi functions as an organizer inducer by inhibition of BMP activity in cooperation with chordin. Dev Cell 7:347–358

Ohta K, Ito A, Kuriyama S, Lupo G, Kosaka M, Ohnuma S, Nakagawa S, Tanaka H (2011) Tsukushi functions as a Wnt signaling inhibitor by competing with Wnt2b for binding to transmembrane protein Frizzled4. Proc Natl Acad Sci U S A 108:14962–14967

Onnerfjord P, Heathfield TF, Heinegard D (2004) Identification of tyrosine sulfation in extracellular leucine-rich repeat proteins using mass spectrometry. J Biol Chem 279:26–33

Orsini G, Ruggeri A Jr, Mazzoni A, Papa V, Mazzotti G, Di Lenarda R, Breschi L (2007) Immunohistochemical identification of decorin and biglycan in human dentin: a correlative field emission scanning electron microscopy/transmission electron microscopy study. Calcif Tissue Int 81:39–45

Parisuthiman D, Mochida Y, Duarte WR, Yamauchi M (2005) Biglycan modulates osteoblast differentiation and matrix mineralization. J Bone Miner Res 20:1878–1886

Paschalis EP, Gamsjaeger S, Klaushofer K (2017) Vibrational spectroscopic techniques to assess bone quality. Osteoporos Int 28:2275–2291

Peplow PV (2005) Glycosaminoglycan: a candidate to stimulate the repair of chronic wounds. Thromb Haemost 94:4–16

Pogany G, Hernandez DJ, Vogel KG (1994) The in vitro interaction of proteoglycans with type I collagen is modulated by phosphate. Arch Biochem Biophys 313:102–111

Quan BD, Sone ED (2015) Structural changes in collagen fibrils across a mineralized interface revealed by cryo-TEM. Bone 77:42–49

Rada JA, Cornuet PK, Hassell JR (1993) Regulation of corneal collagen fibrillogenesis in vitro by corneal proteoglycan (lumican and decorin) core proteins. Exp Eye Res 56:635–648

Rafii MS, Hagiwara H, Mercado ML, Seo NS, Xu T, Dugan T, Owens RT, Hook M, Mcquillan DJ, Young MF, Fallon JR (2006) Biglycan binds to alpha- and gamma-sarcoglycan and regulates their expression during development. J Cell Physiol 209:439–447

Randilini A, Fujikawa K, Shibata S (2020) Expression, localization and synthesis of small leucine-rich proteoglycans in developing mouse molar tooth germ. Eur J Histochem 64

Raspanti M, Viola M, Forlino A, Tenni R, Gruppi C, Tira ME (2008) Glycosaminoglycans show a specific periodic interaction with type I collagen fibrils. J Struct Biol 164:134–139

Reed CC, Iozzo RV (2002) The role of decorin in collagen fibrillogenesis and skin homeostasis. Glycoconj J 19:249–255

Rehn AP, Cerny R, Sugars RV, Kaukua N, Wendel M (2008) Osteoadherin is upregulated by mature osteoblasts and enhances their in vitro differentiation and mineralization. Calcif Tissue Int 82:454–464

Roch C, Kuhn J, Kleesiek K, Gotting C (2010) Differences in gene expression of human xylosyltransferases and determination of acceptor specificities for various proteoglycans. Biochem Biophys Res Commun 391:685–691

Rodahl E, Van Ginderdeuren R, Knappskog PM, Bredrup C, Boman H (2006) A second decorin frame shift mutation in a family with congenital stromal corneal dystrophy. Am J Ophthalmol 142:520–521

Roughley PJ, White RJ, Magny MC, Liu J, Pearce RH, Mort JS (1993) Non-proteoglycan forms of biglycan increase with age in human articular cartilage. Biochem J 295(Pt 2):421–426

Ruehle MA, Li MA, Cheng A, Krishnan L, Willett NJ, Guldberg RE (2019) Decorin-supplemented collagen hydrogels for the co-delivery of bone morphogenetic protein-2 and microvascular fragments to a composite bone-muscle injury model with impaired vascularization. Acta Biomater 93:210–221

Ruhland C, Schonherr E, Robenek H, Hansen U, Iozzo RV, Bruckner P, Seidler DG (2007) The glycosaminoglycan chain of decorin plays an important role in collagen fibril formation at the early stages of fibrillogenesis. FEBS J 274:4246–4255

Sarathchandra P, Cassella JP, Ali SY (2002) Ultrastructural localization of proteoglycans in bone in osteogenesis imperfecta as demonstrated by Cuprolinic Blue staining. J Bone Miner Metab 20:288–293

Schaefer L, Babelova A, Kiss E, Hausser HJ, Baliova M, Krzyzankova M, Marsche G, Young MF, Mihalik D, Gotte M, Malle E, Schaefer RM, Grone HJ (2005) The matrix component biglycan is proinflammatory and signals through Toll-like receptors 4 and 2 in macrophages. J Clin Invest 115:2223–2233

Schonherr E, Witsch-Prehm P, Harrach B, Robenek H, Rauterberg J, Kresse H (1995) Interaction of biglycan with type I collagen. J Biol Chem 270:2776–2783

Schwarcz HP (2015) The ultrastructure of bone as revealed in electron microscopy of ion-milled sections. Semin Cell Dev Biol 46:44–50

Scott JE (1972) Histochemistry of Alcian blue. 3. The molecular biological basis of staining by Alcian blue 8GX and analogous phthalocyanins. Histochemie 32:191–212

Scott IC, Imamura Y, Pappano WN, Troedel JM, Recklies AD, Roughley PJ, Greenspan DS (2000) Bone morphogenetic protein-1 processes probiglycan. J Biol Chem 275:30504–30511

Scott PG, Mcewan PA, Dodd CM, Bergmann EM, Bishop PN, Bella J (2004) Crystal structure of the dimeric protein core of decorin, the archetypal small leucine-rich repeat proteoglycan. Proc Natl Acad Sci U S A 101:15633–15638

Seo NS, Hocking AM, Hook M, Mcquillan DJ (2005) Decorin core protein secretion is regulated by N-linked oligosaccharide and glycosaminoglycan additions. J Biol Chem 280:42774–42784

Septier D, Hall RC, Embery G, Goldberg M (2001) Immunoelectron microscopic visualization of pro- and secreted forms of decorin and biglycan in the predentin and during dentin formation in the rat incisor. Calcif Tissue Int 69:38–45

Shao H, Lee S, Gae-Scott S, Nakata C, Chen S, Hamad AR, Chakravarti S (2012) Extracellular matrix lumican promotes bacterial phagocytosis, and Lum-/- mice show increased Pseudomonas aeruginosa lung infection severity. J Biol Chem 287:35860–35872

Silver FH, Landis WJ (2011) Deposition of apatite in mineralizing vertebrate extracellular matrices: a model of possible nucleation sites on type I collagen. Connect Tissue Res 52:242–254

Sudo H, Kodama HA, Amagai Y, Yamamoto S, Kasai S (1983) In vitro differentiation and calcification in a new clonal osteogenic cell line derived from newborn mouse calvaria. J Cell Biol 96:191–198

Svensson L, Heinegard D, Oldberg A (1995) Decorin-binding sites for collagen type I are mainly located in leucine-rich repeats 4-5. J Biol Chem 270:20712–20716

Svensson L, Aszodi A, Reinholt FP, Fassler R, Heinegard D, Oldberg A (1999) Fibromodulin-null mice have abnormal collagen fibrils, tissue organization, and altered lumican deposition in tendon. J Biol Chem 274:9636–9647

Svensson L, Narlid I, Oldberg A (2000) Fibromodulin and lumican bind to the same region on collagen type I fibrils. FEBS Lett 470:178–182

Takagi M, Hishikawa H, Hosokawa Y, Kagami A, Rahemtulla F (1990) Immunohistochemical localization of glycosaminoglycans and proteoglycans in predentin and dentin of rat incisors. J Histochem Cytochem 38:319–324

Tashima T, Nagatoishi S, Sagara H, Ohnuma S, Tsumoto K (2015) Osteomodulin regulates diameter and alters shape of collagen fibrils. Biochem Biophys Res Commun 463:292–296

Tashima T, Nagatoishi S, Caaveiro JMM, Nakakido M, Sagara H, Kusano-Arai O, Iwanari H, Mimuro H, Hamakubo T, Ohnuma SI, Tsumoto K (2018) Molecular basis for governing the morphology of type-I collagen fibrils by Osteomodulin. Commun Biol 1:33

Tice LW, Barrnett RJ (1965) Diazophthalocyanins as reagents for fine structural cytochemistry. J Cell Biol 25:23–41

Tiedemann K, Olander B, Eklund E, Todorova L, Bengtsson M, Maccarana M, Westergren-Thorsson G, Malmstrom A (2005) Regulation of the chondroitin/dermatan fine structure by transforming growth factor-beta1 through effects on polymer-modifying enzymes. Glycobiology 15:1277–1285

Traub W, Arad T, Weiner S (1992) Origin of mineral crystal growth in collagen fibrils. Matrix 12:251–255

Uzel MI, Scott IC, Babakhanlou-Chase H, Palamakumbura AH, Pappano WN, Hong HH, Greenspan DS, Trackman PC (2001) Multiple bone morphogenetic protein 1-related mammalian metalloproteinases process pro-lysyl oxidase at the correct physiological site and control lysyl oxidase activation in mouse embryo fibroblast cultures. J Biol Chem 276:22537–22543

Vallet SD, Clerc O, Ricard-Blum S (2020) Glycosaminoglycan-protein interactions: the first draft of the glycosaminoglycan interactome. J Histochem Cytochem 69(2):93–104

Von Marschall Z, Fisher LW (2010) Decorin is processed by three isoforms of bone morphogenetic protein-1 (BMP1). Biochem Biophys Res Commun 391:1374–1378

Waddington RJ, Hall RC, Embery G, Lloyd DM (2003) Changing profiles of proteoglycans in the transition of predentine to dentine. Matrix Biol 22:153–161

Wang D, Christensen K, Chawla K, Xiao G, Krebsbach PH, Franceschi RT (1999) Isolation and characterization of MC3T3-E1 preosteoblast subclones with distinct in vitro and in vivo differentiation/mineralization potential. J Bone Miner Res 14:893–903

Wang X, Harimoto K, Xie S, Cheng H, Liu J, Wang Z (2010) Matrix protein biglycan induces osteoblast differentiation through extracellular signal-regulated kinase and Smad pathways. Biol Pharm Bull 33:1891–1897

Wang Y, Azais T, Robin M, Vallee A, Catania C, Legriel P, Pehau-Arnaudet G, Babonneau F, Giraud-Guille MM, Nassif N (2012) The predominant role of collagen in the nucleation, growth, structure and orientation of bone apatite. Nat Mater 11:724–733

Wang L, Foster BL, Kram V, Nociti FH Jr, Zerfas PM, Tran AB, Young MF, Somerman MJ (2014) Fibromodulin and biglycan modulate periodontium through TGFbeta/BMP signaling. J Dent Res 93:780–787

Weiner S, Traub W (1986) Organization of hydroxyapatite crystals within collagen fibrils. FEBS Lett 206:262–266

Whinna HC, Choi HU, Rosenberg LC, Church FC (1993) Interaction of heparin cofactor II with biglycan and decorin. J Biol Chem 268:3920–3924

Xu T, Bianco P, Fisher LW, Longenecker G, Smith E, Goldstein S, Bonadio J, Boskey A, Heegaard AM, Sommer B, Satomura K, Dominguez P, Zhao C, Kulkarni AB, Robey PG, Young MF (1998) Targeted disruption of the biglycan gene leads to an osteoporosis-like phenotype in mice. Nat Genet 20:78–82

Xu Y, Nudelman F, Eren ED, Wirix MJM, Cantaert B, Nijhuis WH, Hermida-Merino D, Portale G, Bomans PHH, Ottmann C, Friedrich H, Bras W, Akiva A, Orgel J, Meldrum FC, Sommerdijk N (2020) Intermolecular channels direct crystal orientation in mineralized collagen. Nat Commun 11:5068

Yamaguchi Y, Mann DM, Ruoslahti E (1990) Negative regulation of transforming growth factor-beta by the proteoglycan decorin. Nature 346:281–284

Yamanaka O, Yuan Y, Coulson-Thomas VJ, Gesteira TF, Call MK, Zhang Y, Zhang J, Chang SH, Xie C, Liu CY, Saika S, Jester JV, Kao WW (2013) Lumican binds ALK5 to promote epithelium wound healing. PLoS One 8:e82730

Yamauchi M, Katz EP (1993) The post-translational chemistry and molecular packing of mineralizing tendon collagens. Connect Tissue Res 29:81–98

Yamauchi M, Uzawa K, Katz EP, Lopes M, Verdelis K, Cheng H (2000) Distribution of small keratan sulfate proteoglycans in predentin and dentin. In: Goldberg M, Boskey A, Robinson C (eds) Proceedings of the interantonal conference on chemistry and biology of mineralized tissues. American Academy of Orthopaedic Surgeons, pp 305–310

Yan W, Wang P, Zhao CX, Tang J, Xiao X, Wang DW (2009) Decorin gene delivery inhibits cardiac fibrosis in spontaneously hypertensive rats by modulation of transforming growth factor-beta/Smad and p38 mitogen-activated protein kinase signaling pathways. Hum Gene Ther 20:1190–1200

Yano K, Washio K, Tsumanuma Y, Yamato M, Ohta K, Okano T, Izumi Y (2017) The role of Tsukushi (TSK), a small leucine-rich repeat proteoglycan, in bone growth. Regen Ther 7:98–107

Yeh LK, Chen WL, Li W, Espana EM, Ouyang J, Kawakita T, Kao WW, Tseng SC, Liu CY (2005) Soluble lumican glycoprotein purified from human amniotic membrane promotes corneal epithelial wound healing. Invest Ophthalmol Vis Sci 46:479–486

Yoon JH, Halper J (2005) Tendon proteoglycans: biochemistry and function. J Musculoskelet Neuronal Interact 5:22–34

Young MF, Fallon JR (2012) Biglycan: a promising new therapeutic for neuromuscular and musculoskeletal diseases. Curr Opin Genet Dev 22:398–400

Zanotti S, Negri T, Cappelletti C, Bernasconi P, Canioni E, Di Blasi C, Pegoraro E, Angelini C, Ciscato P, Prelle A, Mantegazza R, Morandi L, Mora M (2005) Decorin and biglycan expression is differentially altered in several muscular dystrophies. Brain 128:2546–2555
Zappia J, Joiret M, Sanchez C, Lambert C, Geris L, Muller M, Henrotin Y (2020) From translation to protein degradation as mechanisms for regulating biological functions: a review on the SLRP family in skeletal tissues. Biomolecules 10
Zhao Y, Li X, Xu X, He Z, Cui L, Lv X (2016) Lumican alleviates hypertrophic scarring by suppressing integrin-FAK signaling. Biochem Biophys Res Commun 480:153–159
Zheng Z, Jian J, Velasco O, Hsu CY, Zhang K, Levin A, Murphy M, Zhang X, Ting K, Soo C (2014) Fibromodulin enhances angiogenesis during cutaneous wound healing. Plast Reconstr Surg Glob Open 2:e275
Zhou HW, Burger C, Wang H, Hsiao BS, Chu B, Graham L (2016) The supramolecular structure of bone: X-ray scattering analysis and lateral structure modeling. Acta Crystallogr D Struct Biol 72:986–996

Chapter 7
Cementum Proteins Beyond Cementum

Higinio Arzate and Margarita Zeichner-David

Abstract Dental cementum is a specialized complex structural component of teeth, covering its roots. It is avascular, non-innervated, does not remodel but some of it can grow continuously through life (Bosshardt and Selvig. Dental cementum: the dynamic tissue covering of the root. Periodontol 2000 13:41–75, 1997). Comprehensive knowledge of this tissue has been slow to gather due to its low representation as compared to other mineralized tissues like enamel, dentin, and alveolar bone, constituting the teeth. Because of its sparse presence (about 20–50 μm thick around the cementoenamel junction (CEJ), and about 150–200 μm thick at the root apex), initial studies on cementum were for a long time mostly morphological; however, the last two decades have seen tremendous progress on our understanding of the developmental, cellular, and molecular biology of cementum. We now know that this tissue is highly unique, it is organized into different structural varieties, has a unique composition and function, plays a big role in the periodontal apparatus homeostasis and it is a determinant factor for the regeneration of periodontal tissues destroyed by trauma, genetic diseases, and infections.

7.1 Introduction

Periodontal disease (gingivitis and periodontitis) is one of those infectious oral diseases that affects the structures providing support to teeth (gingiva, periodontal ligament, cementum, and alveolar bone), functionally known as the periodontium. The Center for Disease Control (CDC) reports that in the US, 47.2% of adults aged

H. Arzate (✉)
Laboratorio de Biología Periodontal, Facultad de Odontología, Universidad Nacional Autónoma de México, Mexico City, Mexico
e-mail: harzate@unam.mx

M. Zeichner-David
Herman Ostrow School of Dentistry, University of Southern California, Los Angeles, CA, USA
e-mail: Zeichner@usc.edu

M. Goldberg, P. Den Besten (eds.), *Extracellular Matrix Biomineralization of Dental Tissue Structures*, Biology of Extracellular Matrix 10,
https://doi.org/10.1007/978-3-030-76283-4_7

30 years and older have some form of periodontal disease, which increases with age to 70.1% of adults in the US and about 20–50% around the world, having periodontitis (Eke et al. 2012; Sanz et al. 2010). Furthermore, according to the Global Burden of Disease Study (2017), severe periodontal disease was the 11th most prevalent condition in the world. Periodontitis is one of the major causes of tooth loss which can have a negative effect on eating (resulting in malnutrition), talking, esthetics, self-confidence, and overall quality of life (Tonetti et al. 2017; Reynolds and Duane 2018). Furthermore, periodontal disease is associated with several systemic conditions like diabetes, premature delivery, rheumatoid arthritis, cardiovascular disease, and pulmonary disease (Nazir et al. 2020).

As with any other disease, the best treatment is prevention. Unfortunately, not all types of periodontal disease can be prevented, and when that is the case, then the treatment is focused on removing the bacteria and induce the repair, ideally the regeneration, of the tissues of the periodontium. However, complete regeneration of all these tissues is harder to achieve and remains the ultimate goal of periodontal therapeutics; reestablish a complete, integrated, and functional periodontium. A multitude of studies has focused on regenerating the different components of the periodontium with mixed success. In this chapter, we focus on one of the periodontium components: Cementum. In order to be able to regenerate this tissue, we first need to have a better understanding of the formation, structure, composition, and function of this tissue to be able to mimic these events therapeutically.

7.2 The Origins of Cementum

The name "cementum" comes from the Latin word *caementum: "stone from the quarry."* Cementum precursors appeared 310–330 million years ago in the dermal skeletons of Odontostraci armored fish. These denticles contain structures of enamel, dentin, and a "bony" lining supportive tissue that attaches to the dermal bone. These structures are derived from the mesoderm, and the acellular supporting lining "bony" tissue, named Aspidin, is proposed to be the precursor of cementum (Kozawa et al. 2005). Paleontological studies have shown that the differentiation of cementum, with inserted Sharpey's fibers, was also present in lower amniotes such as Diatectomorpha or Diadectidae, the first herbivorous tetrapods, 323 Ma. In mosasaurs, 168–165 Ma, there is the induction of extensive trabeculation of cementum though nothing is known on the phylogenetic temporospatial evolution of cementum before Diadectidae and stem mammals (For a thorough review of the origins of cementum look at Ripamonti 2019).

Mammalian cementum develops into various forms (Muraki 1958; Schmidt and Keil 1971) concurring with mammalian tooth evolution associated and with the food habit development. In a recent publication on the discovery of cementum, Foster (2017) provides a detailed narrative of the events leading to the acceptance of cementum as a constituent of human teeth. Briefly, acellular cementum formed in the human tooth cervical area, although mammalian root cementum is cellular, was thought to be a new arising characteristic of phylogeny (Kozawa et al. 2005). Initial studies on dental cementum go back to the seventeenth century with Marcelo Malpighi being the first to recognize another substance covering the root; the "*substantia tartarea*," which appears as the first name by which cementum was known (Malpighi 1700). Georges Cuvier then coined the term "cement" to describe the material joining the plates in the compound crown of the elephant molar. Eighteenth-century anatomists still believed that human teeth were composed of enamel and dentin only, although it was recognized that cementum was a constitutive part of animal teeth as demonstrated by Tenon's studies in Horses and Blake's on elephant teeth. However, the first crucial description of the presence of cementum in human teeth was described by Ringelmann, 1824, who was the first to confirm the presence of cementum as a normal constituent on roots of human teeth as a "horny substance present on all human teeth, which partially covered the roots." In 1835, Purkinje described the outer surface of the human tooth root as a true osseous substance; "the *substantia ossea* is drawn thin at the neck of the tooth and increases in size near the apices, wherein one can see *osseous corpuscula*, or bony bodies," which were referred to as corpuscles of Purkinje then and later understood to be the cementocytes of cellular cementum (Fränkel 1835; Raschkow 1835). Retzius in 1837 also described the cortical *substantia* covering roots of human teeth, and he recognized that cementum grows throughout life, being relatively thicker in the elderly. Retzius description of cementum referred to zones of acellular and cellular cementum with embedded cementocytes and their cellular processes (Åhrén 2014; Retzius 1837; Pindborg 1962).

Anthropologist, archeologists, and forensic experts have adopted a procedure using the cementum annulations present in several mammalian species to provide age estimation. In humans, cementum increases in thickness with age, and new layers are deposited on the outside of the dentin throughout the life of the individual. Cementum usually does not undergo wear and tear like enamel and dentin; however, there is continuous apposition in order to maintain a healthy periodontium (Gupta et al. 2014; Zander and Hürzeler 1958).

7.3 Cementum Structure

As previously stated, cementum can be defined as a connective mineralized tissue that covers the entire surface of the root, forming the interface between the dentin and the periodontal ligament. Cementum serves primarily for the anchorage of the collagen fiber bundles of the periodontal ligament, and the supra-alveolar gingival fiber system. Historically, cementum was classified into two major types, acellular and cellular cementum, depending on the presence or absence of cells (cementocytes) in its mineralized matrix. Further classification of these types of cementum was based on the presence of intrinsic fibers secreted by cementoblasts or extrinsic collagen fibers secreted by fibroblasts. The coronal cementum is constituted by mineralized substance which does not contain collagen fibers, named for this reason afibrillar cementum in contrast to fibrillar cementum, which contains collagen fibers. However, later it was shown that some of these types of cementum also contained cellular and fibrillar material (Listgarten and Kamin 1969), and it was not until more than a decade later that Jones (1981) proposed a detailed classification of cementum based on both, cellular and fibrillar components. The final cementum classification regarding its structure was established by Schroeder (1986) and five types of cementum are being described:

1. ***Acellular Afibrillar Cementum (AAC).*** This type of cementum is located over the cervical cementoenamel junction (CEJ), often as a prominence, sometimes overlapping the enamel layer (Fig. 7.1a, b). As reflected in its name, AAC does not contain cementocytes or collagen fibers and is morphologically different from the cementum varieties found over the rest of the root surfaces. Its mineralized matrix reflects a fine granular substance with appositional lines parallel to the root surfaces, and because of the lack of collagen fibrils it has been suggested that this cementum type has no function in tooth attachment (Goncalves et al. 2005). The mineralized AAC patches overlap the cervical enamel border and later can be covered by afibrillar extrinsic fiber cementum (AEFC) or may even wrap the Sharpey's fibers of AEFC. AAC is composed of glycosaminoglycans (Listgarten 1968; Listgarten and Kamin 1969; Schroeder 1986; Schroeder and Scherle 1988; Cho and Garant 2000), its origin and functional properties are unknown although it has been postulated that it is possibly secreted by the inner layer of the Hertwig's epithelial root sheath (Slavkin and Boyde 1974; Thomas 1995; Zeichner-David 2006). Whether the deposition of the AAC layer is a cell-mediated phenomenon or a consequence of mere physicochemical interactions between tissue fluid and the exposed surfaces of calcified substrata, or the consequence of both, remains to be explored (Yamamoto et al. 2016).
2. ***Acellular Extrinsic Fiber Cementum (AEFC).*** This type of cementum is found in the cervical third of the roots in both permanent and deciduous teeth in humans (Fig. 7.1c, d, f). In the single rooted anterior teeth, it might cover all parts of the root; however, in multirooted teeth, AEFC is confined to cervical region (Schroeder 1986). The major function of the AEFC is to anchor the root to the periodontal ligament. The AEFC is composed of glycosaminoglycans and extrinsic collagen fibers densely packed, branching with a diameter between 3 and 6 μm, and oriented perpendicularly to the root surface. The number and density of AEFC fiber bundles are dependent on functional demands. The collagen fibers present in the AEFC, and in the vicinity with the root surface, are wrapped by cytoplasmic processes from adjacent cementoblasts or fibroblasts and a ground substance similar to the one found in the AAC. Infrequently, collagen fibers oriented parallel, or with the irregular arrangement, have been observed and interpreted as extrinsic fibers to the AEFC surface. However, the innermost AEFC layer, close to the dentin-cementum junction, presents irregular collagen fibers suggesting an interconnection of AEFC with dentin matrix fibers and nonhomogeneous mineralization. This AEFC layer is less mineralized than its outer counterpart, contains a non-collagenous layer which functions as an

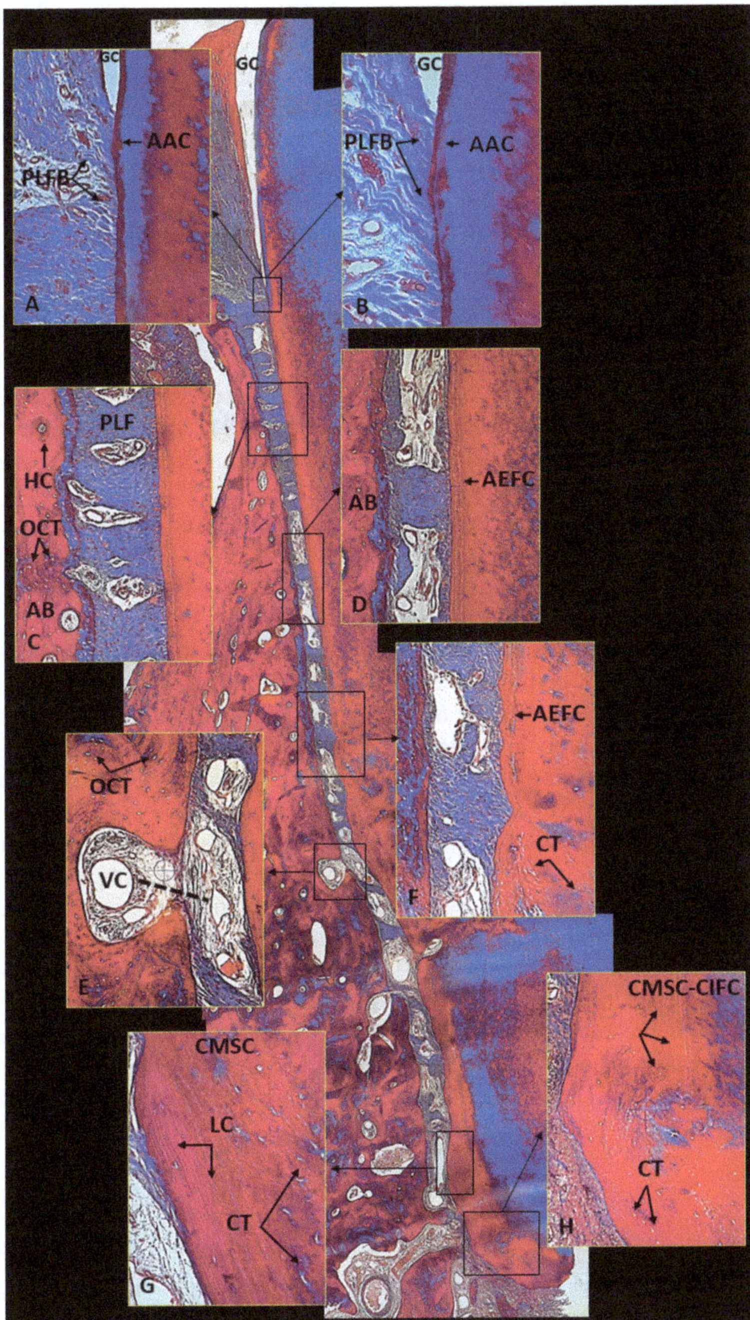

Fig. 7.1 Light micrograph assembly of a panoramic composition of the human periodontal apparatus showing different types of cementum. (**a**) Acellular afibrillar cementum (AAC) beginning at the end of gingival crevice (GC). (**b**) Higher magnification shows the periodontal ligament fiber bundles (PLFB). (**c**) Shows afibrillar extrinsic fiber cementum and at higher magnification lamellar disposition of cementum is observed. (**d**) Bone vascular channel connected to the periodontal

adhesive between dentin and cementum (possibly intermediate cementum). AEFC shows a lamellated pattern containing non-collagenous proteins as organic matrices (resembling finely granular ground substance) are mineralized by the presence of hydroxyapatite crystals and contain embedded fibers from the periodontal ligament (Sharpey's fibers). These alternate lamellae represent variations of periodical activity in matrix deposition and mineralization, with the more profound lamellae representing highly incremental cell lines of cementum showing episodical growth. AEFC shows an alternating bright and dark pattern produced by collagen fibers that change their orientations (Yamamoto et al. 1997, 2010, 2012; Ho et al. 2009; Lieberman 1994; Colard et al. 2016) or crystallite direction (Cape and Kitchin 1930), have different mineralization patterns (Lieberman 1994; Furseth and Johansen 1968; Hals and Selvig 1977; Selvig and Hals 1977; Soni et al. 1962; Colard et al. 2016), or exhibit variations in their organic or inorganic matrices (Kimberly et al. 1994). At the periphery of the AEFC, there is a zone of uncalcified tissue representing "cementoid" tissue which is more obvious during AEFC active growth.

The cells responsible for AEFC deposition are believed to be a particular class of cells resembling fibroblasts (Beertsen and Everts 1990; Bosshardt and Schroeder 1991, 1993) but not necessarily fibroblasts from the periodontal ligament. Fibroblasts of the periodontal ligament are believed to be composed of several subpopulations, whether a single stem cell for cementoblasts and fibroblasts exists in periodontal adult and fetal tissues. Cells that had differentiated into functional fibroblasts and/or cementoblasts could be lineage-restricted (reviewed by McCulloch and Bordin 1991) and are mainly responsible for the modeling and remodeling of the periodontium components including cementum. The possible source for cementoblastic and fibroblastic cell populations also has been attributed to paravascular cells in endosteal spaces of alveolar bone in the rat molar (McCulloch et al. 1987). However, the most accepted hypothesis is that dental follicle cells, identified as pre-cementoblasts, invade the intracellular space of the epithelial root sheath. These cells possess the ability to synthesize matrix components (Cho and Garant 1989) and contribute to the breakup of epithelial root sheath and the formation of Sharpey's fibers (the main component of AEFC). Once these pre-cementoblasts make contact with dentin, they differentiate into cementoblasts. These cells are responsible for the first deposit of AEFC and are characterized by showing a high alkaline phosphatase activity. Accordingly, during this brief period of cementum deposition, cementoblasts detach from the

Fig. 7.1 (continued) ligament providing progenitor cells to the periodontal ligament (red dashed lane). In (**e**) and (**f**) it is observed that the AEFC and the beginning of cellular cementum with the osteocytes (OCT) immersed into the cementum's mineralized matrix. (**g**) Cellular mixed stratified cementum (CMSC) shows cementum lamellae (CL) and cementocytes. (**h**) Cellular mixed stratified cementum (CMSC) and cellular intrinsic fiber cementum where collagen fibers are randomly arranged and in different directions (arrows)

newly formed cementum and acquire a fibroblasts phenotype into de periodontal ligament (Cho and garant 1989, 1996).

3. ***Cellular Mixed Stratified Cementum (CMSC).*** This type of cementum is localized in the interradicular and apical two-third regions of the root (Fig. 7.1e, g). CMSC is occasionally covered by AEFC and is composed of multiple interposed layers of acellular extrinsic fiber cementum; Sharpey's fibers continuing into the periodontal ligament and cellular intrinsic fiber cementum where the intrinsic fibers are completely into CMSC and have no contribution to the maintenance of the tooth into the alveolus. CMSC features are closely related to the lacunae and cementocytes canaliculi, its lamellae form of deposition with sharply delineated incremental lines separating individual lamella and a superficial layer of uncalcified cementum recently secreted; the cementoid layer. The lamellae represent circumscribed patches that cover the perimeter of the apical third which indicate an intermittent deposition pattern which is reflected by a discontinuous increase of CMSC thickness. The increment of CMSC thickness is cyclical and episodic. In CMSC the intrinsic collagen fibers are more voluminous, numerous, and densely packed than the extrinsic collagen fibers. The main components of CMSC, Sharpey's fibers, contain an uncalcified core surrounded by calcified extracellular matrix. However, the intrinsic collagen fibers into CMSC appear to be mineralized (Schroeder 1993; Yamamoto et al. 2016).
4. **Cellular Intrinsic Fiber Cementum (CIFC).** This type of cementum is deposited as a part of the reparative process (Fig. 7.1h), found as a substrate into lacunae following root resorption, represents CMSC without Sharpey's fibers attachment and does not participate in supporting the teeth (Bosshardt and Selvig 1997; Schroeder 1986). However, varieties of CIFC have been demonstrated like CIFC fiber-rich; CIFC fiber-poor; and CIFC fiber-free (Yamamoto et al. 2016). It has also been reported that CIFC shows alternate lines, lamellae-like, with longitudinal and transversal disposition similar to compact bone conforming to the twisted plywood model. According to this model, all collagen fibrils run parallel in a given plane, and their direction rotates from plane to plane. A 180° rotation of fibril arrays corresponds to a period. This periodic change in fibril array orientation appears as the alternating lamellae when observed in histological sections (Yamamoto et al. 1997, 2010, 2016).

   A better appreciation of the cementum structure can be seen in Fig. 7.2a, where Sharpey's fibers (SF) can be seen embedded in the cementum (AEFC) coming from the periodontal ligament (PDL). The presence of the intermediate cementum can be seen in Fig. 7.2b, as well as the epithelial rest of Malassez (ERM) present near the cementum in the PDL. Figure 7.2c shows the PDL fibers inserted in the cementum as well as the presence of cementoblasts (CB) and paravascular cells (PVC) near the cementum layer. Figure 7.2d shows the PDL and alveolar bone, while in Fig. 7.2e the presence of cementocytes can easily be seen in the cellular cementum. Figure 7.2f shows both, CIFC and CMSC and the presence of inserted collagen fibers.
5. **Intermediate cementum (IC).** The presence, origin, and structure of the cementum layer is still controversial (Fig. 7.2b). First described by Bodecker in 1878 as

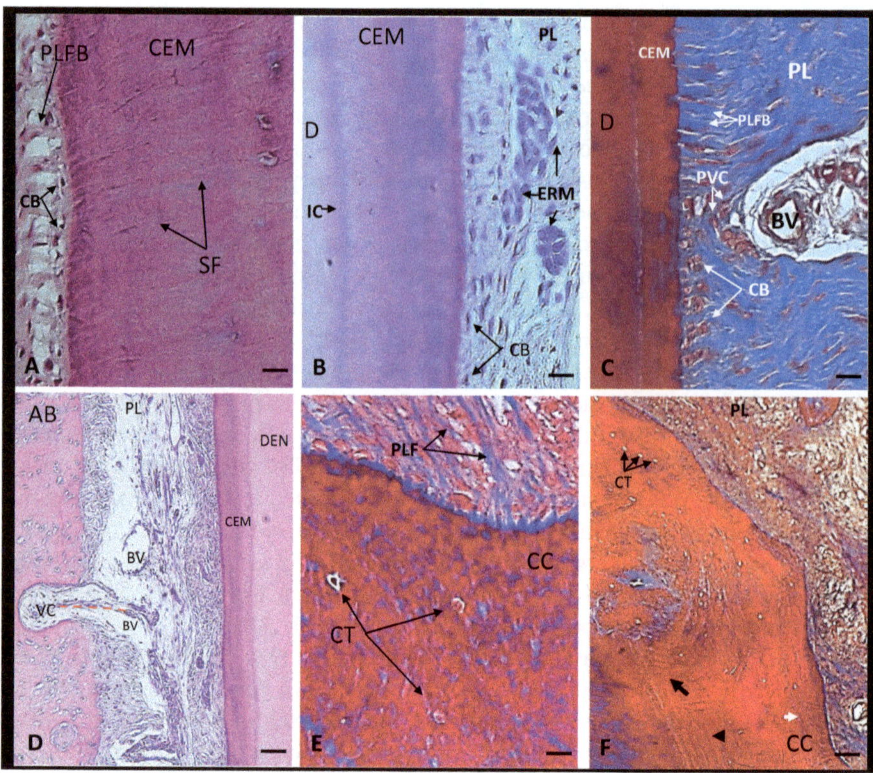

Fig. 7.2 (**a**) Light microphotograph shows Sharpey's Fibers (SF) immersed into AEFC. Cementoblasts, facing cementum surface (CB) and periodontal ligament fiber bundles (PLFB). (**b**) Shows AEFC and epithelial rests of Malassez (ERM) in the vicinity of cementum and a basophilic stained band between cementum and dentin that represents the intermediate cementum (IC). (**c**) Shows periodontal ligament fiber bundles (PLFB) inserted into cementum, and paravascular cells (PVC) adjacent to a periodontal ligament blood vessel (BV) in the vicinity of cementum. (**d**) Vascular channel (VC) and the migration of endosteal progenitor cells to the periodontal ligament (red dashed lane). (**e**) Shows cellular cementum and cementocytes (CT) into its mineralized matrix. (**f**) Cellular mixed stratified cementum (CMSC) and cellular mixed intrinsic cementum (CIFC) were observed in cellular cementum (CC), where collagen fibers could be observed in various directions and randomly arranged (black arrow, black triangle, and white arrow) and cementocytes (CT). Bar = 100 μm

an interzonal layer between dentin and cementum. Bencze in 1927 coined the name "intermediary layer of cementum" to describe an ill-defined layer in the region of the cemento-dentinal junction that does not contain a regular structure. Blackwood in 1957 noticed the presence of protoplasmic bodies and suggested these bodies were odontoblastic processes. Selvig in 1963 demonstrated that a layer of collagen forms from the dental sac, just prior to cementum formation and immediately after the beginning of dentin formation, possibly due to the disintegration of Hertwig's epithelial root sheath (HERS) before the production of any cementum-like tissue thus leading several investigators to suggest that the IC was some form of dentin product of dentinogenesis (Osborn 1965; Owens 1972). Others (el Mostehy and Stallard 1968) concluded that intermediate cementum is not a layer per se, nevertheless, they describe an area between cellular cementum and dentin containing cellular debris trapped in the cementum matrix and pointed out that the cellular inclusions in this layer are epithelial cells (HERS) that failed to separate from the dentin surface and are now surrounded for calcified material. Lester (1969a, b) concluded that intermediate cementum is a distinct form of cementum and its irregularity is a result of the trapping cells on the forming tissue. Notably, in fully mineralized teeth, the junctional zone between the cementum and the dentin is hypercalcified and persists throughout life as a specific layer of mineralized tissue separated from the dental cementum (Lindskog 1982a; Röckert 1958). However, fibrillar material is deposited close to HERS before it disrupts completely and it is proposed that the initial cementum deposition is an epithelial secretory product with higher mineral content than the surrounding cementum and dentin thus suggesting that HERS participates in the formation of intermediate cementum (Schonfeld and Slavkin 1977; Lindskog 1982b; Thomas 1995; Zeichner-David et al. 2003). Later Lindskog and Hammarström (1982) confirmed that the matrix of IC was a non-collagenous matrix, similar in nature to aprismatic enamel matrix (for a more comprehensive review of IC, we recommend the reviews of Harrison and Roda 1995; Bosshardt and Schroeder 1996; Goncalves et al. 2005; Yamamoto et al. 2016).

To this date, the identity and origin of IC continue to be questioned. In later studies, using HERS-derived cells we showed that these cells express molecules related to mineralized tissues such as osteopontin (OPN), bone sialoprotein (BSP), and cementum-derived protein (CEMP-1) at the initial formation of the root and initial deposition of cementum. One possible explanation is that HERS undergo epithelial-mesenchymal transdifferentiation (EMT) and become cementoblasts. Taken together, these findings indicate that IC is a secretory product of the inner layer of HERS (Zeichner-David et al. 2003; Arzate et al. 2015). The function of intermediate cementum remains unknown however potential functions to provide a permeable barrier between cementum and dentin, of being a precursor for cementogenesis during root development, and a precursor for cementogenesis during periodontium wound healing have been suggested (Harrison and Roda 1995). It is also suggested that it serves to join the root dentin and cementum (Yamamoto 1986).

7.4 Cementum Function

As previously stated, cementum is a complex tissue regarding its origin, structure, composition, and function. It has been the interest of periodontists and biologists for centuries. It is accepted that the major role of this tissue is to serve as the anatomical site for the attachment of Sharpey's fibers of the periodontal ligament and therefore provide the attachment of the tooth to the alveolar bone. Black (1890) gave a detailed description of the cells that secrete cementum: the cementoblasts and he pointed out that cementoblasts have no resemblance with osteoblasts in both, form and the matrix they deposit. This statement established the earliest understanding of cementum and its regenerative potential. However, it is in the second decade of the twentieth century when Bernhard Gottlieb (1942) was one of the first to study the function of cementum and theorized about the "vitality of cementum." He suggested that continuous cementum apposition is necessary for maintaining a healthy periodontium. Furthermore, he suggested that it acts as a barrier against the downgrowth of the epithelium. Similar results were seen in the study conducted by Stein and Corcoran (1990) on vital and non-vital teeth.

The main function of the cementum is to participate, together with the principal fibers of the periodontal ligament and alveolar bone, in tooth anchorage into the alveolus. Another function requiring cementum participation is to reposition teeth to maintain opposing teeth in occlusion, as suggested from experimental studies in rats in which when an opposing molar was extracted, the growth of cementum on the remaining molar was less than that of the control contralateral molar in which the occlusion was not modified (Kato et al. 1992), and supported by studies in deer (Pérez-Barbería et al. 2020). Since the biomechanical properties and composition of cementum provide a flexible mechanism for tooth attachment to the alveolus, and a structure to enable repositioning of the occlusal plane; therefore, the rate at which cementum is deposited allows it to be used as an evolutionary life-history trait (Lieberman 1993) has been extensively used in archeology, life history studies in population ecology (Lieberman 1994; Pérez-Barbería et al. 2014, 2015), and as a technique to estimate age (Pérez-Barbería et al. 2014).

From a biomechanics perspective, the intriguing presence of intermediate cementum at the CDJ provides a gradual transition in mechanical properties. Stress and strain singularities are eliminated when occlusal loads are distributed, making the cementum–dentin interphase an efficient biomechanical system. Thus, intermediate cementum can be classified as a functionally graded dental tissue (Ho et al. 2004) and a necessary precursor to cementogenesis in root development. The potential functions of this layer suggest that it can act as a permeability barrier between cementum and dentin; as a barrier to the stimulation of inflammatory root resorption by necrotic tissue or microbial contents of the root canal after trauma (e.g., luxation) to the teeth; a precursor for cementogenesis in root development; and as a precursor for cementogenesis in wound healing (Andreasen 2012). Functions for AEFC and CMSC are directly related to tooth attachment to the dental alveolus. AEFC plays the main role for Sharpey's fibers attachment to cementum, which appears to be solid and long-lasting, and covers the total length of the root in single rooted teeth but not in the same way in multirooted teeth. CMSC begins its deposit before tooth eruption in multirooted teeth, accumulating in the apical region, although the teeth are not functioning, and appears that its deposition is not related to support the external forces on teeth. Only when CMSC is covered with AEFC it contributes to this function therefore its function may be related to the homeostasis process and adaptation of the root in response to occlusal forces. CIFC participates in the repair process of external root resorption lacunae (Ascenzi and Bonucci 1968), compensates tooth movement due to crown wear (Jones 1981; Schroeder 1986, 1992, 1993; Yamamoto et al. 2016), and does not appear to contribute to tooth support. Together, CMSC and CIFC act during the dynamic restructuration of the deeper periodontium, playing a role for the newly apposition of cementum, rather than resorption and replacement.

The biology of cementum seems to be unique in mammals since its morphological characteristics lack Haversian canals, blood, lymphatic vessels, and nerves and cementum deposits continuously throughout the entire life of mammals. Therefore, the next sections of this review will be dedicated to examining both, inorganic and organic constituents of cementum.

7.5 Cementum Composition

7.5.1 Inorganic Components

Cementum mineralization is a cyclic process due to the functional response to mastication. Its mineral density is similar to bone and it varies according to the different types of cementum: AEFC and CMSC contain 45–50% of inorganic mineral composed of amorphous calcium phosphate and crystalline hydroxyapatite $Ca_{10}(PO_4)_6(OH)_2$ by 98.8% and 1.2% ß-TCP (Selvig 1965, 1970). These hydroxyapatite crystals are found between the collagen fibrils, normally arranged with their c-axis parallel to the long axis of collagen fibrils (Yamamoto and Hinrichsen 1993). However, the crystallinity of the cemental inorganic component is lower than other calcified tissues (Bosshardt and Selvig 1997). As expected, AEFC contains more minerals than CIFC and CMSC (Yamamoto and Wakita 1990). This is explained by the presence of uncalcified spaces, such as lacunae and by the uncalcified core of Sharpey's fibers (Goncalves et al. 2005). The inorganic chemical composition of human root cementum and their most common elements such as Ca^{2+}, P, Mg^{2+}, and F have been described (Röckert 1958; Brudevold et al. 1960; Nakata et al. 1972). Using electron probe microanalysis and microradiography, Hals and Selvig (1977) analyzed ground sections of teeth of adult individuals describe the presence of Ca^{2+}, P, F, S, Mg^{2+}, Fe, Cu, Zn, Sn, and Ag restricted to AEFC. However, fluoride concentration in the AEFC outer layer is about 0.5% and decreases at 0.1% at the dentinal-cementum junction. The fluoride content in AEFC is consistently greater than that of CMSF. Ashed cementum contains fluoride in a higher proportion than dentin and enamel, probably by intake of fluorinated drinking water and the high degree of absorption and permeability of cementum. Fluoride is located in superficial layers of cementum with a limited diffusion to the deeper layers. However, cementum possesses a stronger anti-absorption capacity compared to the alveolar bone.

Regarding its hardness, cementum is softer than dentine. The spatial distribution of the physicochemical parameters assessed from the Raman analysis indicates that the mineral/organic and type-B carbonatation ratio reproduced the alternating bright and dark lines of AEFC type of cementum. Dark lines show the highest mineral/

organic ratio, suggesting that they correspond to more mineralized incremental layers, but also more readily decalcifies in the presence of acidic conditions (Bosshardt and Selvig 1997). Zoologists and zooarcheologists called the dark lines annuli, winter lines or rest lines, as they were thought to correspond to slowed growth periods (Ho et al. 2004, 2005, 2009). The dark lines correspond to smaller, more mineralized, and better organized lines produced during a slower period of AEFC growth. Raman spectra analysis has shown that in AEFC the mineral crystals and collagen fibers are always oriented in the same direction both in dark and bright lines. Energy dispersive X-ray microanalysis of cementum determined the mineral content of calcium, phosphorous, and magnesium from the cementoenamel junction to the root apex on the external surface of the cementum. The Ca/P ratio at the cementoenamel junction showed to be 1.8 to 2.2; however, in the area corresponding to the AEFC, Ca/P media value of 1.65 was found corresponding to biological hydroxyapatite. In the area representing the fulcrum of the root, there is an abrupt change of the Ca/P ratio, which decreases to 1.3 indicating the presence of amorphous calcium phosphate, probably the result of occlusal resistance function and concentration increase of Na, K, Cl, and Mg^{2+} ions in this area. In the apical one-third of the root, where the CMSC and CIFC are present, the Ca/P ratio decreases linearly to the root apex from a 1.65 ratio to a 1.3 ratio which corresponds to amorphous calcium phosphate. These findings revealed that on the coronal one-third of the root where AEFC is located, the mineralized tissue is densely and homogeneously mineralized. In contrast, the apical one-third of the root is less mineralized which contributes to its being heterogeneously mineralized, porous, and having soft characteristics, intimately related to the main function of CMSC during adaptation and dynamic reshaping of the root surface as the tooth shifts and drifts in its socket. These differences could be partially accounted by the nonmineralized structures present in CIFC, which include cementocytes lacunae as well as larger inclusions of cellular elements (Selvig 1965). Magnesium is, following calcium, potassium, and sodium, the fourth highest concentrated cation in the human body (Posner 1969) and its distribution throughout the length of human cementum shows a Mg/Ca ratio value of 1.3–1.4% around the fulcrum of the root and an average value of 0.03% as reported elsewhere (Nakata et al. 1972; Neiders et al. 1972; Hals and Selvig 1977). The Mg/Ca ratio pattern distribution showed that in the region where the Ca/P ratio decreased the Mg/Ca ratio reached its maximum value, showing a negative correlation, which supports the fact that Mg^{2+} replaces Ca^{2+}, which is related to a higher crystal growth. It has been suggested that Mg^{2+} controls bone calcification in general as well as the crystallization processes (Serre et al. 1998); therefore, it could be expected that Mg^{2+} incorporation can contribute to the root's mineralization and control growth of hydroxyapatite crystals, particularly on the fulcrum area. In summary, clear compositional differences exist between AEFC and CMSC cementum varieties and the Mg^{2+} distribution suggests a role in regulating the metabolism of cementum, influencing mineralization, especially crystal growth (Wiesmann et al. 1997), stabilizing calcium phosphate, and reducing the degradation rate of calcium phosphate (Serre et al. 1998) at the fulcrum level. Therefore, Mg^{2+} is an important factor in supporting the masticatory forces and in controlling cementum metabolism because Mg^{2+} acts on deposition and resorption of mineralized tissues (Alvarez-Pérez et al. 2005). The mineral content of cementum once formed does not seem to change significantly with age.

7.5.2 Organic Components

In addition to hydroxyapatite mineral crystals, cementum contains a complex lattice of insoluble collagen bundles (Glimcher and Krane 1968) immersed within a matrix formed by non-collagenous proteins (NCPs) (reviewed in Weiner 1984; Butler 1991; Robey et al. 1993; Butler and Ritchie 1995). These macromolecules include plasma-derived proteins (Dickson et al. 1975), proteoglycans (Fisher et al. 1987; Embery et al. 2001), and many other molecules such as growth factors and cell attachment factors (reviewed in Hauschka 1990; Mundy et al. 1995). These macromolecules are embedded into the mineralized matrix of cementum and have the capacity to bind to hydroxyapatite mineral contents of cementum (Termine et al. 1981; Weiner 1986; Moradian-Oldak et al. 1992; Füredi-Milhofer et al. 1994). The association of non-collagenous proteins with structural proteins and mineral lattice determine the structure, biomechanical and biological properties of cementum.

7.5.3 Collagens

Cementum contains 50–55% organic material and water. The organic matrix of cementum consists primarily of Type I collagen. In bovine cementum, type I collagen accounts for more than 90% of the organic matrix and type III collagen approximately 5% (Birkedal-Hansen et al. 1977), while in human cementum type I collagen appears to be the only type of collagen (Christner et al. 1977); however, Sharpey's fibers are coated with type III collagen (constituting approximately 2.5% of the total collagen in human cementum). It was suggested that the presence of type III collagen coating the Sharpey's fiber may prevent mineralization of the type I collagen (Selvig 1965). On the other hand, type III collagen forms cross-striated fibrils in cementum and induce biological mineralization serving as a scaffold for the mineral crystals and maintain the structural integrity of cementum after mineralization (Birkedal-Hansen et al. 1977; Wang et al. 1980). Recently, proteomic analysis of cementum identified the presence of COL5A2, COL6A3, COL6A1, and COL11A1 and an isoform of collagen alpha-1 (XXV) chain; however, they were present in low abundance and their role in cementum mineralization is unknown. Type I collagen plays a structural function in cementum during the mineralization process providing heterogeneous nucleation (in holes and pores) where the inorganic nuclei are formed and develop into intrafibrillar apatite crystals of the collagen fibers (Glimcher 1989).

7.5.4 Proteoglycans/Glycosaminoglycans

The class of glycosylated proteins known as proteoglycans (PG) is represented by its carbohydrate constituents. These are polysaccharides known as glycosaminoglycans (GAG), such as heparin (Hp) (Mulloy et al. 2016) and chondroitin sulfate (CS) (Mantovani et al. 2016). When attached to their native protein cores, these polysaccharides form the glycoconjugates known as proteoglycans (Pomin and Wang 2018). GAGs are linear and heterogeneous sulfated glycans. Although they are structurally complex, the backbones of these polysaccharides are simply made up of repeating disaccharide building blocks. PGs and their constituent glycosaminoglycan (GAG) chains also bind to hydroxyapatite (Boskey et al. 1997; Rees et al. 2001; Coulson-Thomas et al. 2015). Within the tissues that predominantly contain type I collagen such as cementum, the higher molecular weight PGs retain hydrostatic pressure and gross shape characteristics (Seog et al. 2002; Scott 1988; Scott and Haigh 1988), while lower molecular weight PGs within the cementum and their adjacent interfaces are thought to maintain site-specific biophysical and biochemical environment for cells like cementoblasts and cementocytes. The presence of glycosaminoglycans in cementum was first shown using histochemical staining with alcian blue (Vidal et al. 1974). Bartold et al. (1988), using electrophoresis demonstrated the presence of hyaluronan, chondroitin sulfate, and dermatan sulfate in healthy human cementum with chondroitin sulfate being the predominant glycosaminoglycan species.

Different types of proteoglycans have also been identified; versican (large proteoglycan) and decorin, biglycan, and lumican (small proteoglycans) have been identified (Qian et al. 2004). Their presence is exclusively in cellular cementum, (Ababneh et al. 1988, 1999) and keratan sulfate is located around cementocytes (Cheng et al. 1996). Nonetheless, osteoadherin, a keratin sulfate-containing proteoglycan, is associated with the initial phase of cementum formation because Hertwig's epithelial root sheath cells express this proteoglycan during root development (Petersson et al. 2003), although acellular cementum does not contain proteoglycans, initial acellular cementum formation requires a dense accumulation of proteoglycans (Yamamoto et al. 2004). Glycosaminoglycans present in human cementum are chondroitin-4-sulfate/dermatan sulfate, chondroitin-6-sulfate, and keratan sulfate, and localized to the lumina of the lacunae and canaliculi of cellular cementum, as well as on periodontal ligament fibers attached to both types of cementum: AEFC and CMSC. Chondroitin 4- and -6-sulfate (CS), keratan sulfate (KS), and other native chondroitin sulfate epitopes are expressed in cementocytes (Ababneh et al. 1988) and it has been suggested that the expression of proteoglycan/glycosaminoglycan around cementocytes plays a role in inhibiting cell mineralization (Cheng et al. 1996). It is proposed that proteoglycans inhibit mineralization of collagen fibrils by attaching to specific sites on collagen fibrils normally destined to be filled with hydroxyapatite; biglycan has been shown to have a high affinity to collagen (Ababneh et al. 1999). Thus, the proteoglycan content becomes lower after mineralization (Ababneh et al. 1988; Yamamoto et al. 2016). Additionally, the presence of chondroitin sulfated GAGs was confirmed at an interface between cementum and dentin (Ho et al. 2004; Yamamoto et al. 1999). Cementum lamellae contain biglycan, fibromodulin, and small leucine-rich proteoglycans at the interfaces between PDL-cementum and dentin-cementum. Biglycan and fibromodulin, containing higher and lower molecular weight CS- and KS-GAGs, interconnect with collagen and these interconnecting elements maintain needed tissue resilience for tooth movement within the alveolar socket. Digestion of matrices with enzymes confirmed the presence of KS and CS GAGs at these interphases. KS PGs are found to be located almost exclusively in nonmineralized portions of cementum such as pre-cementum and the peri-cementocytes area. This biochemical and immunohistochemical data suggest that the major KS PGs of cementum, lumican, and fibromodulin have a specific tissue distribution and may have regulatory roles in cementum mineralization (Cheng et al. 1996).

From a physical perspective, the presence of GAGs provides tissue resilience and consequently a conserved tooth movement in the socket in response to functional loads. The innate polyanionic nature of GAGs is pivotal in maintaining free-charge density, electrostatic interactions, steric hindrance, water retention, hydrostatic pressure, lubrication, nutrient transport, cell migration, cell differentiation, matrix structure, and mechanical integrity (Seog et al. 2002) all of which are necessary to conserve the biomechanical integrity of the ligament–cementum complex. The cementoid layer contains higher ratios of organic to inorganic matter due to the presence of GAGs. This interphase loss in fibrillar structure in the predominantly hygroscopic regions of the PDL-cementum and cementum-dentin interfaces shows higher concentrations of PGs. It has been shown that GAGs in the cementum–dentin interface retain the structure and suggest that their retention could provide shape memory to tissues (Ho et al. 2005).

7.5.5 Tissue Nonspecific Alkaline Phosphatase (TNALP)

TNALP, also known in humans as ALPL, is critical for proper skeletal mineralization (Zweifler et al. 2015). ALPL is a membrane-bound glycoprotein enzyme that hydrolyses phosphate groups at alkaline pH, inhibits pyrophosphatase, ATPase, and protein phosphatase activity at neutral pH (Hui and Tenenbaum 1998). It belongs to a family of four homologous human alkaline phosphatase genes (Buchet et al. 2013) and plays a role in bone and cementum matrix mineralization. In humans, mutations in the ALPL gene cause infantile hypophosphatasia, a disease marked by poor bone mineralization and defects in the deposition of acellular cementum, Rickets disease, and Osteomalacia (Foster et al. 2014). Ablation of the homologous mouse ALPL gene results in increased levels of inhibitory pyrophosphate and osteopontin leading to the suppression of hydroxyapatite crystals nucleation and growth (Pujari-Palmer et al. 2016). Furthermore, ALPL appears to play a key role in the development of the central nervous system and adult neurogenesis (Chiquoine 1954; Narisawa et al. 1994; Langer et al. 2007; Foster et al. 2012; Sebastián-Serrano et al. 2016), modulates the function of T and B lymphocytes function in adult life (Hernández-Chirlaque et al. 2017), and maintains Blood-Brain-Barrier integrity (Deracinois et al. 2015; Nwafor and Brown 2021). Along with these functions, ALPL is also associated with the regulation of tissue remodeling, cell proliferation, differentiation, and maturation.

As far as cementum is related, in vitro assays demonstrated that ALPL was capable to induce the deposit of a tissue resembling acellular afibrillar cementum on calcified dentin in vivo. This matrix had a granular or filamentous appearance with the presence of incremental lines (Beertsen and Van den Bos 1991). Additional support for the view that ALPL plays a significant role in cementogenesis comes from a naturally occurring condition, i.e., hypophosphatasia, an inherited disorder of osteogenesis and cementogenesis, characterized by a defect in the synthesis and expression of the liver/bone/kidney (L/B/K) isoenzyme of ALPL. Almost no cementum formation is observed in children suffering from this disorder, as a result of which teeth are lost prematurely (Sobel et al. 1953; Bruckner et al. 1962). ALPL deficient mice showed the defective formation of acellular cementum along the molar roots. Cementum was very thin and found as irregularly shaped patches around insertion of the periodontal ligament Sharpey's fibers which were short and poorly developed. Interestingly, ALPL deficiency did not affect alveolar bone, periodontal ligament, or cellular cementum formation thus one can deduce that ALPL is a crucial factor for the formation of acellular cementum (Beertsen et al. 1999). However, in humans affected by hypophosphatasia, both acellular and cellular cementum formation was disturbed (van den Bos et al. 2005). In addition to cementum, the periodontal ligament is also rich in ALPL activity (Yamashita et al. 1987), with very high activity as compared to kidney, liver, small intestine, alveolar bone, periosteum, and dental pulp. The activity of ALPL is heterogeneously distributed, perhaps reflecting local variations in phosphate ALPL activity in the periodontium, and this activity is associated with regional variations in cementum thickness (Groeneveld et al. 1994), which is higher adjacent to the cellular cementum compared with acellular cementum (Groeneveld et al. 1995).

The role of ALPL in calcification came from studies of hypophosphatasia patients, whose disease results from missense mutations in the gene coding for ALPL leading to decreased or absent ALP activity. One of the main functions of ALPL is the hydrolysis of inorganic pyrophosphate, an inhibitor of hydroxyapatite crystal growth (Foster et al. 2008, 2014; McKee et al. 2013; Murshed et al. 2005) creating a low PPi environment promoting acellular cementum initiation. Nevertheless, alterations in PPi have little effect on cellular cementum formation, though cementum matrix mineralization is affected (Zweifler et al. 2015). At enzymatic and molecular levels, two phosphatases, NPP1 and TNAP have antagonistic effects on mineral formation due to their opposing activities: production of Inorganic pyrophosphate (PPi) by NPP1 or its hydrolysis by TNAP (Thouverey et al. 2009). ALPL plays a role during the initial cementogenesis process in both acellular and cellular cementum. However, loss of TNAP blocked acellular cementum formation and increased the formation of hypo-mineralized cellular cementum. NPP1 was detectable in cementoblasts only after cementum apposition and was not consistently found around cellular cementum. Loss of NPP1 activity resulted in a rapidly growing acellular cementum whereas cellular cementum remained unaltered (Zweifler et al. 2015). Cells responsible for cementum deposition (cementoblasts) are specifically sensitive to levels of PPi/Pi within the extracellular matrix as compared to cells involved in the formation of the surrounding alveolar bone (osteoblasts); therefore, PPi prevents mineralization in osteoblast by direct binding to growing crystals and inhibition of ALPL activity. Recent findings focused on understanding the role of inorganic phosphate (Pi) and pyrophosphate (PPi) suggest that the progressive ankylosis gene (ANKH in humans; Ank in mice) regulates the transport of PPi, and mice that lack ANK function feature decreased extracellular PPi, promoting a rapidly forming and thick cementum (Rodrigues et al. 2011). The results showed higher cementum formation and greater new cementum appositional activity in Ank KO mice than controls and elevated phosphatase NPP1 in cementoblasts, suggesting that reduced local levels of PPi could promote increased cementum regeneration, indicating that the local modulation of Pi/PPi may be a potential therapeutic approach for achieving better cementum regeneration, a key for periodontal regeneration, as reported recently that local administration of rhALPL reestablished periodontal attachment and OPN deposition on root surfaces and cementum formation (Nagasaki et al. 2020).

7.6 Phosphoproteins

Cementum contains several phosphoproteins that have been implicated in mineral deposition and cell and matrix–matrix interaction. These proteins belong to the "small integrin binding ligand *N*-glycosylated" known as SIBLING proteins (Boskey 1995; Fisher and Fedarko 2003). These proteins are all located on the same chromosome, all have RGD-cell binding domains, all are anionic, and all are subject to posttranslational modifications including phosphorylation and dephosphorylation, cleavage, and glycosylation. These proteins appear to be positive regulators of hydroxyapatite crystal nucleation and/or growth and the following are present in cementum:

Osteopontin (OPN), *a*lso known as secreted phosphoprotein 1 (SPP1) or bone sialoprotein 1 (BSP1), is abundant at sites of calcification in human atherosclerotic plaques and in calcified aortic valves (Mohler III et al. 1997). OPN is a highly phosphorylated and sulfated sialoprotein which contains attachment RGD motifs that recognize the vitronectin type of the integrin receptor. OPN is expressed by periodontium cells in close contact with acellular cementum as well as cementocytes (Bronckers et al. 1994). It has been suggested that OPN is not only derived from local cellular sources but may be also imported from outside the local environment via circulation. It has been found in such areas as in the cementum laminae, the AEFC layer, and the CIFC layer, which is concordant with OPN's role in cell–matrix interactions during cementum remodeling

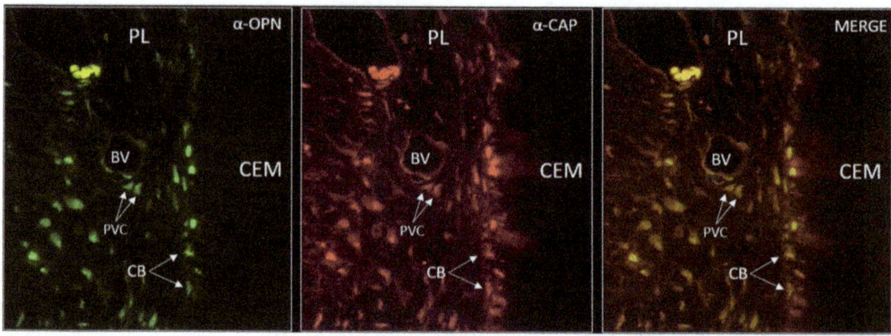

Fig. 7.3 Double staining and colocalization of OPN and CAP expression. It can be seen that OPN is localized in cell populations of the periodontal ligament, paravascular cells (PVC), and some cementoblast cells facing the cementum surface. CAP is strongly expressed by cementoblasts and paravascular cells (PVC). Merged image revealed colocalization of OPN and CAP in same locations with a light evidence that cementoblasts becoming embedded into cementum matrix strongly express CAP. Bar = 100 μm

and suggested that OPN is critical for the cementogenesis process (Foster 2012) since cementoblasts along the root surface showed intracellular labeling for OPN both in elements of their Golgi apparatus and in secretory granules (McKee et al. 1996). OPN was substantially more concentrated in acellular cementum and is distributed as a dense network of organic material closely enveloping the collagen fibrils of acellular, extrinsic fiber cementum (McKee et al. 1996). OPN was also located at the dentinal-cementum junction, suggesting that OPN, in addition to binding to mineral, may participate in maintaining tissue cohesion/adhesion by binding to OCN and type I collagen (Ritter et al. 1992; Gorski et al. 1995). This dentino-cementum junction may be related to the intermediate layer of cementum (Bosshardt and Schroeder 1996) and may thus correspond to this relatively collagen-free zone rich in non-collagenous glycoproteins such as OPN. Our studies showed co-expression of OPN with the cementum-specific protein CAP (describe in detail later) in cementoblast cells lining the cementum layer as well as paravascular cells presumed to be cementoblasts progenitors (Fig. 7.3).

OPN has a poly-Asp domain, whose repetitive sequences are known to bind calcium to mineral surfaces and is related to the initial growth of hydroxyapatite crystals (Alvarez Pérez et al. 2003). However, OPN is a multifunctional protein and can also act as an inhibitor of hydroxyapatite formation (Boskey et al. 1993) and calcium oxalate crystals in the kidney and act principally on crystal growth (Hunter et al. 1996). More recent studies using an OPN knockout (Spp1−/−) mice showed an increase in cellular cementum formation, volume, and mineral density of dentin/cementum and alveolar bone and increase in tissue densities while decreasing the volumes of both PDL and dental pulp. However, acellular cementum apposition was not altered (Foster et al. 2018). It was concluded that OPN has an important and non-redundant role in regulating mineralization in the periodontium, though its importance lies with cellular cementum and at the

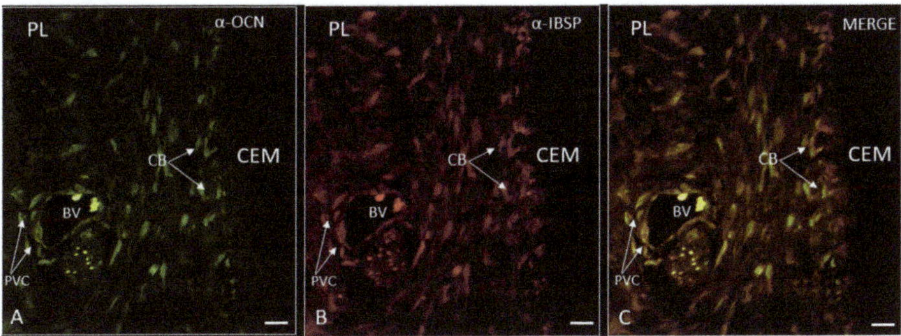

Fig. 7.4 Double staining and colocalization of Osteocalcin (OCN) with Integrin Binding sialoprotein (IBSP). (**a**) OCN is expressed by cementoblasts (CB) and paravascular cells (PVC) and periodontal ligament cell populations. (**b**) IBSP is strongly expressed by cementoblasts (CB), paravascular progenitor cells (PVC), and lightly by periodontal ligament cell populations. (**c**) Merged image shows a predominant expression of IBSP on cementoblasts (CB) periodontal ligament cell populations and colocalizes with OPN mainly in paravascular cells (PVC) and pre-cementoblasts

PDL-bone border and influencing tissue density properties of the PDL and bone, and not by directly controlling acellular cementum. (Foster et al. 2018).

Bone sialophosphoprotein (BSP), also known as integrin-binding sialoprotein (IBSP), is an acidic phosphorylated glycoprotein, it also has an RGD cell attachment motif and repeated sequences of acidic amino acids. It is present in mineralizing tissues like bone, dentin, cementum, and calcified cartilage; therefore, it is implicated in the nucleation and regulation of hydroxyapatite crystals in these tissues (Chen et al. 1991; Chen et al. 1992; Ganss et al. 1999; Fisher et al. 2001). BSP is believed to play a critical regulatory role during the process of cementogenesis beginning with the processes of pre-cementoblasts chemoattraction, adhesion, and differentiation (MacNeil et al. 1995a). BSP is strongly expressed by cementoblasts and paravascular cells in the PDL, (Fig. 7.4b) which suggested it contributes to cementum formation and mineralization (Somerman et al. 1990; MacNeil et al. 1995b). BSP is present in the cementum extracellular matrix in both, the acellular and cellular cementum layers (MacNeil et al. 1994; McKee et al. 1996). BSP is critical for acellular cementum formation promoting mineralization at the root surface in order to provide the insertion of Sharpey's fibers for a strong attachment of the tooth to the alveolar bone and AEFC.

Both glycoproteins, OPN and BSP, participate in tissue remodeling on resorbed roots and form part of the normal protein secretory sequence of cementoblasts to deposit new cementum. The deposition of the collagen fibrils in the repair cementum occurs almost concomitantly with the secretion of BSP and OPN which is consistent with the interfibrillar accumulation of BSP and OPN (Nanci 1999; Bosshardt and Nanci 2000; Bosshardt et al. 2005). BSP is specifically localized to the cemental surface which indicates that this protein is involved in cementoblasts differentiation and early mineralization of the cementum matrix. Localization of OPN to nonmineralized tissues further suggests that OPN functions as an inhibitor of mineralization during periodontal ligament formation. These findings collectively suggest that BSP and OPN are intimately involved in the sequence of cellular and molecular events accompanying cementogenesis. Foster et al. (2015), using knockout mice for BSP (Bsp−/−) and histological, SEM, and TEM methods, demonstrated that molars in these mice lacked a functional acellular cementum and subsequently presented with loss of Sharpey's collagen fiber insertion into the tooth root structure. Furthermore, Bsp−/− mouse presented with alveolar and mandibular bone with fewer osteoclasts at early ages; however, increased RANKL immunostaining and mRNA, and significantly increased number of osteoclast-like cells were found at later ages, corresponding to periodontal breakdown and severe alveolar bone resorption observed following molar teeth entering occlusion. Studies by Ao et al. (2017) showed that BSP−/− mice had increase PPI circulation, increased mRNA expression of Alpl, Spp1, Ank, and increased OPN amounts in the periodontia (Ao et al. 2017). Interestingly, the defective cementum phenotypes between Bsp−/− mice and Alpl−/− mice (the latter featuring elevated PPi and OPN) are quite similar.

Dentin Matrix Protein 1 (DMP1), also known as dentin matrix acidic phosphoprotein 1 and an acidic phosphorylated extracellular matrix protein, was first isolated from dentin (George et al. 1993), and later found also in cementum (Toyosawa et al. 2004) and other tissues like bone, cartilage, brain, pancreas, kidney, and salivary glands (MacDougall et al. 1996; Feng et al. 2003). DMP1 is an acidic phosphoprotein rich in aspartic, glutamic, and serine residues whose functions have been postulated to play a significant role in biomineralization (Ye et al. 2004) ranging from hydroxyapatite nucleator (He et al. 2003) to control of mineral propagation rather than initiation (Qin et al. 2007). It has also been postulated that DMP1 control cell differentiation by targeting the nucleus and/or interacting with cell-surface integrin/CD44 receptors (Narayanan et al. 2001; Kalajzic et al. 2004), regulate cell attachment to the ECM (Kulkarni et al. 2000), and activate matrix metalloproteinase-9 (Fedarko et al. 2004).

Although DMP1 is expressed in all mineralized tissues, the highest expression appears in osteocytes (Toyosawa et al. 2004). DMP1 may provide the molecular design through self-assembly of acidic clusters into a beta-sheet template necessary for controlling the formation of oriented calcium phosphate crystals, likely required for its role in biomineral induction (Feng et al. 2002; He et al. 2003, 2005). Studies localizing DMP1 during root development showed that DMP1 mRNA was expressed in cementoblasts on the surface of the acellular cementum and in cementocytes in the cellular cementum, but not in cementoblasts on the surface of the cellular cementum. The DMP1 protein was localized in the acellular cementum and in the pericellular matrix of cementocytes including their processes in the cellular cementum (Toyosawa et al. 2004). Previously, using Dmp1-lacZ knock-in mice, it was shown that DMP1 is expressed in cementum (Feng et al. 2003), in both, acellular and cellular cementum

(Ye et al. 2008). DMP1 was observed in AEFC, particularly in Sharpey's fibers. Electron immunomicroscopy revealed the close proximity of DMP1 to collagen Sharpey's fibers at the mineralization front. These suggest that DMP1 contributes to Sharpey's fiber biomineralization suggesting that DMP1 functions as a regulatory molecule in the mineralization of acellular extrinsic fiber cementum. This concept is supported by findings in DMP 1 null mice (Ye et al. 2008), where AEFC shows abnormal mineralization and was much thinner than in wild-type mice (Ye et al. 2008; Sawada et al. 2012).

7.7 Vitamin K-Dependent Proteins

Although there are about 17 different Vitamin K-dependent proteins, only Matrix GLA (MGP) and Osteocalcin (OCN), also known as bone gamma-carboxyglutamate protein (BGPLA), are present in cementum (Price et al. 1983). Both have a high affinity for Ca^{2+} and hydroxyapatite through interaction with the Gla residues, but they have different functions in different tissues. For example, it has been shown that carboxylated OCN aids the deposition of calcium into the bone matrix while carboxylated MGP protects blood vessels and may prevent calcification within the vascular wall (Wen et al. 2018).

Matrix GLA (MGP) is an insoluble small protein, with a molecular weight of only 14.7 kD, secreted and localized in the extracellular matrix of chondrocytes or endothelial cells and they may be important in the regulation of mineral apposition in calcified tissues (Hauschka and Reddi 1980; Price et al. 1981; Otawara and Price 1986). Unlike the limited distribution of osteocalcin, MGP is expressed in a wide variety of tissues such as cartilage, kidney, lung, and aorta, besides bone (Hale et al. 1988; Fraser and Price 1988). Genetically engineered, MGP-deficient mice exhibit more severe phenotypic disorders of calcification than osteocalcin-deficient mice (Luo et al. 1997; Ducy et al. 1996). The MGP-deficient mice have abnormal calcification. The mutant mice showed calcification of the aortic walls and valves, which should not normally be calcified under physiological conditions. Thus, both osteocalcin and MGP seem to regulate calcification in the body by acting mainly as negative regulators, but to different extents. The presence of MGP in both acellular and cellular matrix cementum has been reported (Glimcher 1979). Acellular cementum showed more prominent immunoreactivity than cellular cementum. MGP is secreted locally by cementum-forming cells and is incorporated at the mineralization front showing that acellular cementum contains abundant MGP and less osteocalcin. The accumulation of MGP on the acellular and the calcified cellular cementum may be required to prevent hyper calcification of the outer root surface (Hashimoto et al. 2001).

Studies on the inhibition of MGP using transgenic mice (Col1a1-Mgp mice) revealed almost complete abrogation of mineralization of the tooth root ECM. Cellular cementum ECM appeared to assemble into a well-ordered and structured matrix; however, it failed to mineralize in the presence of abundant MGP. Since apical cellular cementum assembled seemingly appropriately as an ECM with occluded cementocytes (but without mineralization), a lack of acellular cementum, which is more occlusally situated, implies that mineralization within root mantle dentin is required for at least the initiation of acellular cementogenesis. These observations also suggest that mineralization at the surface of root mantle dentin may precede the assembly of cementum proteins into an ECM (Kaipatur et al. 2008). The expression of MGP in the tooth root might be associated with the resistance to potential root resorption by physiological distal drift (Bronckers et al. 1993).

Osteocalcin (OCN) is a unique small protein of 46–50 amino acid residues that appears abundantly in the mineralized extracellular matrices of bone, dentin, and cementum. The presence of two or three residues of gamma-carboxyglutamic acid (Gla) in OCN is known to enhance its affinity for Ca^{2+} ions (Hauschka and Carr 1982) and promote the adsorption of the protein to hydroxyapatite. The γ-carboxylated Glu amino acid residues in OCN have a calcium-binding site that attracts calcium ions, binds them, and incorporates them in the hydroxyapatite crystals that form the mineralized matrix while simultaneously promoting bone mineral density (Zoch et al. 2016). The distribution of osteocalcin in the mammalian body is fairly limited to mineralized tissues like bone, dentin, and cementum (Cole and Hanley 1991). In these calcified hard tissues, OCN might act as a negative regulator for mineral apposition (Ducy et al. 1996). This proposition is supported by the observation that purified OCN inhibits the spontaneous conversion of brushite to hydroxyapatite and inhibits the formation of hydroxyapatite crystals in supersaturated calcium/phosphate solutions in vitro (Hauschka et al. 1989; Price 1989). Expression of OCN in cementum and its associated cells has been reported (Bronckers et al. 1994), and our studies show its presence in CFM, expressed by cementoblasts and paravascular cells in the PDL (Fig. 7.4a) showing remarkable co-expression with BSP (Fig. 7.4b, c). Cementoblasts express OCN mRNA transcripts intensely during early cementogenesis and the protein is immunolocalized in both cellular and acellular cementum as well as cementoblasts and cementocytes (Tenorio et al. 1993). OCN could be a common, and possibly fundamental, extracellular matrix component for cellular and acellular types of cementum (Sasano et al. 2001).

7.8 Enamel-Related Proteins

Enamel proteins: Amelogenin (AMEL), Ameloblastin (AMBN), Enamelin (ENAM), and Tuftelin (TUFT), once thought to be specific for enamel, appear to be present in other tissues as well, and some of them, intimately related to cementogenesis, homeostasis, and the regenerative process, although their exact role remains still unclear. In 1974, Slavkin and Boyde proposed the hypothesis that cementum is an epithelial secretory product. They hypothesized that the inner layer of Hertwig's epithelial root sheath (HERS) secreted proteins, which might be related to enamel proteins, and these proteins could initiate the deposition of a thin layer of acellular afibrillar cementum tissue or the so-called intermediate cementum (Slavkin and Boyde 1974; Lindskog and Hammarström 1982; Thomas 1995; Hammarström 1997; Zeichner-David et al. 2003). It was found that human and mouse cementum share enamel-related material (Slavkin et al. 1989a). Both human and mouse cementum tissues contained enamel protein species with a molecular mass of Mr. = 72,000 (enamelin) and Mr. = 26,000 (amelogenin) and the amino acid composition for human and mouse cementum proteins were very similar to one another (Slavkin et al. 1989a). However, compared to amelogenins extracted from enamel, the cementum proteins are low in proline, histidine, and methionine residues, suggesting that cementum contains a unique class of enamel-related proteins possibly produced by HERS during the course of human and mouse root formation (Slavkin et al. 1989b). It was suggested that the production of typical cementum proteins may start before the epithelial cells switch to a mesenchymal phenotype, and therefore epithelial cells are still associated with the first cementum matrix deposited on dentin. It was also postulated that HERS undergo epithelial–mesenchymal transformation (EMT) to produce the initial cementum matrix and that could also explain the apparent disappearance of the majority of HERS cells (Bosshardt and Selvig 1997; Bosshardt and Nanci 1998; Zeichner-David et al. 2003).

7.9 Amelogenin (AMEL)

The presence of AMEL in cementum has been debated; some investigators looking at the presence of AMEL transcripts failed to detect the expression of amelogenin in HERS in vivo or in vitro (Luo et al. 1991; Zeichner-David et al. 2003), while others, using immunohistochemistry report the expression of amelogenins by HERS (Hamamoto et al. 1996; Fong and Hammarström 2000; Fukae et al. 2001; Hatakeyama et al. 2003; Janones et al. 2005: Sonoyama et al. 2007; Arzate et al. 2015). Our own studies show the presence of the AMEL protein on the epithelial rests of Malassez (ERM) originating from HERS, and in cementoblasts cells of the PDL as can be seen in Fig. 7.5a. These cells also express CEMP1, a cementum-specific protein which will be described later, as can be seen in Fig. 7.5b, and their colocalization with AMEL can be seen in the merged Fig. 7.5c thus confirming the identity of these cells as cementogenic. The mesenchymal nature of the cells expressing CEMP1 was demonstrated by colocalization with STRO1 as can be seen in Fig. 7.5d–f. Further support for the expression of AMEL in cementum-associated cells comes from studies using the Amelx null mice where the roots of these animals showed increased formation of cementicles and irregular calcified globules on the root surface and increased osteoclastogenesis as compared to the wild type. Furthermore, increased areas of root resorption were observed in older mice, indicating that loss of AMEL during cementum formation leads to cementum defects therefore AMEL must be involved in preventing idiopathic root resorption (Haruyama et al. 2011). The role of AMEL as a "signal molecule" or "growth factor" was initially suggested by Hammarström (1997) using an extract of developing

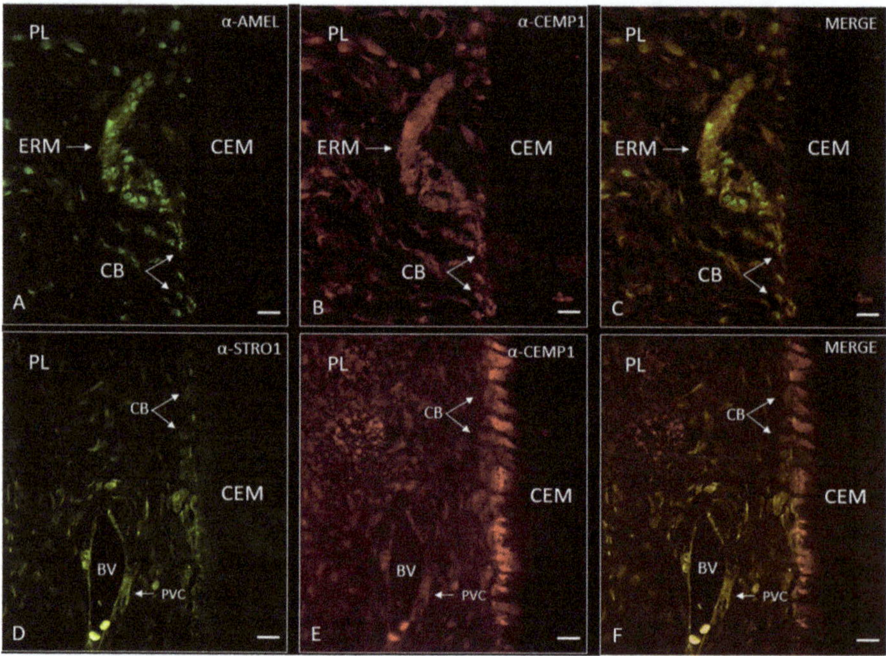

Fig. 7.5 Double staining and colocalization of Amelogenin (AMEL) with CEMP1. (**a**) AMEL is strongly expressed by some cells of the epithelial rests of Malassez (ERM) and cementoblasts (CB). (**b**) CEMP1 is expressed in a similar way and in cell populations of the periodontal ligament. (**c**) Merged image shows that there are some cells in the ERM that express only CEMP1. (**d**) Stro1 is strongly expressed by paravascular cells (PVC) and weakly by cementoblasts. (**e**) CEMP1 is strongly expressed by cementoblasts and paravascular cells (PVC). (**f**) Merged image shows that CEMP1 and STRO1 colocalized in paravascular progenitor cells of the periodontium. Bar = 100 μm

porcine enamel proteins. Veis et al. (2000) demonstrate signaling functions for AMEL peptides and Zeichner-David (2001) showed that PDL cells grown in vitro in the presence of purified recombinant AMEL, increase their rate of proliferation in a dose-dependent manner, while no similar effect was found with HERS cells.

In vitro studies using the Leucine-rich amelogenin peptide (LRAP) showed that this peptide alone can also function as signaling molecules capable of inducing cell differentiation in cementoblasts (Boabaid et al. 2004) possibly using a plasminogen signaling pathway (Martins et al. 2020). One 22-amino acid long peptide region of AMGN referred to as amelogenin-derived peptide 5 (ADP5) has shown to facilitate cell-free formation of a cementum-like hydroxyapatite mineral layer on demineralized human root dentin that, in turn, supported attachment of periodontal ligament cells in vitro. The cementomimetic layer formed by ADP5 has potential clinical applications to repair diseased root surfaces, caused by periodontal disease, and to promote regeneration of periodontal tissue (Gungormus et al. 2012).

7.10 Ameloblastin (AMBN)

At one time also known as Amelin and Sheathlin, AMBN is another protein believed originally to be specific for enamel and later on demonstrated to be expressed in many other tissues including odontoblasts (Fong et al. 1998), osteoblast-like cells (Spahr et al. 2006), and mesenchymal cells where it has been associated with early cell differentiation and repair (Tamburstuen et al. 2011). AMBN is a well-conserved gene among species which was originally identified as an enamel-specific protein and its presence appeared to be critical for proper enamel prism formation (Krebsbach et al. 1996; Cerný et al. 1996; Fong et al. 1996; Hu et al. 1997). Ameloblastin is suggested to be a two domain, intrinsically unstructured protein (Vymetal et al. 2008) that binds calcium (Yamakoshi et al. 2001) and is subjected to intensive proteolysis by the matrix proteases enamelysin and kallikrein (Iwata et al. 2007). AMBN has been found to be expressed by HERS during root development and is associated with cell clusters in the vicinity of acellular cementum, progenitor cells close to the periodontal ligament blood vessels, and cementum deposited in resorption areas (Fong et al. 1996). It was also shown that HERS cells maintained in vitro express AMBN (Zeichner-David et al. 2003) as well as cementoblast cells (Nuñez et al. 2010). As can be seen in Fig. 7.6, expression of AMEL and AMBN colocalizes in cementoblasts as well as paravascular cells which are believed to be cementoblast progenitors. However, AMBN is also present in the cementum layer while AMEL is not. The precise role of AMBN in cementogenesis is unknown; however, it has been shown that AMBN can also act as a growth factor or signaling molecule during tooth development. In Vitro studies using HERS cells grown in the presence of purified recombinant AMBN showed an increase in proliferation of these cells while recombinant AMGN had no effect on these cells. On the other hand, AMBN had no apparent effect on PDL cell proliferation while AMGN did increase PDL cell proliferation (Zeichner-David 2001). Interestingly, studies by Hirose et al.

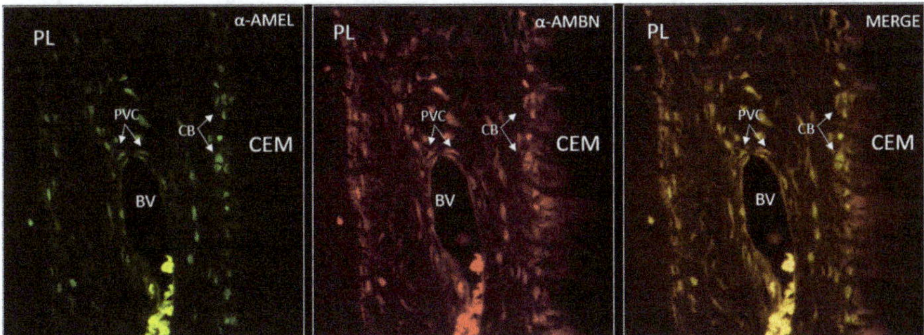

Fig. 7.6 Double staining and colocalization of Amelogenin with Ameloblastin. (**a**) AMEL is expressed by cementoblasts (CB) and paravascular cells (PVC). (**b**) AMBN is highly expressed by cementoblasts (CB) and para vascular cells (PVC). (**c**) Merged image shows the colocalization of both enamel-related proteins; however, there is an expression of AMBN in cementoblast becoming embedded into the matrix they produce. Bar = 100 μm

(2003) blocking expression of AMBN only in roots using siRNA resulted in shorter roots, abnormal HERS cell proliferation, and irregular root dentin formation. These studies support the role of AMBN in HERS proliferation.

7.11 Cementum-Specific Proteins

7.11.1 Cementum-Derived Growth Factor (CGF)

It is well known that mineralized ECM is an outstanding reservoir for growth factors that are actively involved in the process of tissue regeneration. Many growth factors are sequestered in the ECM where ECM interactions can modify the binding of these growth factors to their cell-surface receptors (Schönherr and Hausser 2000). Since cementum is a mineralized ECM, it is also a rich source of many growth factors which influence the activities of various periodontal cell types and can promote cell adhesion, migration, mitogenesis, and cell differentiation (Narayanan and Bartold 1996; Saygin et al. 2000). In 1987, Miki and colleagues reported the presence of a mitogenic factor isolated from human cementum constitutes and later on Nakae et al. (1991) reported a similar discovery in bovine cementum. Part of this mitogenic activity was due to the presence of fibroblast growth factor (FGF); however, about 70% of the mitogenic activity of the cementum extracts was due to another growth factor, that in contrast to FGF, had a low affinity for heparin. This new factor was named cementum growth factor (CGF) because it acts in synergy with epidermal growth factor (EGF) inducing signaling pathways associated with mitogenesis (Yonemura et al. 1992, 1993; Arzate et al. 1992a). CGF is a small molecule (14 kDa protein) which by itself is weakly mitogenic to fibroblasts and smooth muscle cells. It induces mitogenic signaling events, which include Ca^{2+} mobilization, inositol phosphate hydrolysis, activation of phosphokinase C (PKC), and transcription of cellular protooncogenes c-fos and jun-B (Narayanan et al. 1995). The magnitude and pattern of activation of signaling events and their susceptibility to PKC inhibitors and pertussis toxin suggested that CGF may be a distinct molecular species, with some homology with insulin-like growth factor I (IGF-1); however, it is larger than IGF-I therefore implying that CGF factor could be an IGF-I isoform (Ikezawa et al. 1997). Up to date, the gene coding for CGF has not been isolated and therefore CGF has not been fully characterized and the role of this factor in cementum formation remains unclear.

Although not specific for cementum, it is important to mention that there are several other growth factors secreted by cementoblasts and sequestered in the cementum ECM such as BMP-2, 4, 5, and 6 and they might have a role in cementum homeostasis (Grzesik and Narayanan 2002; Saygin et al. 2000). Several polypeptide growth factors with the ability to promote the proliferation and differentiation of putative cementoblasts are also sequestered in the cementum matrix. These include BMP-3 and 4, PDGF, a- and b-FGFs, TGF-b, and IGF-I (Cochran and Wozney 1999; MacNeil and Somerman 1999; Saygin et al. 2000). Although the selection of cells can be achieved at the level of adhesion, it could also be by preferential proliferation. In summary, the cementum ECM has a great potential to regulate the surrounding extracellular matrix formation since it is a reservoir of molecules that might promote periodontal homeostasis and regeneration.

7.11.2 Cementum Attachment Protein (CAP)

Human and bovine cementum extracts also displayed the presence of a 55 kDa protein species which acts as a mediator of cell attachment (McAllister et al. 1990; Olson et al. 1991). This protein was named cementum attachment protein (CAP). Characterization of this protein by amino acid sequencing revealed the presence of four sequences containing Gly-X-Y repeats, which are typical of collagen. Furthermore, a 17-amino acid peptide had 82% homology with a domain present in type XII collagen, and another peptide had 95% homology with collagen type Iα1. Additionally, the attachment activity of CGF was lost after treatment with bacterial collagenase. All of these facts suggest that CAP was some contamination with collagen. However, CAP did not cross-react with antibodies specific to type I, type V, type XII, and type XIV collagens. Moreover, a CAP-specific monoclonal antibody reacted only with proteins present in cementum (Arzate et al. 1992b). These findings suggest that CAP might be a collagenous-attachment protein localized exclusively in cementum (Wu et al. 1996).

The cloning and characterization of a CAP-cDNA clone disclose alternative splicing and homonym of 3-hydroxyacyl-CoA dehydratase 1 (HACD1). CAP encodes a 140-amino acid protein that is identical to the first 125 N-terminal amino acids of a truncated isoform of 3-hydroxyacyl-CoA dehydratase-1/protein-tyrosine phosphatase-like (proline instead of catalytic arginine), known as PTPLA (Valdés De Hoyos et al. 2012), which is 288 amino acids long (isoform 1). The remainder of the C-terminus of CAP is encoded by a read-through of the splice donor site of exon 2 of the HACD1/PTPLA isoform into the adjacent intron. This protein is predicted to have 2 transmembrane spanning domains, and it truncates immediately after the second transmembrane domain. The truncation eliminates the HACD1/PTPLA sequence IVHCLIGIVPT, which has a signature phosphatase active site motif (Uwanogho et al. 1999; Li et al. 2000). A search of the NCBI EST database for the presence of corresponding EST sequences detected 1 human and 1 rat EST showing the same alternate splice product. The human EST (CN305861) was derived from undifferentiated human embryonic stem cell lines. PTPLA mRNA is widely expressed in many tissues (Uwanogho et al. 1999; Li et al. 2000), the CAP/PTPLA mRNA is expressed only in cementum cells and marginally in some periodontal ligament cells in human teeth; however, it has been reported that CAP/PTPLA is not expressed in the cementum of rat teeth (Schild et al. 2009) suggesting that there are some species differences in its distribution. CAP/PTPLA mRNA codes for a protein of 15 kDa, without collagen sequences; however, it is expressed as 39 kDa species in cementoblasts and periodontal ligament-derived cells

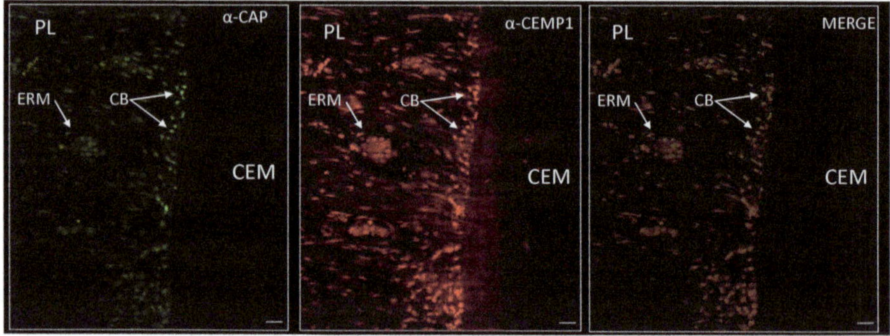

Fig. 7.7 Double staining and colocalization of cementum attachment protein (CAP) with CEMP1. Expression of CAP can be seen in cementoblasts (CB), lightly in epithelial rests of Malassez (ERM). CEMP1 shows strong expression in cementoblasts, pre-cementoblasts, and ERM. Merged image shows that CEMP1 is strongly expressed by cementoblast cells (CB) and ERM and some cementoblast cells showed stronger CAP expression than CEMP1. Bar = 100 μm

in vitro (Valdés De Hoyos et al. 2012) which cross-reacts with the bovine anti-CAP antibody. CAP protein is expressed by cementoblasts, cells with paravascular location into the periodontal ligament, and cells in endosteal spaces of alveolar bone as can be seen in Fig. 7.7, which indicates that CAP possible plays a role promoting the migration of progenitor cells to the periodontal ligament space to differentiate toward the cementoblasts phenotype (Melcher 1976; McCulloch et al. 1987).

Characterization of the secondary structure of CAP/PTPLA using AFM and FE-SEM studies show a protein with a high percentage of random-coil structure (structural disorder), that it is an acidic protein that self-assembles into nanospheres which then form aggregates resembling nanostrings (Montoya et al. 2014). CAP/PTPLA aggregates creating larger particles and multilayers acquiring a supra-molecular mesh-like structure which adopts a functional role that facilitates the nucleation, growth, and direction of hydroxyapatite crystals through guidance for the highly anisotropic growth of apatite crystals in cementum/bone (Montoya et al. 2014). The exact mechanism through which CAP/PTPLA regulates this process is not completely clear; however, it is known that amino acid segments of this protein are of hydrophobic nature, which can interact directly with hydroxyapatite (Weiner and Addadi 1997). CAP/PTPLA is predominantly composed of random-coil and alpha-helix structures therefore this protein can be considered an intrinsically disordered protein (Forman-Kay and Mittag 2013). This class of proteins is multifunctional and has diverse binding properties including those associated with biomineral formation (Weiner and Addadi 1991; He et al. 2003; Delak et al. 2009; Peysselon et al. 2011). Bioinformatics studies have shown that intrinsically disordered and aggregation-prone domains exist within the diverse set of human extracellular matrix protein sequences (Peysselon et al. 2011). These domains are believed to be responsible for observed matrix assembly and hierarchal ordering of the extracellular matrix. In addition, CAP/PTPLA shows a theoretical isoelectric point of 6.37 and acidic proteins perform different roles in crystal formation. These features suggest a role for CAP/PTPLA during cementum/bone formation as a regulator of mineral formation by nucleation of hydroxyapatite and/or control crystal growth, suggesting that the protein itself may provide a dynamic scaffold to form elongated structures directing the growth of mineral as shown for other self-associative proteins involved in the process of biomineralization (Amos et al. 2010, 2011; Marie et al. 2010; Ponce and Evans 2011).

Montoya et al. (2014) demonstrated that CAP/PTPLA possesses functional biological activity of self-assembly and hydroxyapatite nucleation. These studies strongly suggest that the hydroxyapatite crystal formation by this protein might be assembled through an immature mineral phase; therefore, it appears that CAP/PTPLA is responsible for regulating the transformation of an immature mineral phase to crystalline hydroxyapatite. Using an established craniofacial defect model, CAP/PTPLA showed significantly greater bone fill capacity as compared to controls, indicating the potential of CAP in craniofacial bone regeneration/augmentation. CAP/PTPLA showed 73% ± 2.19% and 87% ± 1.97% new bone formation at 4 and 8 weeks, respectively when compared with the absorbable gelatin sponge and sham-surgery controls (Montoya et al. 2014). It is conceivable CAP/PTPLA-induced bone regeneration is due to the induction of neovascularization and the subsequent recruitment of mesenchymal stem cells (MSCs), since this protein has been shown to be expressed by paravascular stem cells in the periodontal ligament, and these cells are the suspected progenitors of osteoblasts and cementoblasts (McCulloch et al. 1987; Arzate et al. 1992b). This statement supports both CAP and STRO-1 co-expression by cells located in the bone marrow spaces. Therefore, this protein may promote cell differentiation and cell commitment toward different phenotypes and make this protein a strong candidate to regulate the mineralization process during periodontal and bone regeneration.

Additional biological properties attributed to CAP include promotion of cell attachment of gingival fibroblasts, endothelial cells, and smooth muscle cells while inhibiting oral sulcular epithelial attachment (Olson et al. 1991); binding selectively to periodontal ligament cells to support their attachment to root surfaces (Pitaru et al. 1994); and CAP mediates cell attachment, through α5β1 integrin, that may play an important role in cementogenesis through interaction with CAP (Ivanovski et al. 1999; Montoya et al. 2014). It was shown that cell attachment to CAP induces immediate-early G1 phase events and increase in cyclin D1 levels in the cells adhered to CAP without growth factors. The expression of cyclin D1 is normally regulated by adhesion in the presence of growth factors, and signal reactions generated by binding CAP to CAP receptors, induce expression of cyclin D1 (Yokokoji and Narayanan 2001). CAP appears to affect the cell cycle progression through mechanisms common to other molecules. However, differences occur in the type and degree of induction of these events (Saito and Narayanan 1999). Cells differing in the capacity to bind CAP also differ in their ability to form mineralized tissue in culture and to express CAP (Liu et al. 1997; Bar-Kana et al. 1998). Furthermore, it was demonstrated that CAP bound to dentin root slices enhances the recruitment of putative cementoblasts in vitro (Liu et al. 1997;

Arzate et al. 1992a). These observations indicate that substances, such as CAP, present in the local cementum environment, could determine which cells are recruited and how they differentiate during normal homeostasis and wound healing, and whether the healing response is repair or regeneration. Since CAP poses selective chemotaxis and cell attachment activity, this might define the progenitor cells pathway from the periodontal ligament and those cells coming from endosteal spaces of alveolar bone and influence the regeneration of cementum. The attachment capacity of clonogenic periodontal ligament cell populations to CAP (15%) is associated with alkaline phosphatase expression and mineralized-like tissue formation, indicating that this protein is associated with the recruitment of cementoblasts progenitors into the periodontal ligament (Liu et al. 1997; Arzate et al. 1996). These findings imply that CAP promotes the differentiation of progenitor cells toward the cementoblastic phenotype.

7.11.3 Cementum Protein 1 (CEMP1)

This protein was first isolated from human cementum and conditioning-media from a human cementoblastoma-derived cell line and was known as cementum protein-23 (CP-23) originally (Arzate et al. 1996, 1998, 2002). It is expressed from a single-copy gene as a 25.9 kDa nascent protein that is extensively subjected to posttranslational modifications. The primary sequence of CEMP1 showed a 98% homology with a predicted protein sequence from *Pan Troglodytes* in chromosome 16 (Alvarez Pérez et al. 2006). Nowadays, CEMP1 orthologs have been identified in 21 different primate species, indicating that CEMP1 descends from a single gene of the last common ancestor: *Pan troglodytes* (chimpanzee). Recently it has been discovered that CEMP1 polymorphisms, including the variants R15Q, K55E, and R80H were detected in cementum obtained from Denisovan man and have been described as associated with a particular phenotype. It can be noticed that the Denisovan individual carries 3 CEMP1 substitutions (R15Q, K55E, R80H) that could be considered as a haplotype. K55E and R80H are located in the domain of the protein showing similarity with collagen a I (I) chain (Arzate et al. 2015) and all three mutations cause a change in the charge of the residue that could impact functional interactions. Noteworthy, K55E corresponds to the ancestral variant, whereas the major substitution K55 in living humans is derived and nearly fixed out of Africa, and likely represents a polymorphism specific to the modern lineage (Zanolli et al. 2017).

CEMP1 is composed of 247 amino acids, with a 25.9 kDa nascent molecular weight and is an alkaline protein with a theoretical isoelectric point of 9.73. Its protein sequence shows that between amino acids 30–110 there is a homology by 48% with human collagen a I (I) chain, as well a similarity with type XI and type X collagens by 46% and 40%, respectively (Alvarez Pérez et al. 2006). Recombinant human CEMP-1 strongly cross-reacts with antibodies against collagen type X but not with antibodies against type I. Nevertheless, CEMP1 does not have the characteristic collagen-like gly-x–y repeats, which suggests that CEMP1 is not a true collagen (Alvarez-Pérez et al. 2006). The original CEMP1 protein is subjected to posttranslational modifications: *N*-glycosylated and phosphorylated. These post-translational modifications are 43.5% that changes its molecular mass to 50 kDa (Villarreal-Ramírez et al. 2009). This species is a glycosylated, phosphorylated, and thermostable protein and is a C-type lectin since it promotes the agglutination of formalized rabbit erythrocytes in the presence of Ca^{2+} (Carmona-Rodríguez et al. 2007). Circular dichroism analyses revealed that its secondary structure is composed of 28.6% alpha-helix, 9.9% of beta-sheet, and 61.5% of random-coil forms (Montoya et al. 2019). Therefore, since CEMP1 contains a high percentage of random-coil secondary structures it belongs to the intrinsically disordered proteins (IDP) (Villarreal-Ramírez et al. 2009; Romo-Arévalo et al. 2016). These domains are believed to be responsible for matrix assembly and hierarchal ordering of the extracellular matrix, because of the plasticity of this intrinsically disordered region and its property to link to hydroxyapatite during the biomineralization process (Evans 2008; Nudelman et al. 2007). This feature might explain how CEMP1 regulates the crystal growth and composition of hydroxyapatite crystals (Villarreal-Ramírez et al. 2009). CEMP1 also induces the expression of proteins related to mineralization and promotes in vitro osteoblastic and/or cementoblastic cell differentiation of HGF (Carmona-Rodríguez et al. 2007). CEMP1 may play a key role in biomineralization, particularly important in the formation of calcium phosphate minerals (Hunter and Goldberg 1994; Weiner and Addadi 1997; Hartgerink et al. 2001). In vitro studies using human recombinant (hr)CEMP-1 (Fig. 7.8) demonstrated that it promotes deposition of hydroxyapatite crystals and organizes them into parallel arrays, through protein–protein and protein–crystal interaction (Romo-Arévalo et al. 2016) and self-assembles into nanospheres that form aggregates resembling nanostrings. This structural framework organization suggests the formation of a platform that facilitates hydroxyapatite crystal formation and orientation (Correa et al. 2019). CEMP1 has shown *to* possess intrinsic properties to bind to hydroxyapatite even without posttranslational modifications and plays a role during the biomineralization process, required for the synthesis of needle-shaped OCP crystals and responsible for OCP crystal nucleation activity. CEMP1's transfection into cells that do not usually mineralize such as human gingival fibroblasts (HGF), promotes proliferation and differentiation of adult HGF to a "mineralizing-like" cell phenotype, and the ability to induce the formation of mineralized nodules and calcium deposition in non-osteogenic cells. Furthermore, CEMP1 induced the expression of OPN, BSP, and OCN proteins as well as the transcription factor RUNX2 in these cells, all of which are related to mineralization of cementum, bone, and cartilage (Bermúdez et al. 2015). Other studies have demonstrated HrCEMP1 promotes up to 97% regeneration of critical-size defects in rat calvaria, all of which indicates that CEMP1 has a function in cementum and bone formation (Serrano et al. 2013).

In vivo studies demonstrated that CEMP1 and STRO-1-positive cells were colocalized adjacent to the root surface in areas where new cementum was being deposited thus implying that the cells that produce the reparative cementum were of mesenchymal origin. In vitro treatment of PDL cells with CEMP1 protein stimulated

Fig. 7.8 SEM images of crystals formed in vitro by CEMP1-p1 where crystals are seen irradiating from an amorphous core (**a**). Higher magnification shows hydroxyapatite crystals with a spear-like morphology (**b**). The crystals induced by CEMP1-p1 are hydroxyapatite crystals revealing a 3-Dvetrtex (120° angle) corresponding to the c-axis of the HA crystal (**c**). EDS analysis of CEMP1-p1-induced crystal showed a Ca/P ratio: 1.8 which corresponds to hydroxyapatite (**d**)

the proliferation and migration of PDL cells, with the migration front comprised of STRO-1-positive cells. Since STRO-1-positive cells are located next to cementum (Barrera-Ortega et al. 2017), it suggests that CEMP1 is likely a mediator in wound healing and periodontal regeneration since it stimulates PDL cell proliferation and migration. CEMP1 also leads to the migration of STRO-1-positive cells, suggesting a possible mechanism for the recruitment of mesenchymal stem cells toward a CEMP1 signal (Paula-Silva et al. 2010). The expression and presence of CEMP1 strongly support these hypotheses; CEMP1 is expressed by cementoblasts lining acellular and cellular cementum (Alvarez Pérez et al. 2006), also expressed by subpopulations in the periodontal ligament of the regenerated periodontium, present in the cementoid and PDL space (Gauthier et al. 2017) and sites of root resorption (Nuñez et al. 2010) but not in osteoblasts. These studies also indicate that CEMP1 is expressed only in cementoblasts cells and their progenitors (Fig. 7.7). CEMP1 expression declines when cells are committed to osteoblastic or chondroblastic lineages, as was shown using an in vivo diffusion chamber assay, suggesting of CEMP1 expression is differentially regulated in precursor mesenchymal stem cells. Furthermore, CEMP1 decreases the expression of osteoblastic and periodontal ligament cell markers while inducing the expression of cementoblastic markers in vitro and in vivo among cells of the periodontium (Komaki et al. 2012).

The role of CEMP1 as chemoattractant and promoter of mineralization is further supported by the findings that mineralization was reduced upon blocking the function of CEMP1 in vitro. Although, both osteoblasts and cementoblasts express BSP, blocking the function of CEMP1 with a CEMP1-specific antibody, resulted in suppression of the expression of BSP in cementoblasts but not osteoblasts, indicating that CEMP1 is associated with regulation of BSP expression in cementoblasts (Alvarez Pérez et al. 2003). On the other hand, over-expression of CEMP1 in periodontal ligament cells resulted in enhancement of cementoblasts cell differentiation (Popowics et al. 2005; Komaki et al. 2012). Previous studies demonstrated that CEMP1 exerts modulation of a number of genes involved in cellular development, cellular growth, cell death, and cell cycle (Bermúdez et al. 2015) and CEMP-1 protein plays a key role in cementoblastic differentiation of dental follicle cells (DFC) and PDL cells (Kémoun et al. 2007a, b; Hoz et al. 2012; Sowmya et al. 2015) through the OSX-CEMP1 pathway (Sowmya et al. 2015). Blocking the activity of CEMP1, with a specific CEMP1 antibody, in cementoblastoma-derived cells, decreased their ALP activity and BSP and OPN expression without altering cell proliferation (Alvarez Pérez et al. 2003). Although it is not clear if cementoblasts and osteoblasts share a common progenitor, this data indicates that periodontal ligament cells that express CEMP1 could be a novel source of adult cementoblasts progenitors for periodontal regeneration therapy (Komaki et al. 2012). PMS scaffolds containing CEMP1/ACP/PCL/COL enhanced differentiation of seeded PDLCs in vitro into a cementoblastic phenotype by upregulating the expression of CAP and CEMP1, supporting the notion that CEMP1 is a key bioactive compound in cementum formation both in vitro and in vivo (Sowmya et al. 2017). In other studies, Matrigel® matrix alone or associated with human transforming growth factor-$\beta 3$ (TGFβ3) proven that this growth factor induces cementogenesis by upregulation of CEMP1 and that CEMP1 is an indicator for cementogenesis (Ripamonti et al. 2017). This study was the first to describe the possible use of CEMP1 as an early predictor of clinical outcomes for periodontal therapy as a marker for cementogenesis, by confirming its presence in gingival crevicular fluid (GCF) of patients. Increased levels of CEMP1 were found mostly in sites where regeneration of the periodontal complex was expected (Dellavia et al. 2019).

In summary, the cementum-specific proteins CAP and CEMP1 are biological factors which participate in chemoattraction, adhesion, and cell differentiation during the homeostasis process of cementogenesis. Moreover, CAP expression is increased by CEMP1-overexpression in periodontal ligament cells at both the mRNA and the protein level (Komaki et al. 2012). The process of new cementum formation, with newly attached Sharpey's fibers, is the key event to achieve periodontal regeneration. Therefore, it is necessary to find biological factors that can attract progenitor stem cells located in the periodontal ligament and endosteal spaces of alveolar bone to the injured site in order to induce them to differentiate into cementoblasts cells. CAP and CEMP1 appear to possess all of these biological properties necessary for the formation of cementum de novo. These proteins also appear to play a pivotal role during the regeneration of the periodontal structures where new deposition of cementum is a key event for the formation of new attachment de novo (Pitaru et al. 1994; Garrett 1996; Arzate et al. 2015). These characteristics open new venues for periodontal regeneration therapies based on CEMP1.

7.12 Cementum Therapeutics

The elucidation of the inorganic and organic composition of cementum allowed us to understand the process of cementum formation, how the mineralization of this unique tissue is regulated, how this biological association maintains the homeostasis of the periodontal structures, especially during the deposition of cementum de novo, offering new alternatives for therapeutic ways to restore the periodontium lost due to periodontitis. Current therapeutic procedures for periodontal disease include root scaling and planning (SRP) and conditioning of the dentin surface with citric acid to widen the dentinal tubules, then reconnecting cytoplasmatic extensions of the odontoblasts with collagen fibers of the periodontal ligament to promote the migration of cementum stem cells, through the periodontal ligament fibers, resulting in the promotion of cementogenesis with subsequent new connective tissue attachment (Mariotti 2003) osseous surgery (Reynolds et al. 2003) and guided tissue regeneration (Al-Hamdan et al. 2003; Miller and Craddock 1996; Murphy and Gunsolley 2003). Nevertheless, the current treatment strategies have shown to produce only limited repair of the damaged periodontium (Southerland et al. 2006; Aichelmann-Reidy and Reynolds 2008; Elangovan et al. 2009; Hanes 2007), they are not predictable and do not restore the periodontium structures or anatomical architecture and function to the normal healthy periodontium.

New therapeutic procedures being tested for a more predictable regeneration of the periodontium include the use of recombinant human growth factors (rhGFs), which are natural biological mediators that regulate key cellular events during tissue repair and regeneration and have become an important therapeutic approach in order to restore the periodontal apparatus (Nakayama et al. 2020; Khoshkam et al. 2015). The regeneration requires an appropriate biological environment which induces the differentiation of undifferentiated cells to make the required structures. Several clinical studies performed on humans raised the possibility that rhPDGF-BB and rhFGF-2 TGFβ, BMPs, FGF, IGF, and EGF, which are present in the mineralized cementum ECM; however, the efficacy of these factors depend on the quantity used, type of carrier combined with the growth factors and principally the mode of application to avoid the rapid loss of topically applied factors over time (Beertsen et al. 1997; Lynch et al. 1991). Other biological therapeutics used to repair/regenerate periodontal tissues are enamel-derived proteins (EMD) shown to participate in the induction and formation of cementum de novo (Gestrelius et al. 1997, 2000; Hammarström 1997; Hammarström et al. 1997; Heijl 1997; Hirooka 1998). Enamel-derived proteins (Emdogain®) have shown to promote periodontal ligament and cementoblastic cell proliferation (Heijl 1997; Heijl et al. 1997; Esposito et al. 2004; Heden and Wennström 2006; Hammarström 1997; Spahr and Hammarström 1999).

This marketed product has been used as an alternative for periodontal regenerative therapy, influencing periodontal wound healing and regeneration in humans (Sculean et al. 2007), showing significant improvement on bone and cementum regeneration (Cochran et al. 2003; Venezia et al. 2004; Sallum et al. 2019; Kulakauskienė et al. 2020). The normal expression of enamel proteins, AMGN and AMBN (Fig. 7.6), during cementum formation, and their ability to serve as signal molecules was previously addressed in this chapter.

7.13 Novel Prospects

7.13.1 Stem Cells

As we have shown in this chapter, cementum is intimately ligated to the homeostasis and regeneration of the periodontium lost due to injury, trauma, or disease. This tissue has been formally studied from 185 years ago. During this period, the histological, ultrastructural, compositional, functional characteristics and mechanisms that regulate cementogenesis during development, homeostasis, and repair at the cellular and molecular levels have been described. The architectural complexity of the periodontium is one of the reasons why it is so difficult to regenerate these tissues following either trauma or periodontitis. Nevertheless, the major goal of periodontal disease treatment to reconstruct the structure and function of periodontal tissues including cementum, PDL fibers, and bone remains as a major challenge in periodontal therapeutics. In 1976, Melcher developed a general theory on the central role of the PDL in periodontal regeneration; he proposed that the progenitor cells for cementum, bone, and PDL fibroblasts are contained in the PDL. Consequently, it made sense that if periodontal regeneration were to be achieved, then the PDL would have to be regenerated first. Later it was found that the vascularization of the PDL was present in the bone-related half of the PDL as compared to the cementum-related half. However, the cementum-related half of the PDL contains a relatively small but clearly measurable supply of blood vessels and it was suggested that paravascular cells in this region constitute a renewing cell population (Leblond et al. 1959; Leblond 1964). It is evident that homeostatic mechanisms in the PDL must regulate the rate of cell generation, cell death, and cell migration. Proliferating cells in the vicinity of blood vessels have small nuclei suggesting that in both normally functioning and stimulated PDL, the paravascular cells comprise a progenitor cell population (Roberts et al. 1982). It has been shown, using radioactive ^{3}H-Tdr, that the progeny of paravascular dividing cells migrate away from the blood vessels and migrate to the vicinity of bone and cementum (McCulloch and Melcher 1983a, b). These dividing cells are located external to the basal lamina of the blood vessels and probably have an origin or function different from that of classical pericytes and cells found in this area. These cells have a tendency to be relatively undifferentiated, retaining embryonic such as small size, scarcity of intracellular organelles, and a high nuclear/cytoplasmic ratio. It is not known whether a single paravascular progenitor within the PDL gives rise to daughter cells which differentiate into fibroblasts, osteoblasts, and cementoblasts, or whether there are separate progenitors (Roberts and Jee 1974; Melcher 1976; McCulloch and Melcher 1983a; Gould et al. 1977, 1980; Gould 1983). These progenitor cells provide a front of new cells for the healing wounds in the periodontal ligament, their number is inversely related to cell density in the periodontal ligament (McCulloch and Melcher 1983a, b, c), and they appear to be retained in the tissue for long periods of time (Leblond et al. 1959; McCulloch and Melcher 1983b; Davidson and McCulloch 1986). Additionally, there are progenitors residing outside of the PDL, which can migrate into the PDL and contribute to its cell populations. It has been demonstrated that mesenchymal cell populations associated with bone surfaces can migrate and form new bone after injury (Patt and Maloney 1975). Cells obtained from fetal rat calvaria cultured with dental root slices in vitro produced a tissue with ultrastructural characteristics that resemble bone or cellular cementum, acellular cementum, and afibrillar cementum (Melcher et al. 1986). These observations suggest that osteoblast and cementoblasts progenitor cells also originate from the endosteal spaces of the alveolar process and migrate into the PDL through vascular channels, enriched with progenitor cells whose progeny rapidly migrate out of the channels following the collagen fibers as they migrate toward cementum in response to chemo-attractive molecules present in cementum (Somerman et al. 1983; McCulloch et al. 1987). The recruitment of precursor cells seems to be a key issue to achieve cementum formation de novo in order to achieve periodontal regeneration.

In other studies, transmission electron microscopy comparison of the mineralized extracellular matrix deposited by HERS in vitro, with acellular cementum deposited in vivo, suggests that these extracellular matrices are similar (Zeichner-David et al. 2003). HERS cells synthesize CAP and CEMP1, supporting the idea that HERS cells are capable of producing cementum. These cells also present with a high activity of ALKP. These findings provide further evidence that the extracellular matrix deposited by these cells is acellular cementum, corroborating that alkaline phosphatase is a very important component of acellular cementum. HERS cells also express osteocalcin in vitro thus indicating the possibility that disruption of the basement membrane is caused by HERS cells when they start depositing the acellular cementum. Altogether, these studies suggested different cellular origins for acellular (HERS cells) and cellular (cementoblasts) cementum (Zeichner-David et al. 2003: Huang et al. 2009; Bosshardt and Nanci 2004). Furthermore, HERS-derived epithelial rests of Malassez (ERM) are a unique population of epithelial cells in the periodontal ligament and are believed to play a crucial role in cementum repair (Spouge 1980). These cells could differentiate into cementoblasts through epithelial–mesenchymal transformation (Krause et al. 2001; Kucia et al. 2007; Xiong et al. 2012). Recently it was demonstrated that in vitro HERS/ERM contain primitive stem cells that express epithelial stem cell markers such as octamer binding transcription factor 4, homeobox protein NANOG, and stage-specific embryonic antigen 4 (Nam et al. 2011) thus implying that the ERM are stem cells of epithelial origin that maintain the PDL as well as the cementum and alveolar bone associated with the ligament. The tissue niche within which ERM is found extends into the supracrestal areas of collagen fiber-containing tissues of the gingiva, above the bony alveolar crest. It has been suggested that epithelial rests of Malassez cultured ERM cells, as their progenitor HERS cells, produced a cementum ECM and demonstrated high ALKP activity (Farea et al. 2016). ERM cells can be clonally expanded, can grow organoids, and express the markers of pluripotency (OCT4, NANOG, SOX2), subcutaneous co-transplantation with mesenchymal stem cells from dental pulp, on poly-l-lactic acid scaffolds, in nude mice gave rise to heterotopic ossicles-like structures similar to cementum (Athanassiou-Papaefthymiou et al. 2015); therefore, these epithelial cells are regarded as a stem cell niche that can give rise to new cementoblasts (Bosshardt et al. 2015). In adulthood, the ERM cells are the only odontogenic epithelial population in the PDL. Although there is no general agreement on the functions of these cells, accumulating evidence suggests that their role in adult periodontal ligament is to maintain PDL homeostasis, to prevent ankylosis by maintaining the PDL space, to prevent root resorption, to serve as a target during periodontal ligament innervation, and to contribute to cementum repair. Recently, ovine ERM cells have been shown to harbor clonogenic epithelial stem cell populations that demonstrated similar properties to mesenchymal stromal/stem cells and demonstrated stem cell-like properties in their differentiation capacity to form bone, fat, cartilage, and neural cells in vitro. When transplanted into immunocompromised mice, ERM generated bone, cementum-like, and Sharpey's fiber-like structures. These cells have the capacity to differentiate into a mesenchymal phenotype and thus represent a unique stem cell population within the PDL both functionally and phenotypically. Therefore, the epithelial cell rests of Malassez, rather than being "cell rests," as indicated by their name, are an important source of stem cells that might play a pivotal role in periodontal regeneration (Xiong et al. 2012, 2013).

7.13.2 Cementum Proteins

Other very important and novel therapeutic candidates to consider are the cementum proteins; CEMP1 and CAP, which have demonstrated the ability to promote new cementum and bone formation in damaged periodontal tissues. These proteins could induce several signaling pathways associated with mitogenesis, increase the concentration of cytosolic Ca^{2+}, activate the protein kinase C cascade, and promote the migration and preferential adhesion of progenitor cells (Treves-Manusevitz et al. 2013). These actions result in cementoblast and osteoblast differentiation with the subsequent production of a mineralized extracellular matrix resembling cementum (Hoz et al. 2012). CAP promotes attachment of HGF in a dose-dependent manner. The attachment activity was comparable with that of fibronectin. CAP is expressed by cementoblasts facing cementum and the matrix and progenitor cells from the endosteal spaces of the alveolar bone (Arzate et al. 1992b; Valdés De Hoyos et al. 2012), promotes hydroxyapatite crystal nucleation, and in vivo promotes new bone formation showing a clinical effectivity at inducing bone repair and healing. Recently we have shown that CAP may have dual biological functions since a peptide comprising the N-terminal domain (from amino acids 40–53) has the ability to inhibit the mineralization process, suggesting that this protein may participate in regulating de deposition of cementum (Montoya et al. 2020). Therefore, this novel molecule has a great potential to be used for mineralized tissue bioengineering and tissue regeneration.

In other studies, it was shown that the presence of CEMP1 protein in cultures of PDL cells caused the stimulation of proliferation, migration, and mineralization of PDL cells. CEMP1 protein increased ALKP activity in 3D PDL cell cultures, induced the expression of cementogenic and osteogenic markers, and resulted in the formation of new tissues that mimicked bone and cementum (Hoz et al. 2012). In vivo, CEMP1- and STRO-1-positive cells were colocalized adjacent to the root surface in areas where de novo cementum was deposited (Fig. 7.5), implying that the cells that deposited the reparative cementum were of mesenchymal origin. Our findings are consistent with other reports showing that STRO-1-positive cells are located next to the cementum layer (Paula-Silva et al. 2010). The overlap between CEMP1- and STRO-1-expressing cells indicates that CEMP1 has the ability not only to recruit progenitor cells from the periodontium but also can signal the differentiation of these cells toward a cementoblast phenotype and these cells form mineralization nodules and continue to express CEMP1 (Kadokura et al. 2019). In other experiments, STRO-1-positive PDL cells transplanted into artificially created PDL defects in immunocompromised rats induced the deposition of a layer of cementum tissue (Fujii et al. 2008). This data further supports that CEMP1 recruits mesenchymal, STRO-1 and α-CD44-positive cells that promote mesenchymal cell differentiation toward cementogenesis (Nuñez et al. 2012; Paula-Silva et al. 2010). It has also been validated that of the mineralization-related genes, BSP is strongly upregulated when CEMP1 is overexpressed in periodontal ligament cells. Although both osteoblasts and cementoblasts express BSP, knockdown of CEMP1 expression affected only BSP expression in cementoblasts not in osteoblasts, indicating that CEMP1 is associated with regulation of BSP expression in cementoblasts (Komaki et al. 2012). High levels of CEMP1 in the cementum ECM play an important role in cementogenesis, as evidenced by the fact that over-expression of CEMP1 in PDL cells enhances their cementoblast differentiation (Komaki et al. 2012; Qi et al. 2014). Microarray studies to evaluate the potential biological effects of CEMP1 transduced into human gingival fibroblasts (HGF), which are not associated with mineralization, revealed a significant increase in the expression of high numbers of genes associated with cellular development, proliferation, growth, cell death, cell cycle, and mineralization (Bermúdez et al. 2015). Remarkable, CEMP1 transfection into HGFs allows these cells to perform cementoblast-like functions without alteration of the ultrastructure of the nucleolus, evaluated by the presence of the different compartments of this organelle involved in ribosomal biogenesis (Villegas-Mercado et al. 2018). In dental follicle cells (DFC), CEMP1 serves as a source for endogenous proteins. These cell-secreted factors/proteins are beneficial in regulating wound healing, cellular proliferation, angiogenesis, osteoblast differentiation and osteogenesis, fibroblast proliferation and differentiation, and cementogenesis; therefore, CEMP-1, protein is thought to play a key role in cementoblastic differentiation of DFCs and

PDL cells (Sowmya et al. 2015). Up to date, other molecules, which like CEMP1, are responsible for recruiting mesenchymal cells and inducing their differentiation into cementoblasts, have not been identified.

Taken together, these findings signify that periodontal ligament cells that express CEMP1 may well be a novel source of adult progenitors for cementoblasts used for de novo formation of cementum in periodontal regeneration therapy (Choi et al. 2014). Bioactive materials which stimulate the proliferation, differentiation, and osteogenic/cementogenic gene expression of PDL cells for periodontal regeneration, such as $Ca_7Si_2P_2O_{16}$ ceramic powders, also stimulate the deposition of CEMP1 by PDL cells in vitro (Zhou et al. 2012; Wu et al. 2012). Biphasic scaffolds composed of polycaprolactone (PCL, Lactel, USA) containing β-tricalcium Phosphate (β-TCP, 20% wt) induced deposition of mineralized tissue on the surface of dentin in vitro and in vivo, and the expression of CEMP1 by cementoblast precursors (Vaquette et al. 2012; Wolf et al. 2013; Zhang et al. 2014). In a similar manner, Mg–calcium silicate cement stimulates the proliferation of PDL cells in vitro and promotes the secretion of CEMP1, CAP, and angiogenic proteins, indicating that these proteins play a significant role during cementum and bone formation and provide the essential basis for their use as biomaterials in bone substitutes and bone regeneration applications (Chen et al. 2015). Growth factors such as BMP-7 significantly increased thickness and integrated area of a newly formed mineralized tissue layer expressing CEMP1 (Cho et al. 2016) and TGF-b3 up-regulates the expression of CEMP1 in vivo (Ripamonti et al. 2017). These studies suggest that CEMP1 may be the most important molecule in cementogenesis, makes cementum different from other hard tissues, and also regulates cementoblast commitment of PDL cells (Chen et al. 2015).

In vitro nucleation studies have shown the capability of recombinant CEMP1 to produce hydroxyapatite (HA) with different crystallinities thus demonstrating that CEMP1 nucleates and regulates hydroxyapatite crystal growth (Fig. 7.8), which leads to the development of new protein loading calcium phosphate materials for dental tissue repair and regeneration (Chen et al. 2015; Romo-Arévalo et al. 2016). Curiously, recombinant human CEMP1 produced in a prokaryotic system has similar HA nucleation functions as recombinant CEMP1 produced in a eukaryotic system, suggesting that posttranslational modifications are not important for this function. It has also been demonstrated that the CEMP1/ACP/PCL/COL scaffold has the potential of generating cementum-like tissue in vitro and in vivo and hindered bone formation in vivo orthotopically, demonstrating that the CEMP1 is a key bioactive compound in cementum formation both in vitro and in vivo (Chen et al. 2016).

Since nascent or posttranslationally modified CEMP1 have the same functions, in more recent studies we designed CEMP1-derived peptides to determine their active site. Several of these peptides are able to promote proliferation and differentiation of PDL cells toward a cementoblast phenotype, as shown by ALKP activity, Osterix, RUNX-2, IBSP, BMP-2, OCN, and CEMP1 expression, both at the mRNA and protein levels. Standardized critical-size calvaria defects in vivo assays treated with CEMP1-peptide 1 (CEMP1-p1) resulted in the formation of new calvaria bone. This

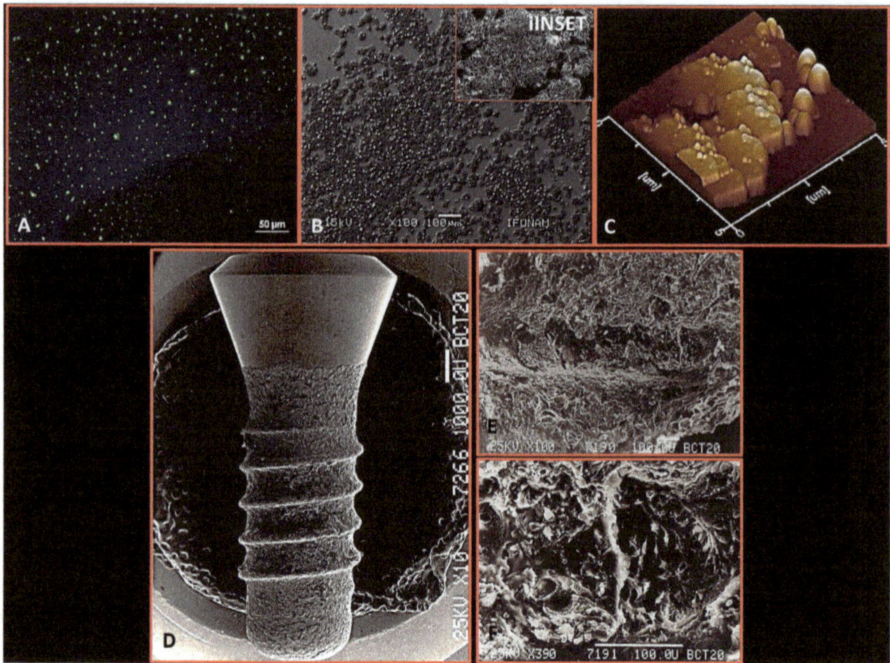

Fig. 7.9 CEMP1-p1 labeled with AF-488 shows adsorption to titanium surface (**a**). Deposition of hydroxyapatite crystals onto the titanium surface treated with CEMP1-p1 shows deposition of hydroxyapatite crystals (clearly observed in the inset) (**b**). AFM shows the topography of HA crystals deposited onto titanium surface (**c**). In vivo experimental assay shows that titanium implant treated with CEMP1-p1 promotes the formation of a fibrous connective tissue (**d**). The imprint of the implant reveals well-organized connective tissue (**e**). Higher magnification of the implant imprint shows the presence of extracellular matrix and cells

data demonstrates that the complete CEMP1 molecule is not necessary for the bone-inductive activity, and the CEMP1-p1 is an effective bioactive peptide for bone tissue regeneration (Correa et al. 2019). Another peptide, CEMP1-p4, activates the Wnt/β-catenin signaling pathway, inducing the differentiation of human oral mucosa stem cells (HOMSCs) toward a cementoblast-like phenotype (Arroyo et al. 2020). These data demonstrate that CEMP1-p1 and CEMP1-p4 are effective bioactive peptides for cementum and bone tissue regeneration. Preliminary studies show that CEMP1-p1 adsorbs to titanium surfaces in vitro and promotes the formation of mineralized tissue in vivo, indicating that this peptide promotes osseointegration and formation of a periodontal ligament-like structure (Fig. 7.9) thus opening new possibilities for implant therapies. The therapeutical application of these bioactive peptides may lead to implementing new strategies for the regeneration of bone and other mineralized tissues. CEMP1-p1 and CEMP1-p4 play a critical role in the modulation of HA crystal formation and somehow mimicking the physiological process that occurs in mammals. In addition, CEMP1-p1 binds other targets to fulfill its function on mineralized tissue formation since it has shown to regulate gene expression, cell differentiation, and regeneration of cementum and bone in vivo thus mimicking CEMP1's function. The advantages of using therapeutic peptides rather than recombinant proteins are numerous; peptides are produced synthetically; their chemical composition can be precisely controlled; and they are free of pathogenic contamination. In conclusion, CEMP1-derived peptides have pharmacological properties, set the basis for novel therapeutic alternatives, and have the potential to be used not only for the regeneration of the periodontium but also beyond cementum. in the field of dental implantology and bone regeneration.

7.14 Conclusion

Cementum is a complex structure involving acellular and cellular cementum. Cementum includes five distinct parts: (1) an acellular afibrillar cementum, (2) an acellular extrinsic fibers, (3) a cellular mixed stratified cementum, (4) a cellular intrinsic cementum, and (5) an intermediate cementum. In addition to inorganic components, organic components include collagens, proteoglycans, and glycosaminoglycans, nonspecific alkaline phosphatase, a series of phosphoproteins (OPN, BSP, DMP1), vitamin K-dependant proteins (MGP, osteocalcin), enamel-related proteins (AMEL, ameloblastin (AMBN), cementum-specific proteins (CAP, CEMP1). Cementum proteins are involved in regeneration therapy and periodontal treatments. Stem cells are also found in cementum, they may contribute to cementum healing and/or regeneration. In summary, the cementum-specific proteins CAP and CEMP1 are biological factors which participate in chemoattraction, adhesion, and cell differentiation during the homeostasis process of cementogenesis.

Acknowledgment This work was partially supported by a grant from DGAPA-PAPIIT-UNAM IN203920.

References

Ababneh KT, Hall RC, Embery G (1988) Immunolocalization of glycosaminoglycans in ageing, healthy and periodontally diseased human cementum. Arch Oral Biol 43:235–246. https://doi.org/10.1016/s0003-9969(98)00001-6

Ababneh KT, Hall RC, Embery G (1999) The proteoglycans of human cementum: immunohistochemical localization in healthy, periodontally involved and ageing teeth. J Periodontal Res 34:87–96. https://doi.org/10.1111/j.1600-0765.999.tb02227.x

Åhrén E (2014) Printing proof of priority: Anders Retzius, Jan Purkinje, and the microscopic anatomy of teeth, 1835–37. 6th international conference of the European society for the history of science, Lisbon, Portugal

Aichelmann-Reidy ME, Reynolds MA (2008) Predictability of clinical outcomes following regenerative therapy in intrabony defects. J Periodontol 79:387–393. https://doi.org/10.1902/jop.2008.060521

Al-Hamdan K, Eber R, Sarment D et al (2003) Guided tissue regeneration-based root coverage: meta-analysis. J Periodontol 74:1520–1533. https://doi.org/10.1902/jop.2003.74.10.1520
Alvarez Pérez MA, Pitaru S, Alvarez Fregoso O et al (2003) Anti-cementoblastoma-derived protein antibody partially inhibits mineralization on a cementoblastic cell line. J Struct Biol 143:1–13. https://doi.org/10.1016/s1047-8477(03)00116-3
Alvarez-Pérez MA, Alvarez-Fregoso O, Ortiz-López J et al (2005) X-ray microanalysis of human cementum. Microsc Microanal 11:313–318. https://doi.org/10.1017/S1431927605050221
Alvarez Pérez MA, Narayanan S, Zeichner-David M et al (2006) Molecular cloning, expression and immunolocalization of a novel human cementum-derived protein (CP-23). Bone 38:409–419. https://doi.org/10.1016/j.bone.2005.09.009
Amos FF, Destine E, Ponce CB et al (2010) The N- and C-terminal regions of the pearl associated EF hand protein, PFMG1, promote the formation of the aragonite polymorph in vitro. Cryst Growth Des 10:4211–4216. https://doi.org/10.1021/cg100363m
Amos FF, Ndao M, Ponce CB et al (2011) A C-RING-like domain participates in protein self-assembly and mineral nucleation. Biochemistry 50:8880–8887. https://doi.org/10.1021/bi201346d
Andreasen JO (2012) Pulp and periodontal tissue repair - regeneration or tissue metaplasia after dental trauma. A review. Dent Traumatol 28:19–24. https://doi.org/10.1111/j.1600-9657.2011.01058.x
Ao M, Chavez MB, Chu EY et al (2017) Overlapping functions of bone sialoprotein and pyrophosphate regulators in directing cementogenesis. Bone 105:134–147. https://doi.org/10.1016/j.bone.2017.08.027
Arroyo R, López S, Romo E et al (2020) Carboxy-terminal cementum protein 1-derived peptide 4 (cemp1-p4) promotes mineralization through wnt/β-catenin signaling in human oral mucosa stem cells. Int J Mol Sci 21:1307. https://doi.org/10.3390/ijms21041307
Arzate H, Olson SW, Page RC et al (1992a) Isolation of human tumor cells that produce cementum proteins in culture. Bone Miner 18:15–30. https://doi.org/10.1016/0169-6009(92)90796-g
Arzate H, Olson SW, Page RC et al (1992b) Production of a monoclonal antibody to an attachment protein derived from human cementum. FASEB J 6:2990–2995. https://doi.org/10.1096/fasebj.6.11.1644261
Arzate H, Chimal-Monroy J, Hernández-Lagunas L et al (1996) Human cementum protein extract promotes chondrogenesis and mineralization in mesenchymal cells. J Periodontal Res 31:144–148. https://doi.org/10.1111/j.1600-0765.1996.tb00476.x
Arzate H, Alvarez-Pérez MA, Aguilar-Mendoza ME et al (1998) Human cementum tumor cells have different features from human osteoblastic cells in vitro. J Periodontal Res 33:249–258. https://doi.org/10.1111/j.1600-0765.1998.tb02197.x
Arzate H, Jiménez-García LF, Alvarez-Pérez MA et al (2002) Immunolocalization of a human cementoblastoma-conditioned medium-derived protein. J Dent Res 8:541–546. https://doi.org/10.1177/154405910208100808
Arzate H, Zeichner-David M, Mercado-Celis G (2015) Cementum proteins: role in cementogenesis, biomineralization, periodontium formation and regeneration. Periodontol 2000 67(1):211–233. https://doi.org/10.1111/prd.12062
Ascenzi A, Bonucci E (1968) The compressive properties of single osteons. Anat Rec 161:377–391. https://doi.org/10.1002/ar.1091610309
Athanassiou-Papaefthymiou M, Papagerakis P, Papagerakis S (2015) Isolation and characterization of human adult epithelial stem cells from the periodontal ligament. J Dent Res 94:1591–1600. https://doi.org/10.1177/0022034515606401
Bar-Kana I, Savion N, Narayanan AS et al (1998) Cementum attachment protein manifestation is restricted to the mineralized tissue forming cells of the periodontium. Eur J Oral Sci 106(Suppl 1):357–364. https://doi.org/10.1111/j.1600-0722.1998.tb02198.x
Barrera-Ortega CC, Hoz-Rodríguez L, Arzate H et al (2017) Comparison of the osteogenic, adipogenic, chondrogenic and cementogenic differentiation potential of periodontal ligament

cells cultured on different biomaterials. Mater Sci Eng C Mater Biol Appl 76:1075–1084. https://doi.org/10.1016/j.msec.2017.03.213
Bartold PM, Miki Y, McAllister B et al (1988) Glycosaminoglycans of human cementum. J Periodontal Res 23:13–17. https://doi.org/10.1111/j.1600-0765.1988.tb01020.x
Beertsen W, Everts V (1990) Formation of acellular root cementum in relation to dental and non-dental hard tissues in the rat. J Dent Res 69:1669–1673. https://doi.org/10.1177/00220345900690100801
Beertsen W, Van den Bos T (1991) Alkaline phosphatase induces the deposition of calcified layers in relation to dentin: an in vitro study to mimic the formation of afibrillar acellular cementum. J Dent Res 70:176–181. https://doi.org/10.1177/00220345910700030401
Beertsen W, McCulloch CA, Sodek J (1997) The periodontal ligament: a unique, multifunctional connective tissue. Periodontol 2000 13:20–40. https://doi.org/10.1111/j.1600-0757.1997.tb00094.x
Beertsen W, VandenBos T, Everts V (1999) Root development in mice lacking functional tissue non-specific alkaline phosphatase gene: inhibition of acellular cementum formation. J Dent Res 78:1221–1229. https://doi.org/10.1177/00220345990780060501
Bencze L (1927) Befunde an der Dentinzementgrenze. Z Stomatol 25:877–896
Bermúdez M, Imaz-Rosshandler I, Rangel-Escareño C et al (2015) CEMP1 Induces Transformation in Human Gingival Fibroblasts. PLoS One 10(5):e0127286. https://doi.org/10.1371/journal.pone.0127286
Birkedal-Hansen H, Butler WT, Taylor RE (1977) Proteins of the periodontium. Characterization of the insoluble collagens of bovine dental cementum. Calcif Tissue Res 23:39–44. https://doi.org/10.1007/BF02012764
Black GV (1890) Descriptive anatomy of the human teeth. Wilmington Dental Mfg. Co, Philadelphia
Blackwood HJ (1957) Intermediate cementum. Br Dent J 102:345–350
Boabaid F, Gibson CW, Kuehl MA et al (2004) Leucine-rich amelogenin peptide: a candidate signaling molecule during cementogenesis. J Periodontol 75:1126–1136. https://doi.org/10.1902/jop.2004.75.8.1126
Bodecker CFW (1878) The distribution of living matter in human dentine, cement, and enamel. Dent Cosmos 20:582–590
Boskey AL (1995) Osteopontin and related phosphorylated sialoproteins: effects on mineralization. Ann N Y Acad Sci 760:249–256. https://doi.org/10.1111/j.1749-6632.1995.tb44635.x
Boskey AL, Maresca M, Ullrich W et al (1993) Osteopontin-hydroxyapatite interactions in vitro: inhibition of hydroxyapatite formation and growth in a gelatin-gel. Bone Miner 22:147–159. https://doi.org/10.1016/s0169-6009(08)80225-5
Boskey AL, Spevak L, Doty SB et al (1997) Effects of bone CS-proteoglycans, DS-decorin, and DS-biglycan on hydroxyapatite formation in a gelatin gel. Calcif Tissue Int 61:298–305. https://doi.org/10.1007/s002239900339
Bosshardt DD, Nanci A (1998) Immunolocalization of epithelial and mesenchymal matrix constituents in association with inner enamel epithelial cells. J Histochem Cytochem 46:135–142. https://doi.org/10.1177/002215549804600201
Bosshardt DD, Nanci A (2000) The pattern of expression of collagen determines the concentration and distribution of noncollagenous proteins along the forming root. In: Goldberg M, Boskey A, Robinson C (eds) Chemistry and biology of mineralized tissues. American Academy of Orthopaedic Surgeons, Vittel, pp 129–136
Bosshardt DD, Nanci A (2004) Hertwig's epithelial root sheath, enamel matrix proteins, and initiation of cementogenesis in porcine teeth. J Clin Periodontol 31:184–192. https://doi.org/10.1111/j.0303-6979.2004.00473.x
Bosshardt DD, Schroeder HE (1991) Initiation of acellular extrinsic fiber cementum on human teeth. A light- and electron-microscopic study. Cell Tissue Res 263:311–324. https://doi.org/10.1007/BF00318773

Bosshardt DD, Schroeder HE (1993) Attempts to label matrix synthesis of human root cementum in vitro. Cell Tissue Res 274(2):343–352. https://doi.org/10.1007/BF00318753
Bosshardt DD, Schroeder HE (1996) Cementogenesis reviewed: a comparison between human premolars and rodent molars. Anat Rec 245:267–292. https://doi.org/10.1002/(SICI)1097-0185(199606)245:2<267::AID-AR12>3.0.CO;2-N
Bosshardt DD, Selvig KA (1997) Dental cementum: the dynamic tissue covering of the root. Periodontol 2000 13:41–75. https://doi.org/10.1111/j.1600-0757.1997.tb00095.x
Bosshardt DD, Degen T, Lang NP (2005) Sequence of protein expression of bone sialoprotein and osteopontin at the developing interface between repair cementum and dentin in human deciduous teeth. Cell Tissue Res 320:399–407. https://doi.org/10.1007/s00441-005-1106-8
Bosshardt DD, Stadlinger B, Terheyden H (2015) Cell-to-cell communication--periodontal regeneration. Clin Oral Implants Res 26:229–239. https://doi.org/10.1111/clr.12543
Bronckers AL, D'Souza RN, Butler WT et al (1993) Dentin sialoprotein: biosynthesis and developmental appearance in rat tooth germs in comparison with amelogenins, osteocalcin and collagen type-I. Cell Tissue Res 272:237–247. https://doi.org/10.1007/BF00302729
Bronckers AL, Farach-Carson MC, Van Waveren E et al (1994) Immunolocalization of osteopontin, osteocalcin, and dentin sialoprotein during dental root formation and early cementogenesis in the rat. J Bone Miner Res 9:833–841. https://doi.org/10.1002/jbmr.5650090609
Bruckner RJ, Rickles NH, Porter DR (1962) Hypophosphatasia with premature shedding of teeth and aplasia of cementum. Oral Surg Oral Med Oral Pathol 15:1351–1369. https://doi.org/10.1016/0030-4220(62)90356-0
Brudevold F, Steadman LT, Smith FA (1960) Inorganic and organic components of tooth structure. Ann N Y Acad Sci 85:110–132. https://doi.org/10.1111/j.1749-6632.1960.tb49951.x
Buchet R, Millán JL, Magne D (2013) Multisystemic functions of alkaline phosphatases. Methods Mol Biol 1053:27–51. https://doi.org/10.1007/978-1-62703-562-0_3
Butler WT (1991) Sialoproteins of bone and dentin. J Biol Buccale 19:83–89
Butler WT, Ritchie H (1995) The nature and functional significance of dentin extracellular matrix proteins. Int J Dev Biol 39:169–179
Cape AT, Kitchin PC (1930) Histologic phenomena of tooth tissues as observed under polarized light; with a note on the roentgen-ray spectra of enamel and dentin. J Am Dent Assoc 17:193–227. https://doi.org/10.14219/jada.archive.1930.0047
Carmona-Rodríguez B, Alvarez-Pérez MA, Narayanan AS et al (2007) Human Cementum Protein 1 induces expression of bone and cementum proteins by human gingival fibroblasts. Biochem Biophys Res Commun 358:763–769. https://doi.org/10.1016/j.bbrc.2007.04.204
Cerný R, Slaby I, Hammarström L et al (1996) A novel gene expressed in rat ameloblasts codes for proteins with cell binding domains. J Bone Miner Res 11:883–891. https://doi.org/10.1002/jbmr.5650110703
Chen JK, Shapiro HS, Wrana JL et al (1991) Localization of bone sialoprotein (BSP) expression to sites of mineralized tissue formation in fetal rat tissues by in situ hybridization. Matrix 11:133–143. https://doi.org/10.1016/s0934-8832(11)80217-9
Chen J, Shapiro HS, Sodek J (1992) Development expression of bone sialoprotein mRNA in rat mineralized connective tissues. J Bone Miner Res 7:987–997. https://doi.org/10.1002/jbmr.5650070816
Chen YW, Yeh CH, Shie MY (2015) Stimulatory effects of the fast setting and suitable degrading Ca-Si-Mg cement on both cementogenesis and angiogenesis differentiation of human periodontal ligament cells. J Mater Chem B 3:7099–7108. https://doi.org/10.1039/c5tb00713e
Chen X, Liu Y, Miao L et al (2016) Controlled release of recombinant human cementum protein 1 from electrospun multiphasic scaffold for cementum regeneration. Int J Nanomedicine 11:3145–3158. https://doi.org/10.2147/IJN.S104324
Cheng H, Caterson B, Neame PJ et al (1996) Differential distribution of lumican and fibromodulin in tooth cementum. Connect Tissue Res 34:87–96. https://doi.org/10.3109/03008209609021494

Chiquoine AD (1954) The identification, origin, and migration of the primordial germ cells in the mouse embryo. Anat Rec 118:135–146. https://doi.org/10.1002/ar.1091180202
Cho MI, Garant PR (1989) Radioautographic study of [3H] mannose utilization during cementoblast differentiation, formation of acellular cementum, and development of periodontal ligament principal fibers. Anat Rec 223:209–222. https://doi.org/10.1002/ar.1092230214
Cho MI, Garant PR (1996) Expression and role of epidermal growth factor receptors during differentiation of cementoblasts, osteoblasts, and periodontal ligament fibroblasts in the rat. Anat Rec 245:342–360. https://doi.org/10.1002/(SICI)1097-0185(199606)245:2<342::AID-AR16>3.0.CO;2-P
Cho MI, Garant PR (2000) Development and general structure of the periodontium. Periodontol 2000 24:9–27. https://doi.org/10.1034/j.1600-0757.2000.2240102.x
Cho H, Tarafder S, Fogge M et al (2016) Periodontal ligament stem/progenitor cells with protein-releasing scaffolds for cementum formation and integration on dentin surface. Connect Tissue Res 57:488–495. https://doi.org/10.1080/03008207.2016.1191478
Choi H, Jin H, Kim JY et al (2014) Hypoxia promotes CEMP1 expression and induces cementoblastic differentiation of human dental stem cells in an HIF-1-dependent manner. Tissue Eng Part A 20:410–423. https://doi.org/10.1089/ten.TEA.2013.0132
Christner P, Robinson P, Clark CC (1977) A preliminary characterization of human cementum collagen. Calcif Tissue Res 23:147–150. https://doi.org/10.1007/BF02012780
Cochran DL, Wozney JM (1999) Biological mediators for periodontal regeneration. Periodontol 2000 19:40–58. https://doi.org/10.1111/j.1600-0757.1999.tb00146.x
Cochran DL, King GN, Schoolfield J et al (2003) The effect of enamel matrix proteins on periodontal regeneration as determined by histological analyses. J Periodontol 74:1043–1055. https://doi.org/10.1902/jop.2003.74.7.1043
Colard T, Falgayrac G, Bertrand B et al (2016) New insights on the composition and the structure of the acellular extrinsic fiber cementum by Raman analysis. PLoS One 11(12):e0167316. https://doi.org/10.1371/journal.pone.0167316
Cole DE, Hanley DA (1991) Osteocalcin. In: Hall BK (ed) Bone matrix and bone specific products. CRC Press, London, pp 239–294
Correa R, Arenas J, Montoya G et al (2019) Synthetic cementum protein 1-derived peptide regulates mineralization in vitro and promotes bone regeneration in vivo. FASEB J 33:1167–1178. https://doi.org/10.1096/fj.201800434RR
Coulson-Thomas YM, Coulson-Thomas VJ, Norton AL et al (2015) The identification of proteoglycans and glycosaminoglycans in archaeological human bones and teeth. PLoS One 10(6): e0131105. https://doi.org/10.1371/journal.pone.0131105
Davidson D, McCulloch CA (1986) Proliferative behavior of periodontal ligament cell populations. J Periodontal Res 21:414–428. https://doi.org/10.1111/j.1600-0765.1986.tb01475.x
Delak K, Harcup C, Lakshminarayanan R et al (2009) The tooth enamel protein, porcine amelogenin, is an intrinsically disordered protein with an extended molecular configuration in the monomeric form. Biochemistry 48:2272–2281. https://doi.org/10.1021/bi802175a
Dellavia C, Canciani E, Rasperini G et al (2019) CEMP-1 levels in periodontal wound fluid during the early phase of healing: prospective. Clinical Trial. Mediators Inflamm 24:1737306. https://doi.org/10.1155/2019/1737306
Deracinois B, Lenfant AM, Dehouck MP et al (2015) Tissue non-specific alkaline phosphatase (TNAP) in vessels of the brain. Subcell Biochem 76:125–151. https://doi.org/10.1007/978-94-017-7197-9_7
Dickson IR, Poole AR, Veis A (1975) Localisation of plasma alpha2HS glycoprotein in mineralising human bone. Nature 256:430–432. https://doi.org/10.1038/256430a0
Ducy P, Desbois C, Boyce B et al (1996) Increased bone formation in osteocalcin-deficient mice. Nature 382:448–452. https://doi.org/10.1038/382448a0
Eke PI, Dye B, Wei L et al (2012) Prevalence of periodontitis in adults in the United States: 2009 and 2010. J Dent Res 91(10):914–920. https://doi.org/10.1177/0022034512457373
el Mostehy MR, Stallard RE (1968) Intermediate cementum. J Periodontal Res 3:24–29

Elangovan S, Srinivasan S, Ayilavarapu S (2009) Novel regenerative strategies to enhance periodontal therapy outcome. Expert Opin Biol Ther 9:399–410. https://doi.org/10.1517/14712590902778423

Embery G, Hall R, Waddington R et al (2001) Proteoglycans in dentinogenesis. Crit Rev Oral Biol Med 12:331–349. https://doi.org/10.1177/10454411010120040401

Esposito M, Coulthard P, Thomsen P et al (2004) Enamel matrix derivative for periodontal tissue regeneration in treatment of intrabony defects: a Cochrane systematic review. J Dent Educ 68:834–844

Evans JS (2008) "Tuning in" to mollusk shell nacre- and prismatic associated protein terminal sequences. Implications for biomineralization and the construction of high performance inorganic-organic composites. Chem Rev 108:4455–4462. https://doi.org/10.1021/cr078251e

Farea M, Husein A, Halim AS et al (2016) Cementoblastic lineage formation in the cross-talk between stem cells of human exfoliated deciduous teeth and epithelial rests of Malassez cells. Clin Oral Investig 20:1181–1191. https://doi.org/10.1007/s00784-015-1601-6

Fedarko NS, Jain A, Karadag A et al (2004) Three small integrin binding ligand N-linked glycoproteins (SIBLINGs) bind and activate specific matrix metalloproteinases. FASEB J 18:734–736. https://doi.org/10.1096/fj.03-0966fje

Feng JQ, Zhang J, Dallas SL et al (2002) Dentin matrix protein 1, a target molecule for Cbfa1 in bone, is a unique bone marker gene. J Bone Miner Res 17:1822–1831. https://doi.org/10.1359/jbmr.2002.17.10.1822

Feng JQ, Huang H, Lu Y et al (2003) The dentin matrix protein 1 (Dmp1) is specifically expressed in mineralized, but not soft, tissues during development. J Dent Res 82:776–780. https://doi.org/10.1359/jbmr.2002.17.10.1822

Fisher LW, Fedarko NS (2003) Six genes expressed in bones and teeth encode the current members of the SIBLING family of proteins. Connect Tissue Res 44(Suppl 1):33–40

Fisher LW, Hawkins GR, Tuross N et al (1987) Purification and partial characterization of small proteoglycans I and II, bone sialoproteins I and II, and osteonectin from the mineral compartment of developing human bone. J Biol Chem 262:9702–9708. https://doi.org/10.1359/jbmr.2002.17.10.1822

Fisher LW, Torchia DA, Fohr B et al (2001) Flexible structures of SIBLING proteins, bone sialoprotein, and osteopontin. Biochem Biophys Res Commun 280:460–465. https://doi.org/10.1006/bbrc.2000.4146

Fong CD, Hammarström L (2000) Expression of amelin and amelogenin in epithelial root sheath remnants of fully formed rat molars. Oral Surg Oral Med Oral Pathol Oral Radiol Endod 90:218–223. https://doi.org/10.1067/moe.2000.107052

Fong CD, Slaby I, Hammarström L (1996) Amelin: an enamel-related protein, transcribed in the cells of epithelial root sheath. J Bone Miner Res 11:892–898. https://doi.org/10.1002/jbmr.5650110704

Fong CD, Cerný R, Hammarström L et al (1998) Sequential expression of an amelin gene in mesenchymal and epithelial cells during odontogenesis in rats. Eur J Oral Sci 106(Suppl 1):324–330. https://doi.org/10.1111/j.1600-0722.1998.tb02193.x

Forman-Kay JD, Mittag T (2013) From sequence and forces to structure, function, and evolution of intrinsically disordered proteins. Structure 21:1492–1499. https://doi.org/10.1016/j.str.2013.08.001

Foster BL (2012) Methods for studying tooth root cementum by light microscopy. Int J Oral Sci 4:119–128. https://doi.org/10.1038/ijos.2012.57

Foster BL (2017) On the discovery of cementum. J Periodontal Res 52:666–685. https://doi.org/10.1111/jre.12444

Foster BL, Tompkins KA, Rutherford RB et al (2008) Phosphate: known and potential roles during development and regeneration of teeth and supporting structures. Birth Defects Res C Embryo Today 84:281–314. https://doi.org/10.1002/bdrc.20136

Foster BL, Nagatomo KJ, Nociti FH Jr et al (2012) Central role of pyrophosphate in acellular cementum formation. PLoS One 7(6):e38393. https://doi.org/10.1371/journal.pone.0038393

Foster BL, Nociti FH Jr, Somerman MJ (2014) The rachitic tooth. Endocr Rev 35:1–34. https://doi.org/10.1210/er.2013-1009

Foster BL, Ao M, Willoughby C et al (2015) Mineralization defects in cementum and craniofacial bone from loss of bone sialoprotein. Bone 78:150–164. https://doi.org/10.1016/j.bone.2015.05.007

Foster BL, Ao M, Salmon CR et al (2018) Osteopontin regulates dentin and alveolar bone development and mineralization. Bone 107:196–207. https://doi.org/10.1016/j.bone.2017.12.004

Fränkel M (1835) De penitiori dentium humanorum structura observationes. Vratislavia, p 20

Fraser JD, Price PA (1988) Lung, heart, and kidney express high levels of mRNA for the vitamin K-dependent matrix Gla protein. Implications for the possible functions of matrix Gla protein and for the tissue distribution of the gamma-carboxylase. J Biol Chem 263:11033–11036

Fujii S, Maeda H, Wada N et al (2008) Investigating a clonal human periodontal ligament progenitor/stem cell line in vitro and in vivo. J Cell Physiol 215:743–749. https://doi.org/10.1002/jcp.21359

Fukae M, Tanabe T, Yamakoshi Y et al (2001) Immunoblot detection and expression of enamel proteins at the apical portion of the forming root in porcine permanent incisor tooth germs. J Bone Miner Metab 19:236–243. https://doi.org/10.1007/s007740170026

Füredi-Milhofer H, Moradian-Oldak J, Weiner S et al (1994) Interactions of matrix proteins from mineralized tissues with octacalcium phosphate. Connect Tissue Res 30:251–264. https://doi.org/10.3109/03008209409015041

Furseth R, Johansen EA (1968) microradiographic comparison of sound and carious human dental cementum. Arch Oral Biol 13:1197–1206. https://doi.org/10.1016/0003-9969(68)90075-7

Ganss B, Kim RH, Sodek J (1999) Bone sialoprotein. Crit Rev Oral Biol Med 10:79–98. https://doi.org/10.1177/10454411990100010401

Garrett S (1996) Periodontal regeneration around natural teeth. Ann Periodontol 1996(1):621–666. https://doi.org/10.1902/annals.1996.1.1.621

Gauthier P, Yu Z, Tran QT et al (2017) Cementogenic genes in human periodontal ligament stem cells are downregulated in response to osteogenic stimulation while upregulated by vitamin C treatment. Cell Tissue Res 368:79–92. https://doi.org/10.1007/s00441-016-2513-8

GBD (2017) Disease and Injury Incidence and Prevalence Collaborators, "Global, regional, and national incidence, prevalence, and years lived with disability for 328 diseases and injuries for 195 countries, 1990–2016: a systematic analysis for the Global Burden of Disease Study 2016" Lancet 390:1211–1259. doi: https://doi.org/10.1016/S0140-6736(17)32154-2

George A, Sabsay B, Simonian PA et al (1993) Characterization of a novel dentin matrix acidic phosphoprotein. Implications for induction of biomineralization. J Biol Chem 268:12624–12630

Gestrelius S, Andersson C, Lidström D et al (1997) In vitro studies on periodontal ligament cells and enamel matrix derivative. J Clin Periodontol 24(9 Pt 2):685–692. https://doi.org/10.1111/j.1600-051x.1997.tb00250.x

Gestrelius S, Lyngstadaas SP, Hammarström L (2000) Emdogain--periodontal regeneration based on biomimicry. Clin Oral Investig 4:120–125. https://doi.org/10.1007/s007840050127

Glimcher MJ (1979) Phosphopeptides of enamel matrix. J Dent Res 58(spec Issue B):790–809. https://doi.org/10.1177/00220345790580023101

Glimcher MJ (1989) Mechanism of calcification: role of collagen fibrils and collagen-phosphoprotein complexes in vitro and in vivo. Anat Rec 224:139–153. https://doi.org/10.1002/ar.1092240205

Glimcher MJ, Krane S (1968) In: Ramachandran GN, Gould BS (eds) Treatise on collagen, Chapter 2, vol 2B. Academic, London

Goncalves PF, Sallum EA, Sallum AW et al (2005) Dental cementum reviewed: development, structure, composition, regeneration and potential functions. Braz J Oral Sci 4:651–658

Gorski JP, Kremer E, Ruiz-Perez J et al (1995) Conformational analyses on soluble and surface bound osteopontin. Ann N Y Acad Sci 760:12–23. https://doi.org/10.1111/j.1749-6632.1995.tb44616.x
Gottlieb B (1942) Biology of the cementum. J Periodontol 13:13–19
Gould TR (1983) Ultrastructural characteristics of progenitor cell populations in the periodontal ligament. J Dent Res 62:873–876. https://doi.org/10.1177/00220345830620080401
Gould TR, Melcher AH, Brunette DM (1977) Location of progenitor cells in periodontal ligament of mouse molar stimulated by wounding. Anat Rec 188:133–141. https://doi.org/10.1002/ar.1091880202
Gould TR, Melcher AH, Brunette DM (1980) Migration and division of progenitor cell populations in periodontal ligament after wounding. J Periodontal Res 15:20–42. https://doi.org/10.1111/j.1600-0765.1980.tb00258.x
Groeneveld MC, Everts V, Beertsen W (1994) Formation of afibrillar acellular cementum-like layers induced by alkaline phosphatase activity from periodontal ligament explants maintained in vitro. J Dent Res 73:1588–1592. https://doi.org/10.1177/00220345940730100201
Groeneveld MC, Everts V, Beertsen W (1995) Alkaline phosphatase activity in the periodontal ligament and gingiva of the rat molar: its relation to cementum formation. J Dent Res 74:1374–1381. https://doi.org/10.1177/00220345950740070901
Grzesik WJ, Narayanan AS (2002) Cementum and periodontal wound healing and regeneration. Crit Rev Oral Biol Med 13:474–484. https://doi.org/10.1177/154411130201300605
Gungormus M, Oren E, Horst J et al (2012) Cementomimetics-constructing a cementum-like biomineralized microlayer *via* amelogenin-derived peptides. Int J Oral Sci 4:69–77. https://doi.org/10.1038/ijos.2012.40
Gupta P, Kaur H, Shankari GS et al (2014) Human age estimation from tooth cementum and dentin. J Clin Diagn Res 8:ZC07–ZC10. https://doi.org/10.7860/JCDR/2014/7275.4221
Hale JE, Fraser JD, Price PA (1988) The identification of matrix Gla protein in cartilage. J Biol Chem 263:5820–5824
Hals E, Selvig KA (1977) Correlated electron probe microanalysis and microradiography of carious and normal dental cementum. Caries Res 11:62–75. https://doi.org/10.1159/000260250
Hamamoto Y, Nakajima T, Ozawa H et al (1996) Production of amelogenin by enamel epithelium of Hertwig's root sheath. Oral Surg Oral Med Oral Pathol Oral Radiol Endod 81:703–709. https://doi.org/10.1016/s1079-2104(96)80077-1
Hammarström L (1997) Enamel matrix, cementum development and regeneration. J Clin Periodontol 24:658–668. https://doi.org/10.1111/j.1600-051x.1997.tb00247.x
Hammarström L, Heijl L, Gestrelius S (1997) Periodontal regeneration in a buccal dehiscence model in monkeys after application of enamel matrix proteins. J Clin Periodontol 24(9 Pt 2):669–677. https://doi.org/10.1111/j.1600-051x.1997.tb00248.x
Hanes PJ (2007) Bone replacement grafts for the treatment of periodontal intrabony defects. Oral Maxillofac Surg Clin North Am 19(499–512):vi. https://doi.org/10.1016/j.coms.2007.06.001
Harrison JW, Roda RS (1995) Intermediate cementum. Development, structure, composition, and potential functions. Oral Surg Oral Med Oral Pathol Oral Radiol Endod 79:624–633. https://doi.org/10.1016/s1079-2104(05)80106-4
Hartgerink JD, Beniash E, Stupp SI (2001) Self-assembly and mineralization of peptide-amphiphile nanofibers. Science 294:1684–1688. https://doi.org/10.1126/science.1063187
Haruyama N, Hatakeyama J, Moriyama K et al (2011) Amelogenins: multi-functional enamel matrix proteins and their binding partners. J Oral Biosci 53:257–266
Hashimoto F, Kobayashi Y, Kobayashi ET et al (2001) Expression and localization of MGP in rat tooth cementum. Arch Oral Biol 46:585–592. https://doi.org/10.1016/s0003-9969(01)00022-x
Hatakeyama J, Sreenath T, Hatakeyama Y et al (2003) The receptor activator of nuclear factor-kappa B ligand-mediated osteoclastogenic pathway is elevated in amelogenin-null mice. J Biol Chem 278:35743–35748. https://doi.org/10.1074/jbc.M306284200
Hauschka PV (1990) Growth factor effects in bone-forming cells. In: Hall BK (ed) Bone: the osteoblast and the osteocyte, vol 1. Telford Press, Caldwell, pp 103–170

Hauschka PV, Carr SA (1982) Calcium-dependent alpha-helical structure in osteocalcin. Biochemistry 21:2538–2547. https://doi.org/10.1021/bi00539a038

Hauschka PV, Reddi AH (1980) Correlation of the appearance of gamma-carboxyglutamic acid with the onset of mineralization in developing endochondral bone. Biochem Biophys Res Commun 92:1037–1041. https://doi.org/10.1016/0006-291x(80)90806-2

Hauschka PV, Lian JB, Cole DE et al (1989) Osteocalcin and matrix Gla protein: vitamin K-dependent proteins in bone. Physiol Rev 1989(69):990–1047. https://doi.org/10.1152/physrev.1989.69.3.990

He G, Dahl T, Veis A et al (2003) Nucleation of apatite crystals in vitro by self-assembled dentin matrix protein 1. Nat Mater 2:552–558. https://doi.org/10.1038/nmat945

He G, Gajjeraman S, Schultz D et al (2005) Spatially and temporally controlled biomineralization is facilitated by interaction between self-assembled dentin matrix protein 1 and calcium phosphate nuclei in solution. Biochemistry 44:16140–16148. https://doi.org/10.1021/bi051045l

Heden G, Wennström JL (2006) Five-year follow-up of regenerative periodontal therapy with enamel matrix derivative at sites with angular bone defects. J Periodontol 77:295–301. https://doi.org/10.1902/jop.2006.050071

Heijl L (1997) Periodontal regeneration with enamel matrix derivative in one human experimental defect. A case report. J Clin Periodontol 24(9 Pt 2):693–696. https://doi.org/10.1034/j.1600-051x.1997.00693.x

Heijl L, Heden G, Svärdström G et al (1997) Enamel matrix derivative (EMDOGAIN) in the treatment of intrabony periodontal defects. J Clin Periodontol 24(9 Pt 2):705–714. https://doi.org/10.1111/j.1600-051x.1997.tb00253.x

Hernández-Chirlaque C, Gámez-Belmonte R, Ocón B et al (2017) Tissue non-specific alkaline phosphatase expression is needed for the full stimulation of T cells and T cell-dependent colitis. J Crohns Colitis 11:857–870. https://doi.org/10.1093/ecco-jcc/jjw222

Hirooka H (1998) The biologic concept for the use of enamel matrix protein: true periodontal regeneration. Quintessence Int 29:621–630

Hirose N, Shimazu A, Watanabe M et al (2003) Ameloblastin in Hertwig's epithelial root sheath regulates tooth root formation and development. PLoS One 8(1):e54449. https://doi.org/10.1371/journal.pone.0054449

Ho SP, Balooch M, Marshall SJ et al (2004) Local properties of a functionally graded interphase between cementum and dentin. J Biomed Mater Res A 70:480–489. https://doi.org/10.1002/jbm.a.30105

Ho SP, Sulyanto RM, Marshall SJ et al (2005) The cementum-dentin junction also contains glycosaminoglycans and collagen fibrils. J Struct Biol 151:69–78. https://doi.org/10.1016/j.jsb.2005.05.003

Ho SP, Yu B, Yun W et al (2009) Structure, chemical composition and mechanical properties of human and rat cementum and its interface with root dentin. Acta Biomater 5:707–718. https://doi.org/10.1016/j.actbio.2008.08.013

Hoz L, Romo E, Zeichner-David M et al (2012) Cementum protein 1 (CEMP1) induces differentiation by human periodontal ligament cells under three-dimensional culture conditions. Cell Biol Int 36:129–136. https://doi.org/10.1042/CBI20110168

Hu CC, Fukae M, Uchida T et al (1997) Sheathlin: cloning, cDNA/polypeptide sequences, and immunolocalization of porcine enamel sheath proteins. J Dent Res 76:648–657. https://doi.org/10.1177/00220345970760020501

Huang X, Bringas P Jr, Slavkin HC et al (2009) Fate of HERS during tooth root development. Dev Biol 334:22–30. https://doi.org/10.1016/j.ydbio.2009.06.034

Hui M, Tenenbaum HC (1998) New face of an old enzyme: alkaline phosphatase may contribute to human tissue aging by inducing tissue hardening and calcification. Anat Rec 253:91–94. https://doi.org/10.1002/(SICI)1097-0185(199806)253:3<91::AID-AR5>3.0.CO;2-H

Hunter GK, Goldberg HA (1994) Modulation of crystal formation by bone phosphoproteins: role of glutamic acid-rich sequences in the nucleation of hydroxyapatite by bone sialoprotein. Biochem J 302:175–179. https://doi.org/10.1042/bj3020175

Hunter GK, Hauschka PV, Poole AR et al (1996) Nucleation and inhibition of hydroxyapatite formation by mineralized tissue proteins. Biochem J 317:59–64. https://doi.org/10.1042/bj3170059

Ikezawa K, Hart CE, Williams DC et al (1997) Characterization of cementum derived growth factor as an insulin-like growth factor-I like molecule. Connect Tissue Res 36:309–319. https://doi.org/10.3109/03008209709160230

Ivanovski S, Komaki M, Bartold PM et al (1999) Periodontal-derived cells attach to cementum attachment protein via alpha 5 beta 1 integrin. J Periodontal Res 34:154–159. https://doi.org/10.1111/j.1600-0765.1999.tb02236.x

Iwata T, Yamakoshi Y, Hu JC et al (2007) Processing of ameloblastin by MMP-20. J Dent Res 86:153–157. https://doi.org/10.1177/154405910708600209

Janones DS, Massa LF, Arana-Chavez VE (2005) Immunocytochemical examination of the presence of amelogenin during the root development of rat molars. Arch Oral Biol 50:527–532. https://doi.org/10.1016/j.archoralbio.2004.10.004

Jones SJ (1981) Cement. In: Osborn JW (ed) Dental anatomy and embryology. Blackwell Scientific, Oxford; Edinburgh, Boston, pp 193–205, 286–294

Kadokura H, Yamazaki T, Masuda Y et al (2019) Establishment of a primary culture system of human periodontal ligament cells that differentiate into cementum protein 1-expressing cementoblast-like cells. In Vivo 33:349–352. https://doi.org/10.21873/invivo.11480

Kaipatur NR, Murshed M, McKee MD (2008) Matrix Gla protein inhibition of tooth mineralization. J Dent Res 87:839–844. https://doi.org/10.1177/154405910808700907

Kalajzic I, Braut A, Guo D et al (2004) Dentin matrix protein 1 expression during osteoblastic differentiation, generation of an osteocyte GFP-transgene. Bone 35:74–82. https://doi.org/10.1016/j.bone.2004.03.006

Kato K, Nakagaki H, Okumura H et al (1992) Influence of occlusion on the fluoride distribution in rat molar cementum. Caries Res 26:418–422. https://doi.org/10.1159/000261480

Kémoun P, Laurencin-Dalicieux S, Rue J et al (2007a) Human dental follicle cells acquire cementoblast features under stimulation by BMP-2/−7 and enamel matrix derivatives (EMD) in vitro. Cell Tissue Res 329:283–294. https://doi.org/10.1007/s00441-007-0397-3

Kémoun P, Laurencin-Dalicieux S, Rue J et al (2007b) Localization of STRO-1, BMP-2/−3/−7, BMP receptors and phosphorylated Smad-1 during the formation of mouse periodontium. Tissue Cell 39:257–266. https://doi.org/10.1016/j.tice.2007.06.001

Khoshkam V, Chan HL, Lin GH et al (2015) Outcomes of regenerative treatment with rhPDGF-BB and rhFGF-2 for periodontal intra-bony defects: a systematic review and meta-analysis. J Clin Periodontol 42:272–280. https://doi.org/10.1111/jcpe.12354

Kimberly G, Smith KA, Strother J et al (1994) Chemical ultrastructure of cementum growth-layers of teeth of black bears. J Mammal 75:406–409. https://doi.org/10.2307/1382560

Komaki M, Iwasaki K, Arzate H et al (2012) Cementum protein 1 (CEMP1) induces a cementoblastic phenotype and reduces osteoblastic differentiation in periodontal ligament cells. J Cell Physiol 227:649–657. https://doi.org/10.1002/jcp.22770

Kozawa Y, Chisaka H, Iwasa Y et al (2005) Origin and evolution of cementum as tooth attachment complex. J Oral Biosci 47:25–32. https://doi.org/10.1016/S1349-0079(05)80005-2

Krause DS, Theise ND et al (2001) Multi-organ, multi-lineage engraftment by a single bone marrow-derived stem cell. Cell 105:369–377. https://doi.org/10.1016/s0092-8674(01)00328-2

Krebsbach PH, Lee SK, Matsuki Y et al (1996) Full-length sequence, localization, and chromosomal mapping of ameloblastin. A novel tooth-specific gene. J Biol Chem 271:4431–4435. https://doi.org/10.1074/jbc.271.8.4431

Kucia M, Zuba-Surma EK, Wysoczynski M et al (2007) Adult marrow-derived very small embryonic-like stem cells and tissue engineering. Expert Opin Biol Ther 7:1499–1514. https://doi.org/10.1517/14712598.7.10.1499

Kulakauskienė R, Aukštakalnis R, Šadzevičienė R (2020) Enamel matrix derivate induces periodontal regeneration by activating growth factors: a review. Stomatologija 21:49–53

Kulkarni GV, Chen B, Malone JP et al (2000) Promotion of selective cell attachment by the RGD sequence in dentine matrix protein 1. Arch Oral Biol 45:475–484. https://doi.org/10.1016/s0003-9969(00)00010-8

Langer D, Ikehara Y, Takebayashi H et al (2007) The ectonucleotidases alkaline phosphatase and nucleoside triphosphate diphosphohydrolase 2 are associated with subsets of progenitor cell populations in the mouse embryonic, postnatal and adult neurogenic zones. Neuroscience 150:863–879. https://doi.org/10.1016/j.neuroscience.2007.07.064

Leblond CP (1964) Classification of cell populations on the basis of their proliferative behavior. Natl Cancer Inst Monogr 14:119–150

Leblond CP, Messier B, KOpriwa B (1959) Thymidine-H3 as a tool for the investigation of the renewal of cell populations. Lab Investig 8:296–306. discussion 306–308

Lester KS (1969a) The unusual nature of root formation in molar teeth of the laboratory rat. J Ultrastruct Res 28:481–506. https://doi.org/10.1016/s0022-5320(69)80035-3

Lester KS (1969b) The incorporation of epithelial cells by cementum. J Ultrastruct Res 27:63–87

Li D, Gonzalez O, Bachinski LL, Roberts R (2000) Human protein tyrosine phosphatase-like gene: expression profile, genomic structure, and mutation analysis in families with ARVD. Gene 256:237–243. https://doi.org/10.1016/s0378-1119(00)00347-4

Lieberman DE (1993) Life history variables preserved in dental cementum microstructure. Science 261(5125):1162–1164. https://doi.org/10.1126/science.8356448

Lieberman DE (1994) The biological basis of seasonal increments in dental cementum and their application to archaeological research. J Archaeol Sci 21:525–539. https://doi.org/10.1006/jasc.1994.1052

Lindskog S (1982a) Formation of intermediate cementum. I: early mineralization of aprismatic enamel and intermediate cementum in monkey. J Craniofac Genet Dev Biol 2:147–160

Lindskog S (1982b) Formation of intermediate cementum. II: a scanning electron microscopic study of the epithelial root sheath of Hertwig in monkey. J Craniofac Genet Dev Biol 2:161–169

Lindskog S, Hammarström L (1982) Formation of intermediate cementum: III. 3H-tryptophan and 3H-proline uptake into the epithelial root sheath of Hertwig in vitro. J Craniofac Genet Dev Biol 2:171–177

Listgarten MA (1968) A light and electron microscopic study of coronal cementogenesis. Arch Oral Biol 13:93–114. https://doi.org/10.1016/0003-9969(68)90040-x

Listgarten MA, Kamin A (1969) The development of a cementum layer over the enamel surface of rabbit molars-a light and electron microscopic study. Arch Oral Biol 14:961–985. https://doi.org/10.1016/0003-9969(69)90272-6

Liu HW, Yacobi R, Savion N et al (1997) A collagenous cementum-derived attachment protein is a marker for progenitors of the mineralized tissue-forming cell lineage of the periodontal ligament. J Bone Miner Res 12:1691–1699. https://doi.org/10.1359/jbmr.1997.12.10.1691

Luo W, Slavkin HC, Snead ML (1991) Cells from Hertwig's epithelial root sheath do not transcribe amelogenin. J Periodontal Res 26:42–47. https://doi.org/10.1111/j.1600-0765.1991.tb01624.x

Luo G, Ducy P, McKee MD et al (1997) Spontaneous calcification of arteries and cartilage in mice lacking matrix GLA protein. Nature 386:78–81. https://doi.org/10.1038/386078a0

Lynch SE, de Castilla GR, Williams RC, Kiritsy CP et al (1991) The effects of short-term application of a combination of platelet-derived and insulin-like growth factors on periodontal wound healing. J Periodontol 62:458–467. https://doi.org/10.1902/jop.1991.62.7.458

MacDougall M, DuPont BR, Simmons D et al (1996) Assignment of DMP1 to human chromosome 4 band q21 by in situ hybridization. Cytogenet Cell Genet 74:189. https://doi.org/10.1159/000134410

MacNeil RL, Somerman MJ (1999) Development and regeneration of the periodontium: parallels and contrasts. Periodontol 2000 19:8–20. https://doi.org/10.1111/j.1600-0757.1999.tb00144.x

MacNeil RL, Sheng N, Strayhorn C et al (1994) Bone sialoprotein is localized to the root surface during cementogenesis. J Bone Miner Res 9:1597–1606. https://doi.org/10.1002/jbmr.5650091013

MacNeil RL, Berry J, D'Errico J et al (1995a) Role of two mineral-associated adhesion molecules, osteopontin and bone sialoprotein, during cementogenesis. Connect Tissue Res 33:1–7. https://doi.org/10.3109/03008209509016974

MacNeil RL, Berry J, D'Errico J et al (1995b) Localization and expression of osteopontin in mineralized and nonmineralized tissues of the periodontium. Ann N Y Acad Sci 760:166–176. https://doi.org/10.1111/j.1749-6632.1995.tb44628.x

Malpighi M (1700) Opera posthuma. Amsterdam

Mantovani V, Maccari F, Volpi N (2016) Chondroitin sulfate and glucosamine as disease modifying anti-osteoarthritis drugs (DMOADs). Curr Med Chem 23:1139–1151. https://doi.org/10.2174/0929867323666160316123749

Marie B, Zanella-Cléon I, Le Roy N et al (2010) Proteomic analysis of the acid-soluble nacre matrix of the bivalve Unio pictorum: detection of novel carbonic anhydrase and putative protease inhibitor proteins. Chembiochem 11:2138–2147. https://doi.org/10.1002/cbic.201000276

Mariotti A (2003) Efficacy of chemical root surface modifiers in the treatment of periodontal disease. A systematic review. Ann Periodontol 8:205–226. https://doi.org/10.1902/annals.2003.8.1.205

Martins L, Amorim BR, Salmon CR et al (2020) Novel LRAP-binding partner revealing the plasminogen activation system as a regulator of cementoblast differentiation and mineral nodule formation in vitro. J Cell Physiol 235:4545–4558. https://doi.org/10.1002/jcp.29331

McAllister B, Narayanan AS, Miki Y et al (1990) Isolation of a fibroblast attachment protein from cementum. J Periodontal Res 25:99–105. https://doi.org/10.1155/2019/1737306

McCulloch CA, Bordin S (1991) Role of fibroblast subpopulations in periodontal physiology and pathology. J Periodontal Res 26:144–154. https://doi.org/10.1111/j.1600-0765.1991.tb01638.x

McCulloch CA, Melcher AH (1983a) Cell migration in the periodontal ligament of mice. J Periodontal Res 18:339–352. https://doi.org/10.1111/j.1600-0765.1983.tb00369.x

McCulloch CA, Melcher AH (1983b) Cell density and cell generation in the periodontal ligament of mice. Am J Anat 167:43–58. https://doi.org/10.1002/aja.1001670105

McCulloch CA, Melcher AH (1983c) Continuous labelling of the periodontal ligament of mice. J Periodontal Res 18:231–241. https://doi.org/10.1111/j.1600-0765.1983.tb00357.x

McCulloch CA, Nemeth E, Lowenberg B et al (1987) Paravascular cells in endosteal spaces of alveolar bone contribute to periodontal ligament cell populations. Anat Rec 219:233–242. https://doi.org/10.1002/ar.1092190304

McKee M, Zalzal S, Nanci A (1996) Extracellular matrix in tooth cementum and mantle dentin: localization of osteopontin and other noncollagenous proteins, plasma proteins, and glycoconjugates by electron microscopy. Anat Rec 245:293–312. https://doi.org/10.1002/(SICI)1097-0185(199606)245:2<293::AID-AR13>3.0.CO;2-K

McKee MD, Hoac B, Addison WN et al (2013) Extracellular matrix mineralization in periodontal tissues: noncollagenous matrix proteins, enzymes, and relationship to hypophosphatasia and X-linked hypophosphatemia. Periodontology 2000(63):102–122. https://doi.org/10.1111/prd.12029

Melcher AH (1976) On the repair potential of periodontal tissues. J Periodontol 47:256–260. https://doi.org/10.1902/jop.1976.47.5.256

Melcher AH, Cheong T, Cox J et al (1986) Synthesis of cementum-like tissue in vitro by cells cultured from bone: a light and electron microscope study. J Periodontal Res 21(6):592–612. https://doi.org/10.1111/j.1600-0765.1986.tb01497.x

Miki Y, Narayanan AS, Page RC (1987) Mitogenic activity of cementum components to gingival fibroblasts. J Dent Res 66:1399–1403. https://doi.org/10.1177/00220345870660082301

Miller PD, Craddock RD (1996) Surgical advances in the coverage of exposed roots. Curr Opin Periodontol 3:103–108

Mohler ER III, Adam LP, McClelland P et al (1997) Detection of osteopontin in calcified human aortic valves. Arterioscler Thromb Vasc Biol 17:547–552. https://doi.org/10.1161/01.atv.17.3.547

Montoya G, Arenas J, Romo E et al (2014) Human recombinant cementum attachment protein (hrPTPLa/CAP) promotes hydroxyapatite crystal formation in vitro and bone healing in vivo. Bone 69:154–164. https://doi.org/10.1016/j.bone.2014.09.014

Montoya G, Correa R, Arenas J et al (2019) Cementum protein 1-derived peptide (CEMP 1-p1) modulates hydroxyapatite crystal formation in vitro. J Pept Sci 25:e3211. https://doi.org/10.1002/psc.3211

Montoya G, Lopez K, Arenas J et al (2020) Nucleation and growth inhibition of biological minerals by cementum attachment protein-derived peptide (CAP-pi). J Pept Sci 26:e3282. https://doi.org/10.1002/psc.3282

Moradian-Oldak J, Frolow F, Addadi L et al (1992) Interactions between acidic matrix macromolecules and calcium phosphate ester crystals: Relevance to carbonate apatite formation in biomineralization. Proc Biol Sci 247:47–55. https://doi.org/10.1098/rspb.1992.0008

Mulloy B, Hogwood J, Gray E et al (2016) Pharmacology of heparin and related drugs. Pharmacol Rev 68:76–141. https://doi.org/10.1124/pr.115.011247

Mundy GR, Boyce B, Hughes D et al (1995) The effects of cytokines and growth factors on osteoblastic cells. Bone 17(2 Suppl):71S–75S. https://doi.org/10.1016/8756-3282(95)00182-d

Muraki Y (1958) Comparative-anatomical studies on the cementum of the mammalian teeth. Acta Anat Nipp 33:583–611

Murphy KG, Gunsolley JC (2003) Guided tissue regeneration for the treatment of periodontal intrabony and furcation defects. A systematic review. Ann Periodontol 8:266–302. https://doi.org/10.1902/annals.2003.8.1.266

Murshed M, Harmey D, Millán JL et al (2005) Unique coexpression in osteoblasts of broadly expressed genes accounts for the spatial restriction of ECM mineralization to bone. Genes Dev 19:1093–1104. https://doi.org/10.1101/gad.1276205

Nagasaki A, Tadesse W, Nagasaki K et al (2020) Delivery of alkaline phosphatase promotes cementum and alveolar bone regeneration 2020 IADR/AADR/CADR General Session (Washington, DC) ID: 3515

Nakae H, Narayanan AS, Raines E et al (1991) Isolation and partial characterization of mitogenic factors from cementum. Biochemistry 30:7047–7052. https://doi.org/10.1021/bi00243a002

Nakata TM, Stepnick RI, Zipkin I (1972) Chemistry of human dental cementum: The effect of age and fluoride exposure on the concentration of ash, fluoride, calcium, phosphorous and magnesium. J Periodontol 43:115–124. https://doi.org/10.1902/jop.1972.43.2.115

Nakayama Y, Matsuda H, Itoh S et al (2020) Impact of adjunctive procedures on recombinant human fibroblast growth factor (rhFGF-2) mediated periodontal regeneration therapy: a retrospective study. J Periodontol. https://doi.org/10.1002/JPER.20-0481

Nam H, Kim J, Park J et al (2011) Expression profile of the stem cell markers in human Hertwig's epithelial root sheath/Epithelial rests of Malassez cells. Mol Cells 31:355–360. https://doi.org/10.1007/s10059-011-0045-3

Nanci A (1999) Content and distribution of noncollagenous matrix proteins in bone and cementum: relationship to speed of formation and collagen packing density. J Struct Biol 126:256–269. https://doi.org/10.1006/jsbi.1999.4137

Narayanan AS, Bartold PM (1996) Biochemistry of periodontal connective tissues and their regeneration: a current perspective. Connect Tissue Res 34:191–201. https://doi.org/10.3109/03008209609000698

Narayanan AS, Ikezawa K, Wu D et al (1995) Cementum specific components which influence periodontal connective tissue cells. Connect Tissue Res 33:19–21. https://doi.org/10.3109/03008209509016976

Narayanan K, Srinivas R, Ramachandran A et al (2001) Differentiation of embryonic mesenchymal cells to odontoblast-like cells by overexpression of dentin matrix protein 1. Proc Natl Acad Sci USA 98:4516–4521. https://doi.org/10.1073/pnas.081075198

Narisawa S, Hasegawa H, Watanabe K et al (1994) Stage-specific expression of alkaline phosphatase during neural development in the mouse. Dev Dyn 201:227–235. https://doi.org/10.1002/aja.1002010306

Nazir M, Al-Ansari A, Al-Khalifa K et al (2020) Global prevalence of periodontal disease and lack of its surveillance. Sci World J. https://doi.org/10.1155/2020/2146160
Neiders ME, Eick JD, Miller WA et al (1972) Electron probe microanalysis of cementum and underlying dentin in young permanent teeth. J Dent Res 51:122–130. https://doi.org/10.1177/00220345720510010501
Nudelman F, Chen HH, Goldberg H et al (2007) Spiers Memorial Lecture. Lessons from biomineralization: comparing the growth strategies of mollusc shell prismatic and nacreous layers in Atrina rigida. Faraday Discuss 136:9–25; discussion 107–123. https://doi.org/10.1039/b704418f
Nuñez J, Sanz M, Hoz-Rodríguez L et al (2010) Human cementoblasts express enamel-associated molecules in vitro and in vivo. J Periodontal Res 45:809–814. https://doi.org/10.1111/j.1600-0765.2010.01291.x
Nuñez J, Sanz-Blasco S, Vignoletti F et al (2012) Periodontal regeneration following implantation of cementum and periodontal ligament-derived cells. J Periodontal Res 47:33–44. https://doi.org/10.1111/j.1600-0765.2011.01402.x
Nwafor DC, Brown CM (2021) A novel role for tissue-nonspecific alkaline phosphatase at the blood-brain barrier during sepsis. Neural Regen Res 16:99–100. https://doi.org/10.4103/1673-5374.286958
Olson S, Arzate H, Narayanan AS et al (1991) Cell attachment activity of cementum proteins and mechanism of endotoxin inhibition. J Dent Res 70:1272–1277. https://doi.org/10.1177/00220345910700090801
Osborn J (1965) A histologic and microradiographic study of intermediate cementum. J Dent Res 44(suppl):1163
Otawara Y, Price PA (1986) Developmental appearance of matrix GLA protein during calcification in the rat. J Biol Chem 261:10828–10832
Owens PDA (1972) Light microscopic observations on the formation of the. layer of Hopewell-Smith in human teeth. Arch Oral Biol 17:1785–1788. https://doi.org/10.1016/0003-9969(72)90243-9
Patt HM, Maloney MA (1975) Bone marrow regeneration after local injury: a review. Exp Hematol 3:135–148
Paula-Silva FW, Ghosh A, Arzate H et al (2010) Calcium hydroxide promotes cementogenesis and induces cementoblastic differentiation of mesenchymal periodontal ligament cells in a CEMP1- and ERK-dependent manner. Calcif Tissue Int 87:144–157. https://doi.org/10.1007/s00223-010-9368-x
Pérez-Barbería FJ, Sl R, Hooper RJ et al (2014) The influence of habitat on body size and tooth wear in Scottish red deer (Cervus elaphus). Can J Zool 93:61–70
Pérez-Barbería FJ, Carranza J, Sánchez-Prieto C (2015) Wear Fast, Die Young: More Worn Teeth and Shorter Lives in Iberian Compared to Scottish Red Deer. PLoS One 10:e0134788
Pérez-Barbería FJ, Guinness FE, López-Quintanilla M et al (2020) What do rates of deposition of dental cementum tell us? Functional and evolutionary hypotheses in red deer. PLoS One 15(4): e0231957. https://doi.org/10.1371/journal.pone.0231957
Petersson U, Hultenby K, Wendel M (2003) Identification, distribution and expression of osteoadherin during tooth formation. Eur J Oral Sci 2003(111):128–136. https://doi.org/10.1034/j.1600-0722.2003.00027.x
Peysselon F, Xue B, Uversky VN et al (2011) Intrinsic disorder of the extracellular matrix. Mol BioSyst 7:3353–3365. https://doi.org/10.1039/c1mb05316g
Pindborg JJ (1962) Studies in the history of dental histology. I. Anders Adolf Retzius and Alexander Nasmyth: a correspondence between two great nineteenth century dental histologists. J Hist Med Allied Sci 17:388–392
Pitaru S, McCulloch CA, Narayanan AS (1994) Cellular origins and differentiation control mechanisms during periodontal development and wound healing. J Periodontal Res 29:81–94. https://doi.org/10.1111/j.1600-0765.1994.tb01095.x

Pomin VH, Wang X (2018) Glycosaminoglycan-protein interactions by nuclear magnetic resonance (NMR) spectroscopy. Molecules 23:2314. https://doi.org/10.3390/molecules23092314
Ponce CB, Evans JS (2011) Polymorph crystal selection by n16, an intrinsically disordered nacre framework protein. Cryst Growth Des 11:4690–4696. https://doi.org/10.1021/bm200231c
Popowics T, Foster BL, Swanson EC et al (2005) Defining the roots of cementum formation. Cells Tissues Organs 181:248–257. https://doi.org/10.1159/000091386
Posner AS (1969) Crystal chemistry of bone mineral. Physiol Rev 49:760–792. https://doi.org/10.1152/physrev.1969.49.4.760
Price PA (1989) Gla-containing proteins of bone. Connect Tissue Res 21:51–57.; discussion 57–60. https://doi.org/10.3109/03008208909049995
Price PA, Lothringer JW, Baukol SA et al (1981) Developmental appearance of the vitamin K-dependent protein of bone during calcification. Analysis of mineralizing tissues in human, calf, and rat. J Biol Chem 256:3781–3784
Price PA, Urist MR, Otawara Y (1983) Matrix Gla protein, a new gamma-carboxyglutamic acid-containing protein which is associated with the organic matrix of bone. Biochem Biophys Res Commun 117:765–771. https://doi.org/10.1016/0006-291x(83)91663-7
Pujari-Palmer M, Pujari-Palmer S, Lu X et al (2016) Pyrophosphate Stimulates Differentiation, Matrix Gene Expression and Alkaline Phosphatase Activity in Osteoblasts. PLoS One 11(10): e0163530. https://doi.org/10.1371/journal.pone.0163530
Qi Y, Feng W, Cai J et al (2014) Effects of conservatively treated diseased cementum with or without EMD on in vitro cementoblast differentiation and in vivo cementum-like tissue formation of human periodontal ligament cells. Cell Prolif 47:310–317. https://doi.org/10.1111/cpr.12116
Qian H, Xiao Y, Bartold PM (2004) Immunohistochemical localization and expression of fibromodulin in adult rat periodontium and inflamed human gingiva. Oral Dis 10:233–239. https://doi.org/10.1111/j.1601-0825.2004.00996.x
Qin C, D'Souza R, Feng JQ (2007) Dentin matrix protein 1 (DMP1): new and important roles for biomineralization and phosphate homeostasis. J Dent Res 86:1134–1141. https://doi.org/10.1177/154405910708601202
Raschkow I (1835) Meletemata circa dentium evolutionem. Inaug-Dissertation, Bratislaviae, Presburg
Rees SG, Wassell DT, Waddington RJ et al (2001) Interaction of bone proteoglycans and proteoglycan components with hydroxyapatite. Biochim Biophys Acta 1568:118–128. https://doi.org/10.1016/s0304-4165(01)00209-4
Retzius AA (1837) Mikroskopiska Undersökningar öfver Tändernes, Särdeles Tandbenets, Struktur. P.A. Norstedt & Söner, Stockholm
Reynolds I, Duane B (2018) Periodontal disease has an impact on patients' quality of life. Evid Based Dent 19:14–15. https://doi.org/10.1038/sj.ebd.6401287
Reynolds MA, Aichelmann-Reidy ME, Branch-Mays GL et al (2003) The efficacy of bone replacement grafts in the treatment of periodontal osseous defects. A systematic review. Ann Periodontol 8:227–265. https://doi.org/10.1902/annals.2003.8.1.227
Ripamonti U (2019) Developmental pathways of periodontal tissue regeneration: developmental diversities of tooth morphogenesis do also map capacity of periodontal tissue regeneration? J Periodontal Res 54:10–26. https://doi.org/10.1111/jre.12596
Ripamonti U, Parak R, Klar RM et al (2017) Cementogenesis and osteogenesis in periodontal tissue regeneration by recombinant human transforming growth factor-β_3: a pilot study in Papio ursinus. J Clin Periodontol 44:83–95. https://doi.org/10.1111/jcpe.12642
Ritter NM, Farach-Carson MC, Butler WT (1992) Evidence for the formation of a complex between osteopontin and osteocalcin. J Bone Miner Res 7:877–885. https://doi.org/10.1002/jbmr.5650070804
Roberts WE, Jee WS (1974) Cell kinetics of orthodontically-stimulated and non-stimulated periodontal ligament in the rat. Arch Oral Biol 19:17–21. https://doi.org/10.1016/0003-9969(74)90219-2

Roberts WE, Mozsary PG, Klingler E (1982) Nuclear size as a cell-kinetic marker for osteoblast differentiation. Am J Anat 165:373–384. https://doi.org/10.1002/aja.1001650403
Robey PG, Fedarko NS, Hefferan TE et al (1993) Structure and molecular regulation of bone matrix proteins. J Bone Miner Res 8(Suppl 2):S483–S487. https://doi.org/10.1002/jbmr.5650081310
Röckert H (1958) A quantitative X-ray microscopical study of calcium in the cementum of teeth. Acta Odontol Scand 16(suppl. 25):1–68. http://hdl.handle.net/2077/13930
Rodrigues TL, Nagatomo KJ, Foster BL et al (2011) Modulation of phosphate/pyrophosphate metabolism to regenerate the periodontium: a novel in vivo approach. J Periodontol 82:1757–1766. https://doi.org/10.1902/jop.2011.110103
Romo-Arévalo E, Arzate H, Montoya-Ayala G et al (2016) High-level expression and characterization of a glycosylated human cementum protein 1 with lectin activity. FEBS Lett 590:129–138. https://doi.org/10.1002/1873-3468.12032
Saito M, Narayana AS (1999) Signaling reactions induced in human fibroblasts during adhesion to cementum-derived attachment protein. J Bone Miner Res 14:65–72. https://doi.org/10.1359/jbmr.1999.14.1.65
Sallum EA, Ribeiro FV, Ruiz KS et al (2019) Experimental and clinical studies on regenerative periodontal therapy. Periodontol 2000 79:22–55. https://doi.org/10.1111/prd.12246
Sanz M, D'Aiuto F, Deanfield J et al (2010) European workshop in periodontal health and cardiovascular disease-scientific evidence on the association between periodontal and cardiovascular diseases: a review of the literature. Eur Heart J 12(Suppl B):B3–B12. https://doi.org/10.1093/eurheartj/suq002
Sasano Y, Maruya Y, Sato H et al (2001) Distinctive expression of extracellular matrix molecules at mRNA and protein levels during formation of cellular and acellular cementum in the rat. Histochem J 33:91–99. https://doi.org/10.1023/A:1017948230709
Sawada T, Ishikawa T, Shintani S et al (2012) Ultrastructural immunolocalization of dentin matrix protein 1 on Sharpey's fibers in monkey tooth cementum. Biotech Histochem 87:360–365. https://doi.org/10.3109/10520295.2012.671493
Saygin NE, Giannobile WV, Somerman MJ (2000) Molecular and cell biology of cementum. Periodontol 2000 24:73–98. https://doi.org/10.1034/j.1600-0757.2000.2240105.x
Schild C, Beyeler M, Lang NP et al (2009) Cementum attachment protein/protein-tyrosine phosphatase-like member A is not expressed in teeth. Int J Mol Med 23:293–296
Schmidt WJ, Keil A (1971) Cementum. In: Polarizing microscopy of dental tissues. theory, methods and results from the structural analysis of normal and diseased hard, dental tissues and tissues associated with them in man and other vertebrates. Pergamon Press, Oxford, pp 296–318
Schonfeld SE, Slavkin HC (1977) Demonstration of enamel matrix proteins on root-analogue surfaces of rabbit permanent incisor teeth. Calcif Tissue Res 24:223–229. https://doi.org/10.1007/BF02223320
Schönherr E, Hausser HJ (2000) Extracellular matrix and cytokines: a functional unit. Dev Immunol 7:89–101. https://doi.org/10.1155/2000/31748
Schroeder HE (1986) Handbook of microscopic anatomy, vol 5. The periodontium. Springer, Berlin, pp 23–119
Schroeder HE (1992) Biological problems of regenerative cementogenesis: synthesis and attachment of collagenous matrices on growing and established root surfaces. Int Rev Cytol 142:1–59. https://doi.org/10.1016/s0074-7696(08)62074-4
Schroeder HE (1993) Human cellular mixed stratified cementum: a tissue with alternating layers of acellular extrinsic- and cellular intrinsic fiber cementum. Schweiz Monatsschr Zahnmed 103:550–560
Schroeder HE, Scherle WF (1988) Cemento-enamel junction--revisited. J Periodontal Res 23:53–59. https://doi.org/10.1111/j.1600-0765.1988.tb01027.x
Scott JE (1988) Proteoglycan–fibrillar collagen interactions. Biochem J 252:313–323. https://doi.org/10.1042/bj2520313

Scott JE, Haigh M (1988) Identification of specific binding sites for keratan sulphate proteoglycans and chondroitin–dermatan sulphate proteoglycans on collagen fibrils in cornea by the use of cupromeronic blue in critical-electrolyte-concentration techniques. Biochem J 253:607–610. https://doi.org/10.1042/bj2530607

Sculean A, Windisch P, Döri F et al (2007) Emdogain in regenerative periodontal therapy. A review of the literature. Fogorv Sz 100(220–232):211–219

Sebastián-Serrano Á, Engel T, de Diego-García L et al (2016) Neurodevelopmental alterations and seizures developed by mouse model of infantile hypophosphatasia are associated with purinergic signalling deregulation. Hum Mol Genet 25:4143–4156. https://doi.org/10.1093/hmg/ddw248

Selvig KA (1963) Electron microscopy of Hertwig's epithelial sheath and of early dentin and cementum formation in the mouse incisor. Acta Odontol Scand 21:175–186. https://doi.org/10.3109/00016356308993957

Selvig KA (1965) The fine structure of human cementum. Acta Odontol Scand 23:423–441. https://doi.org/10.3109/00016356509007523

Selvig KA (1970) Periodic lattice images of hydroxyapatite crystals in human bone and dental hard tissues. Calcif Tissue Res 6:227–238. https://doi.org/10.1007/BF02196203

Selvig KA, Hals E (1977) Periodontally diseased cementum studied by correlated microradiography, electron probe analysis and electron microscopy. J Periodontal Res 12:419–429. https://doi.org/10.1111/j.1600-0765.1977.tb00137.x

Seog J, Dean D, Plaas AHK et al (2002) Direct measurement of glycosaminoglycan intermolecular interactions via high-resolution force spectroscopy. Macromolecules 35:5601–5615. https://doi.org/10.1021/ma0121621

Serrano J, Romo E, Bermúdez M et al (2013) Bone regeneration in rat cranium critical-size defects induced by cementum protein 1 (CEMP1). PLoS One 8(11):e78807. https://doi.org/10.1371/journal.pone.0078807

Serre CM, Papillard M, Chavassieux P et al (1998) Influence of magnesium substitution on a collagen-apatite biomaterial on the production of a calcifying matrix by human osteoblasts. J Biomed Mater Res 42:626–633. https://doi.org/10.1002/(sici)1097-4636(19981215)42:4<626::aid-jbm20>3.0.co;2-s

Slavkin HC, Boyde A (1974) Cementum: An epithelial secretory product? J Dent Res 53(special issue):157

Slavkin HC, Bessem C, Fincham AG et al (1989a) Human and mouse cementum proteins immunologically related to enamel proteins. Biochim Biophys Acta 991:12–18. https://doi.org/10.1016/0304-4165(89)90021-4

Slavkin HC, Bringas P Jr, Bessem C et al (1989b) Hertwig's epithelial root sheath differentiation and initial cementum and bone formation during long-term organ culture of mouse mandibular first molars using serumless, chemically-defined medium. J Periodontal Res 24:28–40. https://doi.org/10.1111/j.1600-0765.1989.tb00854.x

Sobel EH, Clark LC Jr, Fox RP et al (1953) Rickets, deficiency of alkaline phosphatase activity and premature loss of teeth in childhood. Pediatrics 11:309–322

Somerman M, Hewitt AT, Varner HH et al (1983) Identification of a bone matrix-derived chemotactic factor. Calcif Tissue Int 35:481–485. https://doi.org/10.1007/BF02405081

Somerman MJ, Shroff B, Agraves WS et al (1990) Expression of attachment proteins during cementogenesis. J Biol Buccale 18:207–214

Soni NN, Van Huysen GV, Swenson HM (1962) A microradiographic and x-ray densitometric study of cementum. J Periodontol 33:372–378. https://doi.org/10.1902/jop.1962.33.4.372

Sonoyama W, Seo BM, Yamaza T et al (2007) Human Hertwig's epithelial root sheath cells play crucial roles in cementum formation. J Dent Res 86:594–599. https://doi.org/10.1177/154405910708600703

Southerland JH, Taylor GW, Moss K et al (2006) Commonality in chronic inflammatory diseases: periodontitis, diabetes, and coronary artery disease. Periodontol 2000 40:130–143. https://doi.org/10.1111/j.1600-0757.2005.00138.x

Sowmya S, Chennazhi KP, Arzate H et al (2015) Periodontal specific differentiation of dental follicle stem cells into osteoblast, fibroblast, and cementoblast. Tissue Eng Part C Methods 21:1044–1058. https://doi.org/10.1089/ten.TEC.2014.0603

Sowmya S, Mony U, Jayachandran P et al (2017) Tri-layered nanocomposite hydrogel scaffold for the concurrent regeneration of cementum, periodontal ligament, and alveolar bone. Adv Healthc Mater 6. https://doi.org/10.1002/adhm.201601251

Spahr A, Hammarström L (1999) Response of dental follicular cells to the exposure of denuded enamel matrix in rat molars. Eur J Oral Sci 107:360–367. https://doi.org/10.1046/j.0909-8836.1999.eos107507.x

Spahr A, Lyngstadaas SP, Slaby I et al (2006) Ameloblastin expression during craniofacial bone formation in rats. Eur J Oral Sci 114:504–511. https://doi.org/10.1111/j.1600-0722.2006.00403.x

Spouge JD (1980) A new look at the rests of Malassez. A review of their embryological origin, anatomy, and possible role in periodontal health and disease. J Periodontol 51:437–444. https://doi.org/10.1902/jop.1980.51.8.437

Stein TJ, Corcoran JF (1990) Anatomy of the root apex and its histologic changes with age. Oral Surg Oral Med Oral Pathol 69:238–242. https://doi.org/10.1016/0030-4220(90)90334-o

Tamburstuen MV, Reseland JE, Spahr A et al (2011) Ameloblastin expression and putative autoregulation in mesenchymal cells suggest a role in early bone formation and repair. Bone 48:406–413. https://doi.org/10.1016/j.bone.2010.09.007

Tenorio D, Cruchley A, Hughes FJ (1993) Immunocytochemical investigation of the rat cementoblast phenotype. J Periodontal Res 28:411–419

Termine JD, Belcourt AB, Conn KM et al (1981) Mineral and collagen-binding proteins of fetal calf bone. J Biol Chem 256:10403–10408

Thomas HF (1995) Root formation. Int J Dev Biol 39:231–237

Thouverey C, Bechkoff G, Pikula S et al (2009) Inorganic pyrophosphate as a regulator of hydroxyapatite or calcium pyrophosphate dihydrate mineral deposition by matrix vesicles. Osteoarthr Cartil 17:64–72. https://doi.org/10.1016/j.joca.2008.05.020

Tonetti MS, Jepsen S, Jin L et al (2017) Impact of the global burden of periodontal diseases on health, nutrition and wellbeing of mankind: a call for global action. J Clin Periodontol 44 (456–462):2017. https://doi.org/10.1111/jcpe.12732

Toyosawa S, Okabayashi K, Komori T et al (2004) mRNA expression and protein localization of dentin matrix protein 1 during dental root formation. Bone 34:124–133. https://doi.org/10.1016/j.bone.2003.08.010

Treves-Manusevitz S, Hoz L, Rachima H et al (2013) Stem cells of the lamina propria of human oral mucosa and gingiva develop into mineralized tissues in vivo. J Clin Periodontol 40:73–81. https://doi.org/10.1111/jcpe.12016

Uwanogho DA, Hardcastle Z, Balogh P et al (1999) Molecular cloning, chromosomal mapping, and developmental expression of a novel protein tyrosine phosphatase-like gene. Genomics 62:406–416. https://doi.org/10.1006/geno.1999.5950

Valdés De Hoyos A, Hoz-Rodríguez L, Arzate H et al (2012) Isolation of protein-tyrosine phosphatase-like member-a variant from cementum. J Dent Res 91:203–209. https://doi.org/10.1177/0022034511428155

van den Bos T, Handoko G, Niehof A et al (2005) Cementum and dentin in hypophosphatasia. J Dent Res 84:1021–1025. https://doi.org/10.1177/154405910508401110

Vaquette C, Fan W, Xiao Y et al (2012) A biphasic scaffold design combined with cell sheet technology for simultaneous regeneration of alveolar bone/periodontal ligament complex. Biomaterials 33:5560–5573. https://doi.org/10.1016/j.biomaterials.2012.04.038

Veis A, Tompkins K, Alvares K et al (2000) Specific amelogenin gene splice products have signaling effects on cells in culture and in implants in vivo. J Biol Chem 275:41263–41272. https://doi.org/10.1074/jbc.M002308200

Venezia E, Goldstein M, Boyan BD et al (2004) The use of enamel matrix derivative in the treatment of periodontal defects: a literature review and meta-analysis. Crit Rev Oral Biol Med 15:382–402. https://doi.org/10.1177/154411130401500605
Vidal BC, Mello MLS, Valdrighi L (1974) Histochemical and anisotropical aspects of the rat cementum. Acta Anat 89:546–559. https://doi.org/10.1159/000144313
Villarreal-Ramírez E, Moreno A, Mas-Oliva J et al (2009) Characterization of recombinant human cementum protein 1 (hrCEMP1): primary role in biomineralization. Biochem Biophys Res Commun 384:49–54. https://doi.org/10.1016/j.bbrc.2009.04.072
Villegas-Mercado CE, Agredano-Moreno LT, Bermúdez M et al (2018) Cementum protein 1 transfection does not lead to ultrastructural changes in nucleolar organization of human gingival fibroblasts. J Periodontal Res 53:636–642. https://doi.org/10.1111/jre.12553
von Ringelmann CJ (1824) Der Organismus Des Mundes. Besonders der Zähne, deren Krankheiten und Entsetzungen: Für Jedermann, insbesondere für Ältern, Erzieher und Lehrer. Reigel und Wierner, Nürnberg
Vymetal J, Slabý I, Spahr A et al (2008) Bioinformatic analysis and molecular modelling of human ameloblastin suggest a two-domain intrinsically unstructured calcium-binding protein. Eur J Oral Sci 116:124–134. https://doi.org/10.1111/j.1600-0722.2008.00526.x
Wang HM, Nanda V, Rao LG et al (1980) Specific immunohistochemical localization of type III collagen in porcine periodontal tissues using the peroxidase-antiperoxidase method. J Histochem Cytochem 28:1215–1223. https://doi.org/10.1177/28.11.7000890
Weiner S (1984) Organization of organic matrix components in mineralized tissues. Amer Zool 24:945–951. https://doi.org/10.1093/icb/24.4.945
Weiner S (1986) Organization of extracellularly mineralized tissues: a comparative study of biological crystal growth. CRC Crit Rev Biochem 20:365–408. https://doi.org/10.3109/10409238609081998
Weiner S, Addadi L (1991) Acidic macromolecules of mineralized tissues: the controllers of crystal formation. Trends Biochem Sci 16:252–256. https://doi.org/10.1016/0968-0004(91)90098-g
Weiner S, Addadi L (1997) Design strategies in mineralized biological materials. J Mater Chem 7:689–702. https://doi.org/10.1039/A604512J
Wen L, Chen J, Duan L et al (2018) Vitamin K-dependent proteins involved in bone and cardiovascular health (Review). Mol Med Rep 18:3–15. https://doi.org/10.3892/mmr.2018.8940
Wiesmann HP, Tkotz T, Joos U et al (1997) Magnesium in newly formed dentin mineral of rat incisor. J Bone Miner Res 12:380–383. https://doi.org/10.1359/jbmr.1997.12.3.380
Wolf M, Lossdörfer S, Abuduwali N et al (2013) In vivo differentiation of human periodontal ligament cells leads to formation of dental hard tissue. J Orofac Orthop 74:494–505. https://doi.org/10.1007/s00056-013-0155-y
Wu D, Ikezawa K, Parker T et al (1996) Characterization of a collagenous cementum-derived attachment protein. J Bone Miner Res 11:686–692. https://doi.org/10.1002/jbmr.5650110517
Wu C, Zhou Y, Lin C et al (2012) Strontium-containing mesoporous bioactive glass scaffolds with improved osteogenic/cementogenic differentiation of periodontal ligament cells for periodontal tissue engineering. Acta Biomater 8:3805–8815. https://doi.org/10.1016/j.actbio.2012.06.023
Xiong J, Mrozik K, Gronthos S et al (2012) Epithelial cell rests of Malassez contain unique stem cell populations capable of undergoing epithelial-mesenchymal transition. Stem Cells Dev 21:2012–2025. https://doi.org/10.1089/scd.2011.0471
Xiong J, Gronthos S, Bartold PM (2013) Role of the epithelial cell rests of Malassez in the development, maintenance and regeneration of periodontal ligament tissues. Periodontol 2000 63:217–233. https://doi.org/10.1111/prd.12023
Yamakoshi Y, Tanabe T, Oida S et al (2001) Calcium binding of enamel proteins and their derivatives with emphasis on the calcium-binding domain of porcine sheathlin. Arch Oral Biol 46:1005–1014. https://doi.org/10.1016/s0003-9969(01)00070-x
Yamamoto T (1986) The innermost layer of cementum in rat molars: its ultrastructure, development, and calcification. Arch Histol Jpn 49:459–481. https://doi.org/10.1679/aohc.49.459

Yamamoto T, Hinrichsen KV (1993) The development of cellular cementum in rat molars, with special reference to the fiber arrangement. Anat Embryol (Berl) 188:537–549. https://doi.org/10.1007/BF00187009

Yamamoto T, Wakita M (1990) Initial attachment of principal fibers to the root dentin surface in rat molars. J Periodontal Res 25:113–119. https://doi.org/10.1111/j.1600-0765.1990.tb00901.x

Yamamoto T, Domon T, Takahashi S et al (1997) Formation of an alternate lamellar pattern in the advanced cellular cementogenesis in human teeth. Anat Embryol 196:115–121. https://doi.org/10.1007/s004290050084

Yamamoto T, Domon T, Takahashi S et al (1999) The structure and function of the cemento–dentinal junction in human teeth. J Periodont Res 34:261–268. https://doi.org/10.1111/j.1600-0765.1999.tb02252.x

Yamamoto T, Domon T, Takahashi S et al (2004) Immunolocation of proteoglycans and bone-related noncollagenous glycoproteins in developing acellular cementum of rat molars. Cell Tissue Res 317:299–312. https://doi.org/10.1007/s00441-004-0896-4

Yamamoto T, Li M, Liu Z et al (2010) Histological review of the human cellular cementum with special reference to an alternating lamellar pattern. Odontology 98:102–109. https://doi.org/10.1007/s10266-010-0134-3

Yamamoto T, Hasegawa T, Sasaki M et al (2012) Structure and formation of the twisted plywood pattern of collagen fibrils in rat lamellar bone. J Electron Microsc (Tokyo) 61:113–121. https://doi.org/10.1093/jmicro/dfs033

Yamamoto T, Hasegawa T, Yamamoto T et al (2016) Histology of human cementum: Its structure, function, and development. Jpn Dent Sci Rev 52:63–74. https://doi.org/10.1016/j.jdsr.2016.04.002

Yamashita Y, Sato M, Nocuchi T (1987) Alkaline phosphatase in the periodontal ligament of the rabbit and macaque monkey. Arch Oral Biol 32:677–678. https://doi.org/10.1016/0003-9969(87)90044-6

Ye L, MacDougall M, Zhang S et al (2004) Deletion of dentin matrix protein-1 leads to a partial failure of maturation of predentin into dentin, hypomineralization, and expanded cavities of pulp and root canal during postnatal tooth development. J Biol Chem 279:19141–19148. https://doi.org/10.1074/jbc.M400490200

Ye L, Zhang S, Ke H et al (2008) Periodontal breakdown in the Dmp1 null mouse model of hypophosphatemic rickets. J Dent Res 87:624–629. https://doi.org/10.1177/154405910808700708

Yokokoji T, Narayanan AS (2001) Role of D1 and E cyclins in cell cycle progression of human fibroblasts adhering to cementum attachment protein. J Bone Miner Res 16:1062–1067. https://doi.org/10.1359/jbmr.2001.16.6.1062

Yonemura K, Narayanan AS, Miki Y et al (1992) Isolation and partial characterization of a growth factor from human cementum. Bone Miner 18:187–198. https://doi.org/10.1016/0169-6009(92)90806-o

Yonemura K, Raines EW, Ahn NG et al (1993) Mitogenic signaling mechanisms of human cementum-derived growth factors. J Biol Chem 268(35):26120–26126

Zander HA, Hürzeler B (1958) Continuous cementum apposition. J Dent Res 37:1035–1044. https://doi.org/10.1177/00220345580370060301

Zanolli C, Hourset M, Esclassan R, Mollereau C (2017) Neanderthal and Denisova tooth protein variants in present-day humans. PLoS One 12(9):e0183802. https://doi.org/10.1371/journal.pone.0183802

Zeichner-David M (2001) Is there more to enamel matrix proteins than biomineralization? Matrix Biol 20:307–316. https://doi.org/10.1016/s0945-053x(01)00155-x

Zeichner-David M (2006) Regeneration of periodontal tissues: cementogenesis revisited. Periodontol 2000 41:196–217. https://doi.org/10.1111/j.1600-0757.2006.00162.x

Zeichner-David M, Oishi K, Su Z et al (2003) Role of Hertwig's epithelial root sheath cells in tooth root development. Dev Dyn 228:651–663. https://doi.org/10.1002/dvdy.10404

Zhang X, Han P, Jaiprakash A et al (2014) A stimulatory effect of $Ca_3ZrSi_2O_9$ bioceramics on cementogenic/osteogenic differentiation of periodontal ligament cells. J Mater Chem B 2:1415–1423. https://doi.org/10.1039/c3tb21663b

Zhou Y, Wu C, Xiao Y (2012) The stimulation of proliferation and differentiation of periodontal ligament cells by the ionic products from Ca7Si2P2O16 bioceramics. Acta Biomater 8:2307–2316. https://doi.org/10.1016/j.actbio.2012.03.012

Zoch ML, Clemens TL, Riddle RC (2016) New insights into the biology of osteocalcin. Bone 82:42–49. https://doi.org/10.1016/j.bone.2015.05.046

Zweifler LE, Patel MK, Nociti FH Jr et al (2015) Counter-regulatory phosphatases TNAP and NPP1 temporally regulate tooth root cementogenesis. Int J Oral Sci 7:27–41. https://doi.org/10.1038/ijos.2014.62

Chapter 8
Biochemistry of Non-collagenous Proteins of Bone

Jeffrey P. Gorski

Abstract Bone is a unique biological composite structure composed of an extracellular phase called osteoid in which is embedded living cells (osteoblasts and osteocytes). The cells synthesize and secrete the organic extracellular matrix of bone. They do so in a manner that reflects and anticipates their biomechanical force environment and provides a viable home (lacunae and associated dendrites) which facilitates nutrition and communication with other embedded osteocytes. Osteoid becomes mineralized, e.g., a process in which crystalline calcium hydroxyapatite crystals are formed and deposited along and among collagen fibrils. In addition to osteoblasts, bone also contains nerves that provide sensation and vascular channels which provide nutrition, oxygen, waste removal, and a source of calcium and phosphorus. As discussed elsewhere in this text, bones serve several key systemic functions: a dynamic reservoir for calcium and phosphorus, a rigid framework that allows for erect stance as well as an attachment of muscles and tendons facilitating movement, and protection for vital internal organs such as the heart, brain, and internal organs. Non-collagenous proteins of bone, as described herein, greatly increase the functionality of bone and directly/indirectly mediate many of its unusual properties.

8.1 Introduction

Osteoid is a highly collagenous matrix composed predominantly of type I collagen fibrils along with minor collagens and a series of non-collagenous proteins and proteoglycans. This chapter describes current information on the biochemistry of

J. P. Gorski (✉)
Department of Oral and Craniofacial Sciences, School of Dentistry, University of Missouri-Kansas City, Kansas City, MO, USA

Center of Excellence in the Study of Dental and Musculoskeletal Tissues, University of Missouri-Kansas City, Kansas City, MO, USA
e-mail: gorskij@umkc.edu

M. Goldberg, P. Den Besten (eds.), *Extracellular Matrix Biomineralization of Dental Tissue Structures*, Biology of Extracellular Matrix 10,
https://doi.org/10.1007/978-3-030-76283-4_8

the most prominent non-collagenous proteins in bone and attempts to provide an understanding of how each functions separately or as a complex. We also provide specific insights into how the methodology can influence the way we think about bone structure and bone formation. The story of the structure and function of non-collagenous proteins in bone is one that historically has been severely restricted by the physical properties of the tissue and secondarily by applicable methodology. As is well appreciated, biochemical approaches have generally lagged behind genetic approaches such as GWAS, genome sequencing, exome sequencing, RNA sequencing. Until the recent application of mass spectrometry, the primary tool for characterization of non-collagenous proteins in bone has been SDS PAGE and Western blotting. While many different sizes can be readily distinguished by electrophoresis, the basis for these differences has been difficult to determine. Specifically, it has been difficult to distinguish between N- and O-linked glycosylation, phosphorylation, sulfation, alternative splicing, or proteolytic fragmentation as the source(s) of molecular weight disparities. This is because of the semiquantitative nature of Western blotting, the 10% variability inherent in electrophoretic methods, and the biological variation existing among different skeletal sites and among different individuals. Our discussion now has the benefit of the precise molecular weight resolution of mass spectrometry and its capacity to identify changes in phosphorylation. We have sought to highlight how new biochemical and immunochemical approaches are gradually providing molecular explanations for the skeletal phenotypes arising from mutations and knockouts of non-collagenous proteins in bone. Finally, we conclude with a section describing how biochemical structure/function information has been applied to successfully treat hyperphosphatemia in children with impressive results.

8.2 Function of Non-collagenous Proteins in Bone

The inherent solid nature of bone not only limits our ability to isolate its composite proteins in a native conformation, it also restricts our ability to investigate the function of bone-derived proteins within a native environment. In general, the function of proteins has been most directly addressed via their enzymatic or receptor activating functions. However, only a few non-collagenous proteins in bone express enzymatic functions: Phospho1, alkaline phosphatase, FAM20C, and PHEX.

8.2.1 Non-collagenous Proteins with Catalytic Functions

Alkaline Phosphatase and Phospho1 Play a Key Role in Mineralization Due to its ease of histological staining and relationship with serum isoforms, tissue nonspecific alkaline phosphatase has long been a marker of osteoblastic cells and, based on its phosphatase nature, assumed to play a role in bone mineralization. This role was

Table 8.1 Human genetic mutations involving bone non-collagenous proteins

Gene name	Disease/syndrome name	Chromosome	Type of inheritance
FAM20A	Amelogenesis imperfecta type IG	17	Autosomal recessive
FAM20C	Raine	7	Autosomal recessive
Col11A1	Stickler, type II; Marshall; Fibrochondrogenesis I	1	Autosomal recessive
PHEX	Hypophosphatemic rickets. XLD	X	X-linked
MGP	Keutel	12	Autosomal recessive
DMP1	Hypophosphatemic rickets	4	Autosomal recessive
Decorin	Cornea dystrophy	12	Autosomal dominant
Biglycan	Meester-Loeys; Spondyloepimetaphyseal dysplasia, X-linked	X	X-linked
Alkaline phosphatase	Hypophosphatemia	1	Autosomal recessive; autosomal dominant
Sclerostin	Craniodiaphyseal dysplasia, autosomal dominant; Sclerosteosis 1; Van Buchem	17	Autosomal recessive; autosomal dominant
MEPE	Hereditary congenital facial paresis family and otosclerosis	4	
ENPP1	Diabetes mellitus non-insulin-dependent; obesity; arterial calcification; Cole disease; hypophosphatemic rickets	6	Autosomal recessive
DMP1 and OPN (digenic)	Hypophosphatemic osteosclerosis, hyperostosis, and enthesopathy	4	

reinforced by an association of mutations in the human *TNAP* gene with skeletal disorders including hypophosphatasia (Nielson et al. 2012; Ozono et al. 2012; Ermakov et al. 2010) (Table 8.1). Also, the *TNAP* null mouse exhibited a phenotype similar to that for infantile hypophosphatasia (Narisawa et al. 1997; Fedde et al. 1999; Anderson et al. 2004) (Table 8.2). Interestingly, although the *TNAP* mouse skeleton was hypo-mineralized, hydroxyapatite crystals were still produced normally (Anderson et al. 2004). Under these conditions, it was apparent that phosphate nucleation was still ongoing and an alternate source of inorganic phosphate was needed.

Phospho1 was known to be a phosphatase enriched in osteoblasts which had a neutral pH optimum and a capacity to cleave phosphoethanolamine (PE) and phosphocholine (PC) generating inorganic phosphate(Roberts et al. 2005). Phospho1 was selected as a potential candidate based on its restricted expression to developing bones and its association with a loss of mineralization in chicks treated with specific inhibitor lansoprazole (Macrae et al. 2010). Phospho1 is now

Table 8.2 Mouse knockouts of non-collagenous proteins yielding skeletal phenotypes

Gene name	Summary of skeletal phenotype	Reference
Irisin	Using an Osx-Cre to ablate in osteoblast cells, *FNDC5/irisin* deficient mice showed a lower bone density and significantly delayed bone development and mineralization from early-stage to adulthood. Our phenotypical analysis exhibited decreased osteoblast-related gene expression and increased osteoclast-related gene expression in bone tissues, and reduced adipose tissue browning due to bone-born irisin deletion. In addition, positive effects of exercise, including bone strength enhancement and body weight loss were remarkably weakened due to irisin deficiency. Interestingly, these changes can be rescued by supplemental administration of recombinant irisin during exercise	Zhu et al. (2021)
Bone sialoprotein	Delayed and reduced cementum deposition and mineralization ; increased root and mandibular resorption after P30; increased osteoclast number in P60 bone; short femur; increased trabecular bone volume; decreased bone mineral density and compact bone thickness at 4 mo of age which resolves by 12 mo of age; delayed intramembranous bone mineralization	Malaval et al. (2008); Foster et al. (2015)
Col11A1 (*cho/cho*)	Much wider long bones, decreased long bone length, trabeculae are much wider in diameter; most bone formation occurs at the periosteum (not endochondral); decreased bone mineralization in the cartilage of sternum and phalanges	Fernandes et al. (2007); Li et al. (1995)
Osteopontin	Despite the absence of an obvious phenotype in *OPN* deficient mice being undetectable by radiographic and histological methods, FTIRM analyses revealed that eh relative amount of mineral in the more mature areas of the bone (central cortical bone) of *OPN* knockout mice was significantly increased. Consistent with a role as an inhibitor of mineral formation and crystal growth/proliferation	Boskey et al. (2002)
Osteonectin	Decreased osteoblast and osteoclast number; decreased bone mineral density; reduced trabecular bone volume; abnormal bone strength with age	Delany et al. (2000)
Periostin	Severe growth retardation; smaller skull and shorter long bones; abnormal tooth morphology; cancellous bone defects grossly rescued by the use of soft powdered diet	Rios et al. (2005)
ENPP1	Fused joints after 4 weeks of age; aortic media display calcification and chondrogenic differentiation	Li et al. (2013)
Phospho1	*Phospho1*(–/–) mice display growth plate abnormalities, spontaneous fractures, bowed long bones, osteomalacia, and scoliosis in early life. Primary cultures of *Phospho1* (–/–) tibial growth plate chondrocytes and chondrocyte-derived matrix vesicles (MVs) show reduced mineralizing ability, and plasma samples from *Phospho1* (–/–) mice show reduced levels of TNAP and elevated plasma	Yadav et al. (2011)

(continued)

Table 8.2 (continued)

Gene name	Summary of skeletal phenotype	Reference
	PP(i) concentrations. However, transgenic overexpression of TNAP does not correct the bone phenotype in *Phospho1*(–/–) mice despite normalization of their plasma PP(i) levels. In contrast, double ablation of *PHOSPHO1* and *TNAP* function leads to the complete absence of skeletal mineralization and perinatal lethality	
Osteocalcin	Mice develop marked by higher bone mass and bones of improved functional quality. After OVX, mice without *BGP* show an increase in bone formation without impairing bone resorption. (A single mutational event was used to target both *Bglap1* and *Bgla2*)	Ducy et al. (1996)
MGP	Deficiency results in extensive mineralization of all arteries and chondrocytes. Recue data support the hypothesis that the arterial calcification, not *MGP* deficiency itself, causes the low bone mass phenotype in *MGP* (–/–) mice	Marulanda et al. (2013); Luo et al. (1997)

appreciated to play a central role in the biomineralization of bone and other hard tissues. At sites of new bone formation, Phospho1 is localized within matrix vesicles where it is believed to cleave its substrates and generate inorganic phosphate (Roberts et al. 2007). Millan and Farquharson and colleagues have hypothesized an elegant phosphate mineralization pathway in which TNAP, ENPP1, and Phospho1 all participate together to produce inorganic phosphate to feed the formation of calcium hydroxyapatite crystals within matrix vesicles (Dillon et al. 2019) (see Fig. 8.1 for a summary of mineralization mechanism). Specifically, they proposed that the protected environment and bilayer membrane of the matrix vesicles represents a barrier to diffusion where continued action on phosphocholine and phosphoethanolamine by Phospho1 raises the concentration of inorganic phosphate above the threshold for calcium hydroxyapatite crystal nucleation. The membranes of matrix vesicles are known to be enriched in PC and PE (Wuthier 1975). Within this model, TNAP's role is envisioned as catalyzing the hydrolysis of pyrophosphate and regulating the concentration of this mineralization inhibitor (Fig. 8.1). Ectonucleotide pyrophosphatase/phosphodiesterase1 (ENPP1) rounds out this mechanistic pathway as a source of pyrophosphate derived from both its inherent 5′-nucleotide phosphodiesterase I and nucleotide pyrophosphohydrolyase catalytic activities. Additional support for a role for ENPP1 in systemic calcification reactions derives from the skeletal phenotype associated with its genetic ablation in mice (Table 8.2). In humans, mutations in the *ENPP1* gene can cause arterial calcification or AR hypophosphatemic rickets 2. The latter phenotype was dependent at least in part on an elevation in plasma FGF23 concentration (Oheim et al. 2020). Elevated FGF23 levels are also associated with *PHEX* and *DMP1* mutations (Ichikawa et al. 2017; Murali et al. 2016). FGF23 is the active shared principle mediating the changes in phosphate metabolism in all these latter cases.

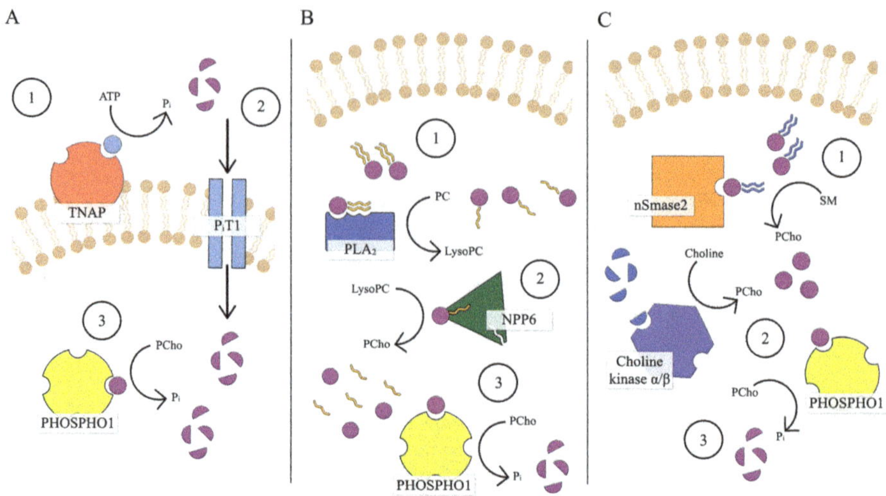

Fig. 8.1 Schematic diagram illustrating the hypothesized mechanism of PHOSPHO1 function within MVs. (**a**) PHOSPHO1 functions synergistically with TNAP: (1) TNAP hydrolyses its substrates to produce P_i extravesicularly; (2) extravesicular P_i is transported into the MV via P_iT1; (3) PHOSPHO1 hydrolyses Pcho intravesicularly to further accumulate P_i. (**b**) Generation of PHOSPHO1 substrates within MVs: (1) an unidentified PLA_2 converts PC from the vesicle membrane to lysoPC; (2) NPP6 subsequently catalyzes the hydrolysis of lysoPC to generate Pcho; (3) PHOSPHO1 liberates P_i from PCho. (**c**) Alternative pathways of PCho generation: (1) nSmase2 breaks down SM from the MV membrane to form Pcho; (2) the α/β choline kinases phosphorylate choline to form PCho; (3) PHOSPHO1 generates P_i from PCho (taken from reference Dillon et al. 2019)

PHEX, MEPE, and FAMC20: A Role in the Regulation of Mineralization Genetic mutations in at least eight genes have linked their function directly or indirectly to the systemic regulation of phosphate concentration in blood (Rowe 1997; Strom and Juppner 2008). One of these is *PHEX*. Mutations in phosphate-regulating endopeptidase homolog, X-linked (*PHEX, PEX*) are the cause of hypophosphatemic rickets, X-linked dominant (Table 8.1). A *Pex* mutation is also the cause for the “*hyp*” mouse skeletal phenotype (hypophosphatasia and elevated FG23 expression (Ruchon et al. 2000). In support, all three transgenic attempts to replace mutant *PHEX* in the *Hyp* mouse were able to bring about at least partial rescue of bone mineralization. Taken together, these data indicate that PHEX has a direct effect on bone mineralization as well as an indirect effect on systemic phosphate hemostasis.

McKee and co-workers have shown that PHEX can cleave and inactivate the mineralization inhibitory activity of some ASARM peptides derived from MEPE (Addison et al. 2008). However, the most phosphorylated forms which are predicted to be the most effective minhibins were not effectively cleaved by PHEX. Minhibins, mineral crystal growth inhibitors, refer to ASARM peptides representing conserved acidic protease-resistant C-terminal regions of SIBLING proteins DMP1 and MEPE (Martin et al. 2008). Interestingly, the release of minhibins is greatly increased in the *HYP* (*PHEX* deficient) mouse skeleton and ASARM peptides are directly responsible for the defective mineralization phenotype (Addison et al. 2008; Martin et al. 2008). Neutral protease cathepsin B seems to cause the accelerated release of ASARM peptides in *HYP* mice since cathepsin inhibitors block this release and rescue the defect (Rowe et al. 2006). ASARM peptides are potent inhibitors of mineralization (Addison et al. 2008; Boskey et al. 2010) and administration in vitro induces increased expression of FGF23 (Liu et al. 2007).

Finally, phosphorylation of the ASARM region of MEPE is catalyzed by FAM20C kinase (Christensen et al. 2020). In all, 31 serine residues in MEPE could be phosphorylated by FAM20C kinase including the nine serine residues located in the C-terminal ASARM region. FAM20C (Table 8.1) is the cause of Raine syndrome an autosomal recessive lethal osteosclerotic bone dysplasia disorder characterized by generalized osteosclerosis with periosteal bone formation, a distinctive facial phenotype, and systemic hypophosphatemia. While not formally proven, it seems likely that the absence of phosphorylated residues on ASARM peptides directly contributes to their capacity to inhibit mineral crystal growth (minihibins) and lead to calcification. Elevated FGF-23 observed in Raine's patients likely contributes to hypophosphatemia (Mameli et al. 2020).

In addition to its C-terminal ASARM peptide region, MEPE also expresses an RGD cell adhesion receptor-binding motif. Recent exome sequencing of rare variant sequences implicated in two distinct bone disorders, hereditary congenital facial paresis (HCFP) and otosclerosis indicate that the RGD and ASARM motifs may contribute separately to these bone phenotypes (Schrauwen et al. 2019) (Table 8.1). Specifically, all variants associated with otosclerosis are predicted to result in nonsense-mediated decay or an ASARM- and RGD-truncated MEPE isoform. The HCFP mutation, in contrast, is predicted to produce an ASARM-truncated MEPE with an intact RGD motif (Table 8.1). Since this difference in effect on the protein correlates with the presumed phenotype of both diseases, Schrauwen et al. suggest the above-noted rationale providing a molecular explanation for the observed distinct phenotypes (Schrauwen et al. 2019).

8.2.2 Non-catalytic Non-collagenous Proteins of Bone

Since the majority of secreted non-collagenous proteins of bone do not display enzymatic activity, and thus cannot self-report, it has been necessary to use other means to determine their functions. The most direct and obvious way to assess function indirectly has been to use genetics to determine whether mutational inactivation or knockout of individual genes causes reproducible phenotypic changes in the skeleton (Tables 8.1 and 8.2).

As a result, it is now evident that a number of non-collagenous proteins are required for normal skeletal function. For example, in addition to proteins described above, type XI collagen, matrix GLA protein, dentin matrix protein1, bone sialoprotein, biglycan, decorin, periostin, irisin, osteonectin, MEPE, sclerostin, osteocalcin, and osteopontin have all been associated with altered mineral content, bone volume, or altered bone structure. However, while these findings demonstrate a functional requirement, the specific mechanism cannot always be rationalized from the phenotype alone. The skeletal phenotype can provide important clues as to the general area affected, e.g., Stickler syndrome (altered growth plate), ENNP1 (abnormal phosphate handling), Sclerostin deficiency (inhibition of bone formation), and matrix GLA protein deficiency (inhibition of mineralization). In addition, knockout mice deficient in either bone sialoprotein, osteonectin, or periostin all exhibit abnormal bone formation (Table 8.2). However, other experimental approaches are necessary to better understand the crucial question of how these proteins perform their required functions. Ultimately, this question of functional mechanism can be distilled down to: who are the binding partners of the protein in question. In this regard, a number of approaches can be used to determine binding partners including immunoprecipitation, co-purification during chromatography, peptide blotting, and co-localization by immunofluorescence microscopy. To illustrate this point, we use the example of type XI collagen and bone sialoprotein. The A1 chain of Col11 is subject to alternative splicing wherein the N-terminal domain (NTD), composed of potentially nine different exons, is expressed as seven major isoforms which display temporal and tissue specificity (Davies et al. 1998; Morris et al. 2000; Warner et al. 2007; Oxford et al. 1995). As shown in Fig. 8.2, immunoprecipitation of bone sialoprotein from osteoblastic cell extracts yields the expected 90 kDa BSP polypeptide which characteristically stains blue with cationic Stains All dye (Gorski

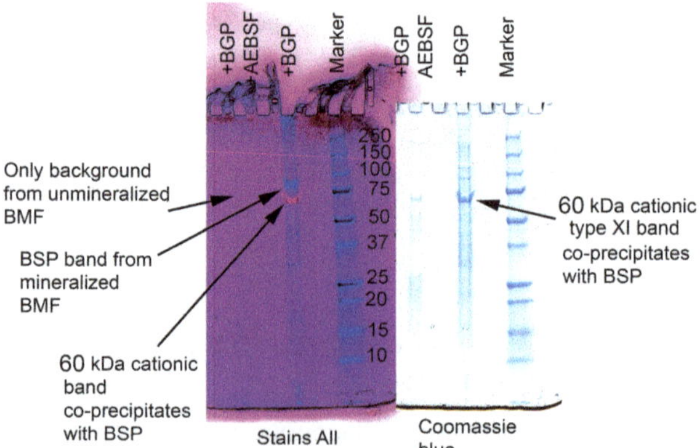

Fig. 8.2 Bone sialoprotein is immunoprecipitated together with 60 kDa cationic type XI N-terminal fragment only from mineralized osteoblastic cultures. Extracts of mineralized and unmineralized osteoblastic cultures were immunoprecipitated with anti-bone sialoprotein antibodies. Immunoprecipitates were subjected to SDS-PAGE and the same gel was stained sequentially with Stains All (left) and then with Coomassie blue dye (right). *Key*: +BGP +AEBSF, extract from a culture where mineralization was inhibited by 4-(2-aminoethyl)-benzenesulfonyl fluoride hydrochloride; +BGP, extract from mineralized culture; marker, molecular weight standards [reprinted from (Gorski 2011)]

Table 8.3 Entire type XI collagen A1 chain NTD 6b alternative splice sequence and two partial peptide sequences used experimentally

KKKSNYT**KKK**RTLATNS**KKK**SKMSTTPKSEKFAS**KKKK**RNQATAKAKLGVQ
KKKSNYT**KKK**RTLATNS**KKK**S**K**M (PEPTIDE 3)
KKKS**K**MSTTP**K**SE**K**FAS**KKKK**R (PEPTIDE 5)

2011). Unexpectedly, a 60 kDa band also co-immunoprecipitates with BSP under mineralizing conditions (+BGP) but not when mineralization is inhibited by AEBSF (Fig. 8.2) (Huffman et al. 2007; Gorski et al. 2011). Importantly, other work shows that this 60 kDa band also reacts with monospecific antibodies against the "6b" exon of the NTD domain of Col11a1 (Gorski 2011; Gorski et al. 2021) (Table 8.3). Also, we have shown that transcription of Col11A1 in osteoblastic cells is blocked by AEBSF, an inhibitor of proprotein convertase SKI-1 (Gorski et al. 2011). As further proof of the presence of the "6b" exon sequence, the 60 kDa band displays a cationic charge character reflected in its capacity to stain red with Stains All and to stain blue with negatively charged Coomassie blue dye (which does not effectively stain BSP) (Fig. 8.2).

Based on the above evidence which focused our attention on the "6b" exon sequence, we immediately identified the cationic character in general and the presence of four lysine triplets within the "6b" exonal sequence as potential binding targets for bone sialoprotein (Gorski et al. 2021). Demonstration of a specific ligand binding interaction with bone sialoprotein was achieved using a modification of the western blotting technique in which proteins are blotted onto PVDF membranes and then probed with fluorescently tagged small peptides representing parts of the whole "6b" sequence (Table 8.3) (Gorski et al. 2021). Specifically, proteins were extracted from UMR106-01 osteoblastic cultures under different mineralizing conditions (plus and minus BGP to induce mineralization and with and without AEBSF to block mineralization) and then blotted with overlapping peptides representing the Col11A1 NTD "6b" sequence (Fig. 8.3). Interestingly, under mineralizing conditions, peptide 3, containing three lysine triplet motif sequences, bound strongly to only two bands at 110 and 90 kDa (arrows, Fig. 8.3) (Gorski et al. 2021). LC-MS/MS on peptide digests of gel slices subsequently established that these bands were nucleolin and bone sialoprotein (data not shown), respectively. When mineralization was prevented, peptide 3 bound strongly to bands at 35 and 18 kDa (Gorski et al. 2021) which other work has shown are fragments of bone sialoprotein and nucleolin (Fig. 8.3) (Huffman et al. 2007).

At the same time, binding with peptide 5, which only contains two lysine triplet sequences (Table 8.3), did not detect the 35 and 18 kDa bands (arrows, Fig. 8.3) (Gorski et al. 2021). These data demonstrate that the 6b exonal sequence, explicitly peptide 3, of the N-terminal domain of type XI collagen binds robustly to bone sialoprotein. In view of the apparent binding preference of bone sialoprotein for peptide 3 versus peptide 5, we believe that the bone sialoprotein binding site involves lysine triplet sequences since peptide 3 contains three such motifs versus

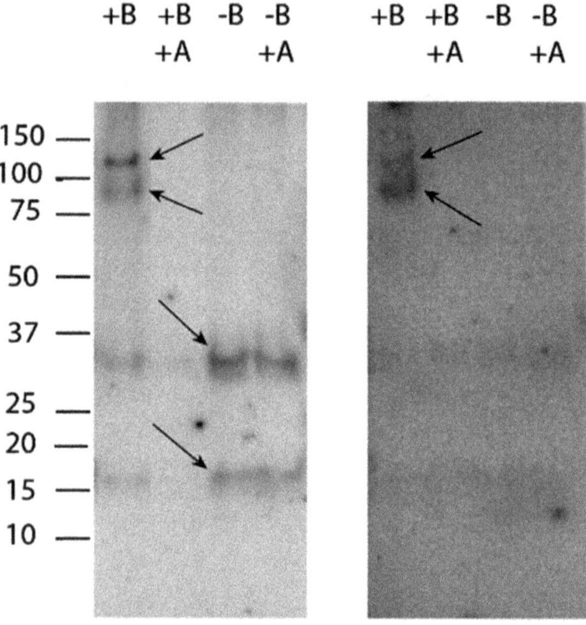

Fig. 8.3 Blotting of extracts of osteoblastic cultures treated with different conditions with FAM-labeled peptides 3 and 5 derived from Col11A1 NTD "6b" exon identifies bone sialoprotein and nucleolin as binding ligands (taken from reference Gorski et al. 2021). *Key*: +B, plus β-glycerolphosphate; +B + A, plus β-glycerolphosphate plus AEBSF; –B, minus β-glycerolphosphate; –B – A, minus β-glycerolphosphate minus AEBSF. See Huffman et al. (2007) for more details

2 for peptide 5 (Table 8.3) (Gorski et al. 2021). Both bone sialoprotein and Col11A1 are secreted proteins found in the osteoid matrix and, in contrast to other short-lived N-terminal collagen propeptides, the NTD of Col11a1 resists degradation and is long-lived in vivo (Rousseau et al. 1996; Moradi-Ameli et al. 1998). As a result, we propose that type XI collagen and bone sialoprotein interact with each other in real bone (Gorski et al. 2021).

To further establish this point and illustrate another method that can be used to demonstrate protein–protein interactions, we carried out double labeling of new bone tissue with a monospecific antibody against bone sialoprotein (LF-83) and with N-terminally biotin-labeled peptide 3 or peptide 5 (see Table 8.3 for sequence). New bone was produced using a bilateral tibial marrow ablation rat model where 4 days after surgery the intramedullary cavity of the tibia is largely filled with a fibrin clot which begins to be reorganized into intramembranous bone. As shown in Fig. 8.4, anti-bone sialoprotein antibodies (green) label spherical "mineralization centers" where pre-osteoblastic and osteoblastic cells produce an osteoid matrix which is deposited in thin concentric layers (yellow arrows). Importantly, peptide 3 binding co-localizes to the same concentric layers of osteoid which are identified by anti-bone sialoprotein (Fig. 8.4) (Gorski et al. 2021). However, it appears also that peptide 3 may preferentially bind to intracellular BSP protein in discrete regions (Fig. 8.4, white arrows) (Gorski et al. 2021). In contrast, controls using N-biotinylated peptide 1, derived from an alternative 6a exonal sequence without

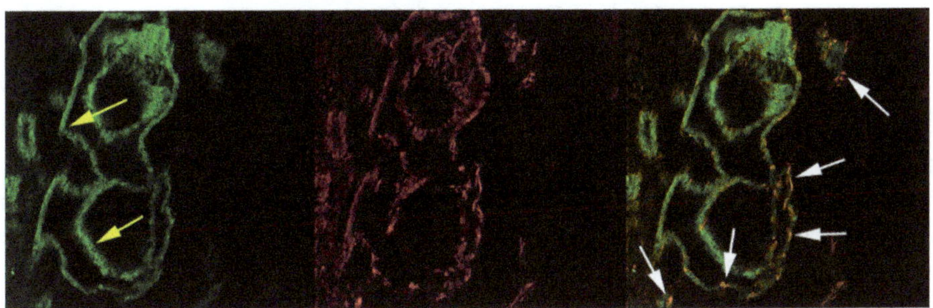

Fig. 8.4 Double immunofluorescence staining of newly forming new bone with anti-bone sialoprotein antibody and N-biotinylated peptide 3. Thin sections of fixed and decalcified paraffin-embedded marrow ablation tissue at day 4 (Magnuson et al. 1997) was incubated with primary antibody and N-biotinylated peptide 3 and then stained with Alexa 488 conjugated secondary antibody (green) and with Rhodamine-conjugated Streptavidin (red). *Key: right panel*, anti-bone sialoprotein antibody; *middle panel*, N-biotinylated peptide 3; and *left panel*, both anti-bone sialoprotein antibody and N-biotinylated peptide 3. See also Gorski et al. (2021)

lysine triplet sequences, yielded only background levels of binding (not shown). As a result, we conclude that the Col11A1 NTD binds bone sialoprotein in vitro (protein blots) as well as in vivo (in forming new bone tissue) (Gorski et al. 2021).

8.2.3 Other Binding Sites Through Which Non-collagenous Proteins Mediate Their Functions

Matrix Gla Protein (MGP) MGP contains up to five potential posttranslational structural modifications which serve as functional motifs. Specifically, these vitamin K-dependent modifications mediate in part its role as an inhibitor of vascular calcification (Nigwekar et al. 2017). In addition, a second modification, serine phosphorylation, is also required to effectively inhibit calcification (Schurgers et al. 2007). When genetically ablated in mice, null mice display denser bones and excessive cartilage mineralization and expire within 2 months of age due to severe vascular and esophageal calcification (Luo et al. 1997). Importantly, ongoing results extend these findings to improve outcomes in kidney transplant and chronic kidney disease patients (Roumeliotis et al. 2020). Briefly, treatment with vitamin K has been found to prolong life and to prevent vascular calcification in these patients which is a significant cause of death (Haroon et al. 2020; Lees et al. 2020). Histomorphometric analysis of femoral and vertebral sites showed significantly decreased bone volume and increased trabecular separation in rats treated with warfarin, a vitamin K antagonist (Fusaro et al. 2015).

While the physiological importance of MGP/BMP-2 complexes in vivo is not yet fully appreciated, it is clear that BMP-2 and BMP-4 display a specific binding for the gamma-carboxylated region(s) of MGP (Sweatt et al. 2003; Kuronuma et al. 2020).

Specifically, BMP-2/MGP complexes were clearly demonstrated within the mucosa of the nasal septum of aged rats using double-label immunofluorescence microscopy (Sweatt et al. 2003). Current functional hypotheses involve a role for MGP by indirectly regulating the amount of biologically active BMP in vivo via competing for its binding to cellular receptors (Zebboudj et al. 2003).

Bone Sialoprotein and Osteopontin Mediated Adhesion of Osteoclasts to Bone Initial reports characterizing the bone phenotype of BSP null mice described a higher bone mass than control littermates along with very low bone formation activity and decreased osteoclast number and OC surface area (Malaval et al. 2008). When osteoclasts were generated using spleen-derived precursors from BSP null mice and control littermates, this resulted in twofold fewer TRAP+ cells and fourfold fewer multinucleated osteoclasts. Interestingly, the addition of RGD-containing proteins (bone sialoprotein or osteopontin) to these cultures restored the osteoclast number to control levels. Also, in vitro resorption pit assays of BSP-deficient osteoclasts on dentine slices were similar to that for controls, whereas resorption of mineral coated glass slides (presumably missing natural adhesive proteins) by BSP null osteoclasts was dramatically reduced. However, resorption of mineral coated slides could be rescued by first coating the mineral surface with recombinant bone sialoprotein (Boudiffa et al. 2010). In addition, recombinant osteopontin was partially effective in restoring resorptive activity with osteoclasts from BSP-deficient mice. These in vitro results, which fully support in vivo findings, strongly suggest that the RGD cell adhesion site of bone sialoprotein plays a direct role in both osteoclast formation and resorptive activity.

In related studies, comparisons of bone healing in OPN- and BSP-deficient mice revealed no reductions in the rate of healing or in mineralization due to osteopontin (Monfoulet et al. 2010). However, BSP-deficient mice did display a modest but significant reduction in healing rate as well as impaired mineralization when measured at 14 and 21 days after surgical insult (Monfoulet et al. 2010). These results indicate that BSP also affects the mineralization of primary bone.

Irisin and Sclerostin FNDC5 (fibronectin type III domain-containing protein 5) is a transmembrane glycoprotein that can be cleaved to yield irisin (Nie et al. 2020). Irisin is a dimer (Schumacher et al. 2013) that acts as a hormone-like myokine produced in abundance by skeletal muscle in response to exercise, both in mice and humans. Serum levels of irisin are dependent upon age, sex, and body weight. Interestingly, levels are raised in response to acute bouts of exercise in adults and children (Loffler et al. 2015). Once released into the circulation, irisin acts on brain, fat, and bone tissues. Irisin effects on white adipocytes include the browning response and subsequently activates non-shivering thermogenesis (Colaianni et al. 2017; Cao et al. 2019). Recently Kim et al. (2019) clarified how irisin mediates its effects on cells (Kim et al. 2019). Specifically, irisin binds to $alpha_V$ class of integrin cell surface receptors since its effects on fat cells and osteocytes can be blocked by chemical inhibition. Functionally, irisin increases both osteocytic survival and production of sclerostin, an inhibitor of the canonical Wnt signaling bone formation pathway. Finally, knockout of *FNDC5* or *irisin* in mice completely blocks osteocytic osteolysis caused by ovariectomy and stimulates bone formation in this model via reduced sclerostin secretion. Thus, irisin's overall effects on bone are to block resorption and inhibit bone formation via sclerostin.

Mechanistically, sclerostin's effects on osteoblastic bone formation appear to be mediated by at least two different pathways. In one, sclerostin blocks the stimulatory actions of BMP-2, BMP-4, BMP-5, BMP-6, and BMP-7 by forming an inhibitory complex with each which competitively blocks the ability of BMPs to bind to type I and type II BMP receptors on osteoblasts or the DAN inhibitor (Winkler et al. 2003). In a second pathway, sclerostin blocks Wnt signaling through the LRP receptors on osteoblasts (Semenov et al. 2005; Leupin et al. 2011). Direct interaction of sclerostin with LRP4 at the level of the membrane was required to mediate this inhibitory effect (Leupin et al. 2011).

Sibling Proteins Modify the Function of MMPs Three members of the SIBLING family of integrin-binding phosphoglycoproteins (bone sialoprotein, BSP; osteopontin, OPN; and dentin matrix protein-1, DMP1) were recently shown to bind with high affinity (nM) and to activate 3 different matrix metalloproteinases (MMP-2, MMP-3, and MMP-9, respectively) in vitro (Fedarko et al. 2004; Ogbureke and Fisher 2004). At this point, dentin sialoprotein and MEPE, the other two SIBLINGs, have not yet been shown to form complexes with any MMPs. At this juncture, the physiological impact of SIBLING-MMP complexes is not fully known. However, Ogbureke and colleagues have hypothesized based on indirect evidence that these complexes may have their greatest functional role in cancer (Nikitakis et al. 2018; Ogbureke et al. 2007).

8.2.4 Methodology Affects the Way We Think About Bone Formation

Mineral-to-matrix ratio. Boskey and colleagues first applied the FTIR-based mineral-to-matrix ratio parameter in 1992 in studies of the mineralization of chick limb bud mesenchymal cell cultures (Boskey et al. 1992). Later applications of this method were used successfully to examine a range of mineralizing systems including the turkey tendon model (Gadaleta et al. 1996) and the *oim* mouse (Camacho et al. 1996). However, in all cases, the method was always meant to be used as a relative measure of mineralization not an absolute measure of mineralization, e.g., since context is important and the ratio can be strongly influenced by sample preparation. This point is emphasized by the work cited above. Specifically, the mineral-to-matrix ratio for *oim/oim* cortical bone was higher than that for either *oim*/+ or +/+ control mice. Rather than ascribe the result to the increased mineral content of the *oim/oim* bones, the authors interpreted their findings in terms of reduced collagen content. Similarly, Kimura-Suda et al. (2018) found that mineral-to-matrix ratio measurements on plastic embedded bone samples were altered by dehydration. Specifically, the M/M for resin embedded bone sections was higher than that for frozen sections taken from the same bone, whereas the carbonate-to-phosphate ratio was lower in resin embedded sections than in frozen sections. The point is that normalization of a relative mineral parameter, e.g., a hydroxyapatite-derived FTIR phosphate signal, with a spectroscopic amide I signal yields a relative value that correlates with the content of calcium hydroxyapatite. However, this value is not a quantitively accurate representation of mineral content and as such, it cannot be used to directly compare with other published values like a BMD or quantitative calcium content value which were determined using standardized methods. While the mineral-to-matrix ratio has continued to be used in mineralization research, it is apparent that recent applications of the method have begun to broaden the meaning of the FTIR-derived mineral-to-matrix ratio to reflect "mineral content" when in fact the data remain relative measures of mineral content and useful only for internal correlations. For example, Faibish et al. (2005) analyzed the mineral-to-matrix ratio of 11 osteomalacic women's bones and then compared it to that for seven control bones. A significant difference was found in the mineral content of the trabecular regions with a lower mean value in osteomalacia bones ($p = 0.01$) than in controls. Support for this conclusion was provided by a separate linear plot of ash weight versus mineral-to-matrix ratio for a large number of control bone samples.

Several criticisms can be raised regarding this latter approach. First, strictly speaking, we do not know whether the collagen content of human osteomalacic bones is similar to that for control bones. The validity of this latter assumption will impact directly the mineral-to-matrix ratio. Second, it is not possible to compare the results from this study directly to another analysis of human osteomalacic bones since the values derived are relative. We only know that the M/M value is less than a small control group of seven individuals. If the control group is not representative, this conclusion may not be valid. In contrast, if the authors had used methods that were quantitative and provided an absolute measure of mineral content, e.g., calcium or phosphate, it would be possible to compare their values with other future and past published reports. Similarly, Wu et al. (2020) documented that subchondral bone collagen in the femoral head weight-bearing area is degraded in osteoarthritis. This implies that the collagen-dependent factor used to calculate the mineral-to-matrix ratio in OA subchondral bone is less than that from controls and would be expected to impact substantially the interpretation of results. Finally, bone porosity is a parameter that can independently alter the mineral-to-matrix ratio determined by two-dimensional FTIR scanning of bone sections. In this regard, Wong et al. (2009) observed reduced porosity in knee specimens taken from patients with advanced knee OA, e.g., suggesting a collapse of the subchondral plate (Wong et al. 2009). Taken together, these findings indicate the mineral-to-matrix ratio is a multi-factorial variable that should be used with caution in the absence of supplemental analyses. In particular, our discussion argues that, in the absence of other data, interpretation of the FTIR-derived M/M ratio as "mineral content" (Zhai et al. 2019) is highly questionable and likely yields misleading conclusions.

8.2.5 How does age affect the composition of bone matrix (or woven bone versus lamellar bone)?

In long bone diaphyses, woven bone forms first and then transitions into a more highly mineralized compact bone tissue. A series of studies have demonstrated that the composition of woven bone is distinct from that of lamellar bone (summarized in Gorski 1998). Direct comparison of the effect of age on the composition of canine cortical bone also demonstrates a decrease with maturity likely reflecting the transition from young (woven) to the older (lamellar) bone. For example, the total protein, collagen, organic phosphate, sialic acid, and organic sulfate content of cortical bone decreases 33–50% going from 4–6 months of age (pups) to 10–12 years of age (old group) (Pinto et al. 1988). Commensurate with the loss of organic content, the mineral content of bone increases dramatically over this same age range going from 8.26 mol/l calcium in pups to 12.96 mol/l in the old group. It is assumed that the loss of water as bone ages provides the necessary internal spatial volume required to accommodate the increased mineral crystal volume, e.g., from about 55% to about 25% water in the old group. Recently, we revisited this topic using a novel method to address changes in cortical bone composition as a function of tissue age (Midura et al. 2011). We developed an in situ approach to isolate the composition of newly formed, most recently deposited, bone from that for more mature, older bone within the same long bone by using sequential extractions with an increasing series of EDTA solutions which progressively extracted proteins from ever deeper levels of cortical bone. These extracts were then examined by Western blotting to make relative comparisons of 7 phosphorylated matrix proteins important for bone physiology and biomineralization. Interestingly, bone sialoprotein, osteopontin, and bone acidic glycoprotein-75 (BAG-75) were enriched in the primary bone as opposed to more mature cortical bone, while osteonectin, fetuin-A, and matrix extracellular phosphoglycoprotein (MEPE) appeared to be equally distributed between these two bone compartments (Midura et al. 2011). Interestingly, intact DMP1 was only detected in extracts of the unmineralized newly deposited matrix. These findings indicate that newly formed bone exhibits a non-collagenous matrix protein composition distinct from that of more mature compact bone even within the same long bone, and suggest that the temporal fate of individual non-collagenous proteins is variable in growing bone. Anecdotally, early work isolating proteins such as bone sialoprotein, decorin, biglycan, bone glycoprotein-75 from bone emphasized the use of young rodent calvarial bone as a starting material because these components were preferentially enriched in this woven bone versus tibial cortical bone. This result needs to be taken into account when considering the functions of non-collagenous proteins in bone since it implicitly suggests that such functionality should be most predominant in younger forming bone.

8.2.6 Replacement of Non-collagenous Proteins to Treat Human Disease

In view of the many functions which non-collagenous proteins play in bone and cartilage, it is now reasonable to question whether diseases that are caused by the loss of a non-collagenous protein may be rescued through replacement. This hypothetical promise was recently fulfilled by the replacement of tissue nonspecific alkaline phosphatase enzyme in children with infantile hypophosphatasia (Whyte et al. 2016) (Table 8.1). The loss of alkaline phosphatase leads to a buildup of pyrophosphate in serum since pyrophosphate is a natural substrate of alkaline phosphatase. The buildup in pyrophosphate, a potent inhibitor of mineralization, leads to osteomalacia or rickets and hypomineralization of teeth. However, the story dates back to the early 1980s when Whyte and colleagues infused platelet-rich plasma from Paget's patients into young children with hereditary hypophosphatasia (Whyte et al. 1982, 1984). While weekly infusions were able to restore circulating levels of alkaline phosphatase to relatively normal values over a short-term period, no changes were noted in bone mineralization. Alkaline phosphatase has a circulatory half-life of about 2 days when administered in this way which is similar to that for the native enzyme. This outcome was interpreted as due to a requirement for the enzyme to function in situ in order to effectively impact bone mineralization (Whyte et al. 1984). We then speed up 20 years to the development of recombinant enzyme asfotase, which is a dimer of two polypeptides linked via disulfide bonds wherein each half consists of a recombinant form of the catalytic domain of tissue nonspecific alkaline phosphatase conjugated to the Fc domain of IgG1 that is linked at the C-terminus to poly-Asp10. It is noteworthy in this context that poly-Asp and poly-Glu peptides have been used for many years as anti-scaling agents to chelate calcium in water systems. As a result, the C-terminal poly-aspartic acid motif acts to localize and bind the catalytic portion to the surfaces of mineral crystals in bone. At these sites, bound asfotase effectively catalyzes the localized removal of pyrophosphate and facilitates mineralization preventing or reversing osteomalacia. Asfotase was first validated as a therapeutic treatment for hypophosphatasia in 2008 in mice deficient in *TNAP* (Millan et al. 2008). Daily subcutaneous injections provided significant improvement in mineralization of long bones and other sites (Millan and Whyte 2016). Following optimization, the current dosing schedule in young children involves weekly subcutaneous injections of asfotase (Whyte et al. 2016).

8.3 Conclusions

1. Non-collagenous proteins comprise a large number of proteins secreted or released from osteoblastic and osteocytic cells in bone. Functionally they represent enzymes as well as binding proteins that assemble to form complexes influencing the growth and turnover of bone, as well as the formation and deposition of hydroxyapatite mineral crystals.

2. Genetic knockout studies in mice and hereditary inactivating mutations in man provide ample evidence that many specific non-collagenous proteins are required for normal bone growth and turnover. Once such a requirement is demonstrated, however, it frequently necessitates a series of detailed biochemical studies to define how a particular protein fulfills this requirement. As a result, we have attempted to provide an update on how a selected group of non-collagenous proteins, previously defined as required, mediate this functionality in bone. In the process, emphasis has been placed upon the biochemical mechanisms by which these non-collagenous proteins mediate their functions. Special attention was also focused on how experimental methodology can influence the way in which think about the structure and turnover of bone tissue.
3. Finally, a recent example is described of how knowledge of the structure/function of a particular non-collagenous protein, alkaline phosphatase, was utilized to rescue the debilitating phenotype of young patients with hereditary infantile hypophosphatasia.

References

Addison WN, Nakano Y, Loisel T, Crine P, McKee MD (2008) MEPE-ASARM peptides control extracellular matrix mineralization by binding to hydroxyapatite: an inhibition regulated by PHEX cleavage of ASARM. J Bone Miner Res 23:1638–1649

Anderson HC, Sipe JB, Hessle L, Dhanyamraju R, Atti E, Camacho NP, Millan JL, Dhamyamraju R (2004) Impaired calcification around matrix vesicles of growth plate and bone in alkaline phosphatase-deficient mice. Am J Pathol 164:841–847

Boskey AL, Camacho NP, Mendelsohn R, Doty SB, Binderman I (1992) FT-IR microscopic mappings of early mineralization in chick limb bud mesenchymal cell cultures. Calcif Tissue Int 51:443–448

Boskey AL, Spevak L, Paschalis E, Doty SB, McKee MD (2002) Osteopontin deficiency increases mineral content and mineral crystallinity in mouse bone. Calcif Tissue Int 71:145–154

Boskey AL, Chiang P, Fermanis A, Brown J, Taleb H, David V, Rowe PS (2010) MEPE's diverse effects on mineralization. Calcif Tissue Int 86:42–46

Boudiffa M, Wade-Gueye NM, Guignandon A, Vanden-Bossche A, Sabido O, Aubin JE, Jurdic P, Vico L, Lafage-Proust MH, Malaval L (2010) Bone sialoprotein deficiency impairs osteoclastogenesis and mineral resorption in vitro. J Bone Miner Res 25:2669–2679

Camacho NP, Landis WJ, Boskey AL (1996) Mineral changes in a mouse model of osteogenesis imperfecta detected by Fourier transform infrared microscopy. Connect Tissue Res 35:259–265

Cao RY, Zheng H, Redfearn D, Yang J (2019) FNDC5: a novel player in metabolism and metabolic syndrome. Biochimie 158:111–116

Christensen B, Schytte GN, Scavenius C, Enghild JJ, McKee MD, Sorensen ES (2020) FAM20C-mediated phosphorylation of mepe and its acidic serine- and aspartate-rich motif. JBMR Plus 4: e10378

Colaianni G, Cinti S, Colucci S, Grano M (2017) Irisin and musculoskeletal health. Ann N Y Acad Sci 1402:5–9

Davies GB, Oxford JT, Hausafus LC, Smoody BF, Morris NP (1998) Temporal and spatial expression of alternative splice-forms of the alpha1(XI) collagen gene in fetal rat cartilage. Dev Dyn 213:12–26

Delany AM, Amling M, Priemel M, Howe C, Baron R, Canalis E (2000) Osteopenia and decreased bone formation in osteonectin-deficient mice. J Clin Invest 105:915–923

Dillon S, Staines KA, Millan JL, Farquharson C (2019) How to build a bone: PHOSPHO1, biomineralization, and beyond. JBMR Plus 3:e10202

Ducy P, Desbois C, Boyce B, Pinero G, Story B, Dunstan C, Smith E, Bonadio J, Goldstein S, Gundberg C, Bradley A, Karsenty G (1996) Increased bone formation in osteocalcin-deficient mice. Nature 382:448–452

Ermakov S, Toliat MR, Cohen Z, Malkin I, Altmuller J, Livshits G, Nurnberg P (2010) Association of ALPL and ENPP1 gene polymorphisms with bone strength related skeletal traits in a Chuvashian population. Bone 46:1244–1250

Faibish D, Gomes A, Boivin G, Binderman I, Boskey A (2005) Infrared imaging of calcified tissue in bone biopsies from adults with osteomalacia. Bone 36:6–12

Fedarko NS, Jain A, Karadag A, Fisher LW (2004) Three small integrin binding ligand N-linked glycoproteins (SIBLINGs) bind and activate specific matrix metalloproteinases. FASEB J 18:734–736

Fedde KN, Blair L, Silverstein J, Coburn SP, Ryan LM, Weinstein RS, Waymire K, Narisawa S, Millan JL, MacGregor GR, Whyte MP (1999) Alkaline phosphatase knock-out mice recapitulate the metabolic and skeletal defects of infantile hypophosphatasia. J Bone Miner Res 14:2015–2026

Fernandes RJ, Weis M, Scott MA, Seegmiller RE, Eyre DR (2007) Collagen XI chain misassembly in cartilage of the chondrodysplasia (cho) mouse. Matrix Biol 26:597–603

Foster BL, Ao M, Willoughby C, Soenjaya Y, Holm E, Lukashova L, Tran AB, Wimer HF, Zerfas PM, Nociti FH Jr, Kantovitz KR, Quan BD, Sone ED, Goldberg HA, Somerman MJ (2015) Mineralization defects in cementum and craniofacial bone from loss of bone sialoprotein. Bone 78:150–164

Fusaro M, Dalle Carbonare L, Dusso A, Arcidiacono MV, Valenti MT, Aghi A, Pasho S, Gallieni M (2015) Differential effects of dabigatran and warfarin on bone volume and structure in rats with normal renal function. PLoS One 10:e0133847

Gadaleta SJ, Camacho NP, Mendelsohn R, Boskey AL (1996) Fourier transform infrared microscopy of calcified turkey leg tendon. Calcif Tissue Int 58:17–23

Gorski JP (1998) Is all bone the same? Distinctive distributions and properties of non-collagenous matrix proteins in lamellar vs. woven bone imply the existence of different underlying osteogenic mechanisms. Crit Rev Oral Biol Med 9:201–223

Gorski JP (2011) Biomineralization of bone: a fresh view of the roles of non-collagenous proteins. Front Biosci (Landmark Ed) 16:2598–2621

Gorski JP, Huffman NT, Chittur S, Midura RJ, Black C, Oxford J, Seidah NG (2011) Inhibition of proprotein convertase SKI-1 blocks transcription of key extracellular matrix genes regulating osteoblastic mineralization. J Biol Chem 286:1836–1849

Gorski JP, Franz NT, Pernoud D, Keightley A, Eyre DR, Oxford JT (2021) A repeated triple lysine motif anchors complexes containing bone sialoprotein and the type XI collagen A1 chain involved in bone mineralization. J Biol Chem 296:100436

Haroon SW, Tai BC, Ling LH, Teo L, Davenport A, Schurgers L, Teo BW, Khatri P, Ong CC, Low S, Yeo XE, Tan JN, Subramanian S, Chua HR, Tan SY, Wong WK, Lau TW (2020) Treatment to reduce vascular calcification in hemodialysis patients using vitamin K (Trevasc-HDK): a study protocol for a randomized controlled trial. Medicine (Baltimore) 99:e21906

Huffman NT, Keightley JA, Chaoying C, Midura RJ, Lovitch D, Veno PA, Dallas SL, Gorski JP (2007) Association of specific proteolytic processing of bone sialoprotein and bone acidic glycoprotein-75 with mineralization within biomineralization foci. J Biol Chem 282:26002–26013

Ichikawa S, Gerard-O'Riley RL, Acton D, McQueen AK, Strobel IE, Witcher PC, Feng JQ, Econs MJ (2017) A mutation in the Dmp1 gene alters phosphate responsiveness in mice. Endocrinology 158:470–476

Kim H, Wrann CD, Jedrychowski M, Vidoni S, Kitase Y, Nagano K, Zhou C, Chou J, Parkman VA, Novick SJ, Strutzenberg TS, Pascal BD, Le PT, Brooks DJ, Roche AM, Gerber KK, Mattheis L, Chen W, Tu H, Bouxsein ML, Griffin PR, Baron R, Rosen CJ, Bonewald LF,

Spiegelman BM (2019) Irisin mediates effects on bone and fat via alphaV integrin receptors. Cell 178:507–508

Kimura-Suda H, Takahata M, Ito T, Shimizu T, Kanazawa K, Ota M, Iwasaki N (2018) Quick and easy sample preparation without resin embedding for the bone quality assessment of fresh calcified bone using Fourier transform infrared imaging. PLoS One 13:e0189650

Kuronuma K, Yokoi A, Fukuoka T, Miyata M, Maekawa A, Tanaka S, Matsubara L, Goto C, Matsuo M, Han HW, Tsuruta M, Murata H, Okamoto H, Hasegawa N, Asano S, Ito M (2020) Matrix Gla protein maintains normal and malignant hematopoietic progenitor cells by interacting with bone morphogenetic protein-4. Heliyon 6:e03743

Lees JS, Mangion K, Rutherford E, Witham MD, Woodward R, Roditi G, Hopkins T, Brooksbank K, Jardine AG, Mark PB (2020) Vitamin K for kidney transplant organ recipients: investigating vessel stiffness (ViKTORIES): study rationale and protocol of a randomised controlled trial. Open Heart 7:e001070

Leupin O, Piters E, Halleux C, Hu S, Kramer I, Morvan F, Bouwmeester T, Schirle M, Bueno-Lozano M, Fuentes FJ, Itin PH, Boudin E, de Freitas F, Jennes K, Brannetti B, Charara N, Ebersbach H, Geisse S, Lu CX, Bauer A, Van Hul W, Kneissel M (2011) Bone overgrowth-associated mutations in the LRP4 gene impair sclerostin facilitator function. J Biol Chem 286:19489–19500

Li Y, Lacerda DA, Warman ML, Beier DR, Yoshioka H, Ninomiya Y, Oxford JT, Morris NP, Andrikopoulos K, Ramirez F et al (1995) A fibrillar collagen gene, Col11a1, is essential for skeletal morphogenesis. Cell 80:423–430

Li Q, Guo H, Chou DW, Berndt A, Sundberg JP, Uitto J (2013) Mutant Enpp1asj mice as a model for generalized arterial calcification of infancy. Dis Model Mech 6:1227–1235

Liu S, Tang W, Zhou J, Vierthaler L, Quarles LD (2007) Distinct roles for intrinsic osteocyte abnormalities and systemic factors in regulation of FGF23 and bone mineralization in Hyp mice. Am J Physiol Endocrinol Metab 293:E1636–E1644

Loffler D, Muller U, Scheuermann K, Friebe D, Gesing J, Bielitz J, Erbs S, Landgraf K, Wagner IV, Kiess W, Korner A (2015) Serum irisin levels are regulated by acute strenuous exercise. J Clin Endocrinol Metab 100:1289–1299

Luo G, Ducy P, McKee MD, Pinero GJ, Loyer E, Behringer RR, Karsenty G (1997) Spontaneous calcification of arteries and cartilage in mice lacking matrix GLA protein. Nature 386:78–81

Macrae VE, Davey MG, McTeir L, Narisawa S, Yadav MC, Millan JL, Farquharson C (2010) Inhibition of PHOSPHO1 activity results in impaired skeletal mineralization during limb development of the chick. Bone 46:1146–1155

Magnuson SK, Booth R, Porter S, Gorski JP (1997) Bilateral tibial marrow ablation in rats induces a rapid hypercalcemia arising from extratibial bone resorption inhibitable by methylprednisolone or deflazacort. J Bone Miner Res 12:200–209

Malaval L, Wade-Gueye NM, Boudiffa M, Fei J, Zirngibl R, Chen F, Laroche N, Roux JP, Burt-Pichat B, Duboeuf F, Boivin G, Jurdic P, Lafage-Proust MH, Amedee J, Vico L, Rossant J, Aubin JE (2008) Bone sialoprotein plays a functional role in bone formation and osteoclastogenesis. J Exp Med 205:1145–1153

Mameli C, Zichichi G, Mahmood N, Elalaoui SC, Mirza A, Dharmaraj P, Burrone M, Cattaneo E, Sheth J, Gandhi A, Kochar GS, Alkuraya FS, Kabra M, Mercurio G, Zuccotti G (2020) Natural history of non-lethal Raine syndrome during childhood. Orphanet J Rare Dis 15:93

Martin A, David V, Laurence JS, Schwarz PM, Lafer EM, Hedge AM, Rowe PS (2008) Degradation of MEPE, DMP1, and release of SIBLING ASARM-peptides (minhibins): ASARM-peptide(s) are directly responsible for defective mineralization in HYP. Endocrinology 149:1757–1772

Marulanda J, Gao C, Roman H, Henderson JE, Murshed M (2013) Prevention of arterial calcification corrects the low bone mass phenotype in MGP-deficient mice. Bone 57:499–508

Midura RJ, Midura SB, Su X, Gorski JP (2011) Separation of newly formed bone from older compact bone reveals clear compositional differences in bone matrix. Bone 49:1365–1374

Millan JL, Whyte MP (2016) Alkaline phosphatase and hypophosphatasia. Calcif Tissue Int 98:398–416

Millan JL, Narisawa S, Lemire I, Loisel TP, Boileau G, Leonard P, Gramatikova S, Terkeltaub R, Camacho NP, McKee MD, Crine P, Whyte MP (2008) Enzyme replacement therapy for murine hypophosphatasia. J Bone Miner Res 23:777–787

Monfoulet L, Malaval L, Aubin JE, Rittling SR, Gadeau AP, Fricain JC, Chassande O (2010) Bone sialoprotein, but not osteopontin, deficiency impairs the mineralization of regenerating bone during cortical defect healing. Bone 46:447–452

Moradi-Ameli M, de Chassey B, Farjanel J, van der Rest M (1998) Different splice variants of cartilage alpha1(XI) collagen chain undergo uniform amino-terminal processing. Matrix Biol 17:393–396

Morris NP, Oxford JT, Davies GB, Smoody BF, Keene DR (2000) Developmentally regulated alternative splicing of the alpha1(XI) collagen chain: spatial and temporal segregation of isoforms in the cartilage of fetal rat long bones. J Histochem Cytochem 48:725–741

Murali SK, Andrukhova O, Clinkenbeard EL, White KE, Erben RG (2016) Excessive osteocytic Fgf23 secretion contributes to pyrophosphate accumulation and mineralization defect in Hyp mice. PLoS Biol 14:e1002427

Narisawa S, Frohlander N, Millan JL (1997) Inactivation of two mouse alkaline phosphatase genes and establishment of a model of infantile hypophosphatasia. Dev Dyn 208:432–446

Nie Y, Dai B, Guo X, Liu D (2020) Cleavage of FNDC5 and insights into its maturation process. Mol Cell Endocrinol 510:110840

Nielson CM, Zmuda JM, Carlos AS, Wagoner WJ, Larson EA, Orwoll ES, Klein RF (2012) Rare coding variants in ALPL are associated with low serum alkaline phosphatase and low bone mineral density. J Bone Miner Res 27:93–103

Nigwekar SU, Bloch DB, Nazarian RM, Vermeer C, Booth SL, Xu D, Thadhani RI, Malhotra R (2017) Vitamin K-dependent carboxylation of matrix Gla protein influences the risk of calciphylaxis. J Am Soc Nephrol 28:1717–1722

Nikitakis NG, Gkouveris I, Aseervatham J, Barahona K, Ogbureke KUE (2018) DSPP-MMP20 gene silencing downregulates cancer stem cell markers in human oral cancer cells. Cell Mol Biol Lett 23:30

Ogbureke KU, Fisher LW (2004) Expression of SIBLINGs and their partner MMPs in salivary glands. J Dent Res 83:664–670

Ogbureke KU, Nikitakis NG, Warburton G, Ord RA, Sauk JJ, Waller JL, Fisher LW (2007) Up-regulation of SIBLING proteins and correlation with cognate MMP expression in oral cancer. Oral Oncol 43:920–932

Oheim R, Zimmerman K, Maulding ND, Sturznickel J, von Kroge S, Kavanagh D, Stabach PR, Kornak U, Tommasini SM, Horowitz MC, Amling M, Thompson D, Schinke T, Busse B, Carpenter TO, Braddock DT (2020) Human heterozygous ENPP1 deficiency is associated with early onset osteoporosis, a phenotype recapitulated in a mouse model of Enpp1 deficiency. J Bone Miner Res 35:528–539

Oxford JT, Doege KJ, Morris NP (1995) Alternative exon splicing within the amino-terminal nontriple-helical domain of the rat pro-alpha 1(XI) collagen chain generates multiple forms of the mRNA transcript which exhibit tissue-dependent variation. J Biol Chem 270:9478–9485

Ozono K, Namba N, Kubota T, Kitaoka T, Miura K, Ohata Y, Fujiwara M, Miyoshi Y, Michigami T (2012) Pediatric aspects of skeletal dysplasia. Pediatr Endocrinol Rev 10(Suppl 1):35–43

Pinto MR, Gorski JP, Penniston JT, Kelly PJ (1988) Age-related changes in composition and Ca2+-binding capacity of canine cortical bone extracts. Am J Physiol 255:H101–H110

Rios H, Koushik SV, Wang H, Wang J, Zhou HM, Lindsley A, Rogers R, Chen Z, Maeda M, Kruzynska-Frejtag A, Feng JQ, Conway SJ (2005) Periostin null mice exhibit dwarfism, incisor enamel defects, and an early-onset periodontal disease-like phenotype. Mol Cell Biol 25:11131–11144

Roberts SJ, Stewart AJ, Schmid R, Blindauer CA, Bond SR, Sadler PJ, Farquharson C (2005) Probing the substrate specificities of human PHOSPHO1 and PHOSPHO2. Biochim Biophys Acta 1752:73–82

Roberts S, Narisawa S, Harmey D, Millan JL, Farquharson C (2007) Functional involvement of PHOSPHO1 in matrix vesicle-mediated skeletal mineralization. J Bone Miner Res 22:617–627

Roumeliotis S, Dounousi E, Salmas M, Eleftheriadis T, Liakopoulos V (2020) Vascular calcification in chronic kidney disease: the role of vitamin K-dependent matrix Gla protein. Front Med (Lausanne) 7:154

Rousseau JC, Farjanel J, Boutillon MM, Hartmann DJ, van der Rest M, Moradi-Ameli M (1996) Processing of type XI collagen. Determination of the matrix forms of the alpha1(XI) chain. J Biol Chem 271:23743–23748

Rowe PS (1997) The PEX gene: its role in X-linked rickets, osteomalacia, and bone mineral metabolism. Exp Nephrol 5:355–363

Rowe PS, Matsumoto N, Jo OD, Shih RN, Oconnor J, Roudier MP, Bain S, Liu S, Harrison J, Yanagawa N (2006) Correction of the mineralization defect in hyp mice treated with protease inhibitors CA074 and pepstatin. Bone 39:773–786

Ruchon AF, Tenenhouse HS, Marcinkiewicz M, Siegfried G, Aubin JE, DesGroseillers L, Crine P, Boileau G (2000) Developmental expression and tissue distribution of Phex protein: effect of the Hyp mutation and relationship to bone markers. J Bone Miner Res 15:1440–1450

Schrauwen I, Valgaeren H, Tomas-Roca L, Sommen M, Altunoglu U, Wesdorp M, Beyens M, Fransen E, Nasir A, Vandeweyer G, Schepers A, Rahmoun M, van Beusekom E, Huentelman MJ, Offeciers E, Dhooghe I, Huber A, Van de Heyning P, Zanetti D, De Leenheer EMR, Gilissen C, Hoischen A, Cremers CW, Verbist B, de Brouwer APM, Padberg GW, Pennings R, Kayserili H, Kremer H, Van Camp G, van Bokhoven H (2019) Variants affecting diverse domains of MEPE are associated with two distinct bone disorders, a craniofacial bone defect and otosclerosis. Genet Med 21:1199–1208

Schumacher MA, Chinnam N, Ohashi T, Shah RS, Erickson HP (2013) The structure of irisin reveals a novel intersubunit beta-sheet fibronectin type III (FNIII) dimer: implications for receptor activation. J Biol Chem 288:33738–33744

Schurgers LJ, Spronk HM, Skepper JN, Hackeng TM, Shanahan CM, Vermeer C, Weissberg PL, Proudfoot D (2007) Post-translational modifications regulate matrix Gla protein function: importance for inhibition of vascular smooth muscle cell calcification. J Thromb Haemost 5:2503–2511

Semenov M, Tamai K, He X (2005) SOST is a ligand for LRP5/LRP6 and a Wnt signaling inhibitor. J Biol Chem 280:26770–26775

Strom TM, Juppner H (2008) PHEX, FGF23, DMP1 and beyond. Curr Opin Nephrol Hypertens 17:357–362

Sweatt A, Sane DC, Hutson SM, Wallin R (2003) Matrix Gla protein (MGP) and bone morphogenetic protein-2 in aortic calcified lesions of aging rats. J Thromb Haemost 1:178–185

Warner LR, Blasick CM, Brown RJ, Oxford JT (2007) Expression, purification, and refolding of recombinant collagen alpha1(XI) amino terminal domain splice variants. Protein Expr Purif 52:403–409

Whyte MP, Valdes R Jr, Ryan LM, McAlister WH (1982) Infantile hypophosphatasia: enzyme replacement therapy by intravenous infusion of alkaline phosphatase-rich plasma from patients with Paget bone disease. J Pediatr 101:379–386

Whyte MP, McAlister WH, Patton LS, Magill HL, Fallon MD, Lorentz WB Jr, Herrod HG (1984) Enzyme replacement therapy for infantile hypophosphatasia attempted by intravenous infusions of alkaline phosphatase-rich Paget plasma: results in three additional patients. J Pediatr 105:926–933

Whyte MP, Rockman-Greenberg C, Ozono K, Riese R, Moseley S, Melian A, Thompson DD, Bishop N, Hofmann C (2016) Asfotase alfa treatment improves survival for perinatal and infantile hypophosphatasia. J Clin Endocrinol Metab 101:334–342

Winkler DG, Sutherland MK, Geoghegan JC, Yu C, Hayes T, Skonier JE, Shpektor D, Jonas M, Kovacevich BR, Staehling-Hampton K, Appleby M, Brunkow ME, Latham JA (2003) Osteocyte control of bone formation via sclerostin, a novel BMP antagonist. EMBO J 22:6267–6276
Wong AK, Beattie KA, Emond PD, Inglis D, Duryea J, Doan A, Ioannidis G, Webber CE, O'Neill J, de Beer J, Adachi JD, Papaioannou A (2009) Quantitative analysis of subchondral sclerosis of the tibia by bone texture parameters in knee radiographs: site-specific relationships with joint space width. Osteoarthritis Cartilage 17:1453–1460
Wu Z, Wang B, Tang J, Bai B, Weng S, Xie Z, Shen Z, Yan D, Chen L, Zhang J, Yang L (2020) Degradation of subchondral bone collagen in the weight-bearing area of femoral head is associated with osteoarthritis and osteonecrosis. J Orthop Surg Res 15:526
Wuthier RE (1975) Lipid composition of isolated epiphyseal cartilage cells, membranes and matrix vesicles. Biochim Biophys Acta 409:128–143
Yadav MC, Simao AM, Narisawa S, Huesa C, McKee MD, Farquharson C, Millan JL (2011) Loss of skeletal mineralization by the simultaneous ablation of PHOSPHO1 and alkaline phosphatase function: a unified model of the mechanisms of initiation of skeletal calcification. J Bone Miner Res 26:286–297
Zebboudj AF, Shin V, Bostrom K (2003) Matrix GLA protein and BMP-2 regulate osteoinduction in calcifying vascular cells. J Cell Biochem 90:756–765
Zhai M, Lu Y, Fu J, Zhu Y, Zhao Y, Shang L, Yin J (2019) Fourier transform infrared spectroscopy research on subchondral bone in osteoarthritis. Spectrochim Acta A Mol Biomol Spectrosc 218:243–247
Zhu X, Li X, Wang X, Chen T, Tao F, Liu C, Tu Q, Shen G, Chen JJ (2021) Irisin deficiency disturbs bone metabolism. J Cell Physiol 236:664–676

Part III
Enamel

Chapter 9
Human Tooth Enamel, a Sophisticated Material

E. F. Brès, J. Reyes-Gasga, and J. Hemmerlé

Abstract Human teeth are sophisticated materials that we use every day.

We attempt here to describe how the mechanical and chemical properties (resistance to carious dissolution), are optimised in the long term by the construction of the tissue at the different hierarchical levels.

Thanks to their hierarchical architecture, under load, teeth distribute stress to their successive sub-elements.

Human tooth enamel, dentine and cementum are characterised by mechanical properties that include: elasticity, hardness, viscoelasticity and fracture behaviour. For enamel, hardness and elastic modulus and crack resistance decrease from the enamel surface to the dentine/enamel junction. A similar phenomenon is observed for dentine with a decrease of hardness and elastic modulus towards the pulp cavity.

Similar to the bone, dental hard tissues show a piezoelectric response. This effect is more important for dentine than for enamel because of the presence of collagen molecules. It is proposed that the strength of the effect is correlated to the remodelling capacity of alveolar bone.

During the carious process -at the nanoscale, tooth enamel crystals are destroyed in a systematic fashion with the formation of a lesion on the basal plane that develops anisotropically through the crystals. Several explanations are put forward: the

In homage to professor R M Frank

E. F. Brès (✉)
Unité des Matériaux et Transformation UMR CNRS 8207, Faculté des Sciences et Technologies, Université de Lille, Villeneuve d'Ascq, France
e-mail: etienne.bres@univ-lille.fr

J. Reyes-Gasga
Instituto de Física, Universidad Nacional Autónoma de México, Circuito de la Investigación s/n, Ciudad Universitaria, Coyoacán, México, DF, México
e-mail: jreyes@fisica.unam.mx

J. Hemmerlé
ITMO Technologies pour la Santé de l'Aviesan, Strasbourg Cedex 2, France
e-mail: hemmerle@unistra.fr

M. Goldberg, P. Den Besten (eds.), *Extracellular Matrix Biomineralization of Dental Tissue Structures*, Biology of Extracellular Matrix 10,
https://doi.org/10.1007/978-3-030-76283-4_9

existence of screw dislocations, grain boundaries, high concentration of Mg^{2+}, Na^{+}, F^{-} and CO_3^{2-} ions.

9.1 Introduction

The human jaw and its sub-constituents (alveolar bone, mandible, teeth, periodontal filament) play a role in food mastication and speech.

As all bones, jawbones possess different structures at different scales of length (hierarchical structure). The interactions between the different sub-elements confer enhanced properties essential for maintaining the long-term functions of the jaw.

Amongst the properties required are the mechanical properties (chewing) and the chemical properties (tissue growth, resistance to the carious dissolution).

9.2 The Hierarchical Structure of Teeth

The structure of natural teeth consists of enamel, dentine, cementum and dental pulp; the first three tissues constitute the hard parts of the human tooth and are characterised by unique mechanical properties that include elasticity, hardness, viscoelasticity and fracture behaviour.

The architecture of human teeth is shown in Fig. 9.1.

9.2.1 Construction of Human Tooth Enamel

9.2.1.1 Microscale

At the microscale, human tooth enamel is formed by the amelogenesis process. Prisms (or rods) of roughly 1 µm diameter perpendicular to the enamel surface are formed, each prism corresponding to one amelogenin cell.

9.2.1.2 Nanoscale

The study of the nanoscale growth control of enamel crystals is an ongoing work. Robinson and Connell (2017) have proposed a model in which 30–50 nm clusters form the crystals.

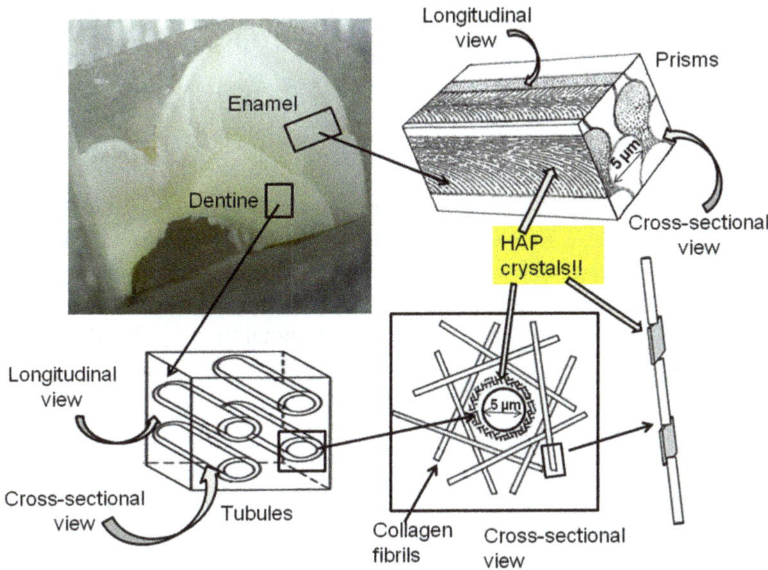

Fig. 9.1 Image of a piece of a human tooth sectioned in longitudinal section in such a way that the enamel, dentine and pulp chamber are observed. The schematic presentation of the dentine tubules and enamel prisms, as well as collagen fibres, is shown

Crystals Morphology

The nanoscale structure of enamel is formed by flat hexagonal shape crystals of rough dimensions 800 nm × 500 nm × several microns length aligned perpendicular to the enamel surface (Matalová et al. 2015; Voegel 1978).

A possible explanation for the elongated shape of the crystals might be the existence of an octacalcium phosphate layer seed oriented along [11–20] (Brown et al. 1979). This has been confirmed experimentally by growing hydroxyapatite on an OCP seed using the constant composition method (Nelson et al. 1986); an OCP layer has also been observed in a dentine crystal in formation (Bodier-Houllé et al. 1998).

Crystals Chemical Composition

Tooth enamel is the most mineralised tissue of the human body; its composition is 96 wt.% inorganic material and 4 wt.% organic material and water. In dentine, the inorganic material represents 70 wt.% which is mainly composed of calcium phosphate related to the hexagonal hydroxyapatite (HA), whose nominal chemical formula is $Ca^{2+}{}_{10}(PO_4{}^{3-})_6(OH^-)_2$ (Weatherell 1975; de Dios Teruel et al. 2015). As a whole, HA crystals have the ability to accommodate a large variety of ions inside their structure (Leroy and Brès 2001) which explains the presence of substituted ions (F^-, Mg^{2+}...) inside enamel crystals.

Crystals Atomic Structure

A full atomic structure determination of human tooth enamel crystals has shown a $P6_3$ space group with a loss of the mirror plane from the $P6_3/m$ hydroxyapatite (HAP) symmetry described by Kay et al. (1964). The loss of the mirror plane in the structure is accompanied by the loss of the crystal centrosymmetry that can explain the piezoelectric property of the whole tissue (Mugnaioli et al. 2014).

Crystals Surface Structure

As for all crystals and most specifically ionic crystals, the structure and chemistry of the apatite crystal surface are always different than those of the bulk (Tasker 1979).

At the neighbourhood of the surface of ionic crystals, the chemical potentials of lattice defects are different from those in the bulk; a space charge is automatically formed in such a way that a positive or negative space-charge layer adjacent to the surface is compensated by a much thicker space-charge layer of opposite sign further away from the surface. Such subsurface modifications have been observed by Brès and Hutchison (2002) in human tooth enamel crystals (Fig. 9.2). Chappell et al. (2008) studying the (100) and (200) bone crystal surfaces using NMR and with first principles calculations have shown that the calculated ^{31}P chemical shifts for the surface slab are found to be significantly different from the bulk crystal. Rotational

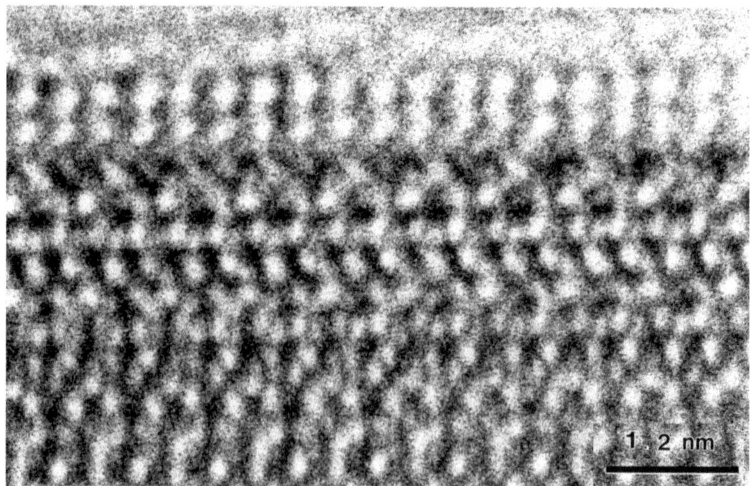

Fig. 9.2 High resolution electron microscopy profile view of a human tooth enamel crystal observed along the [11–20] direction.

relaxations of the surface phosphate ions and the sub-surface displacement of other nearby ions are identified as causing the main differences.

Disordered HA surface structures have also been shown using a combined approach using molecular dynamics and density functional calculations (Wang et al. 2018). In this model hydroxyls ions beneath the surface move outward and cover the exposed calcium ions. The destabilisation caused by hydroxyl vacancy is compensated by the stabilisation of hydroxyl coverage.

Upon adhesion, the apatite crystal surface is modified, in particular by water (Mkhontoa and Leeuw 2002). This has been confirmed experimentally by Rey et al. (2014) who have shown that nanocrystals in bone are covered by a hydrated layer with a composition and structure different from that of crystalline apatite which controls their surface reactivity

Bertinetti et al. (2009) have suggested that the Mg^{2+}/Ca^{2+} exchange process does not affect the morphology and surface topology of the apatite nanocrystals significantly, while a new phase, likely hydrated calcium and/or magnesium phosphate, was formed in a small amount for high Mg concentrations.

At the nanoscale, the large specific surface of HAP nanoparticles (Kolmas et al. 2007) favours the adhesion of a large variety of molecules which justifies their use as nanovectors for gene delivery (Roy et al. 2003; Bouladjine et al. 2009).

9.3 Mechanical Properties

9.3.1 Mechanical Properties of the Jaw

9.3.1.1 Macroscale Properties

The Jaw Bones

The jaw itself is composed of two bones: the upper jawbone (the maxilla) and the lower jawbone (the mandible).

The maxilla itself contains different anatomical regions: the palate, the nasal floor, the alveolus, and the body while the mandible is a U-shaped bone.

During biting and mastication, a combination of bending and rotation occurs. This result is a complex pattern of stresses and strains (compressive, tensile, shear, torsional) on both bones. To be able to resist forces and bending and torsional moments, not only the material properties of the mandible but also its geometrical design is of importance.

Obviously, the mechanical properties of both bones are difficult to analyse on living humans but it must be that the different regions of the maxilla are defined by the material properties: cortical bones from the body of the maxilla tends to be thicker, less dense and less stiff while the maxilla is thinner, denser and stiffer.

Palatal cortical bone is intermediate in some features but overall is more similar to the cortical bone from the alveolar region (Peterson et al. 2006).

A similar phenomenon is observed for the mandible where it was noted that stress calculated from strains and average materials properties vary depending on variations of maximum stiffness and anisotropy (Schwartz-Dabney and Dechow 2003).

Relationship of Bone Piezoelectric Properties and Activity of Bone Cells

The generation of piezoelectricity in bone is a complex process that has been shown to play a key role both in bone adaptation and regeneration (Fukuda and Yasuda 1957; Marino and Becker 1970; Mohammadkhaha et al. 2019).

HA nanoparticles are -at respectively the micro and nanoscale contributors to the bone piezoelectric effect (Ahn and Grodzinsky 2009; Mugnaioli et al. 2014).

This is due to the non-polar nature of apatite crystals which act as transducers (Mugnaioli et al. 2014).

At the different scales of length, the bone strain produced by mechanical loading varies significantly from macroscopic to nanoscopic levels (Verbruggen et al. 2012; Giri et al. 2012).

The types of loads (compressive or tensile) influence the activity of bone cells (osteoblasts and osteoclasts). It has been shown that alveolar bone resorption occurs at the compression side and bone formation at the tension side (Persson 2005). This is contradicted by Shen et al. (2017) who indicate that bone cell activity rather depends on the magnitude of the load and not on its nature.

9.4 Nanoscale Properties

9.4.1 The Teeth

The mechanical properties of whole human teeth are determined by their structure and composition.

9.4.1.1 Human Tooth Dentine and Enamel

Unlike most vital tissues, enamel has no ability to repair and must develop damage tolerance in order to withstand long term use.

Yilmaz et al. (2015) have classified enamel into four hierarchical levels and analysed the implication of each level to the whole tissue fracture behaviour. The result presented indicate a sharp decrease in fracture strength from the HAP nanocrystals to the bulk enamel. The authors also describe a transition from brittle to quasi-ductile deformation behaviour indicating that the damage behaviour is acquired by increasing hierarchy. Toughening mechanisms including microcracking, protein and mineral bridges, crack deflection and branching identified at every hierarchical level.

At the nanoscale level, crystals carry the tensile load while the protein matrix transfers the load between the crystals via shear. The nanometre size of the crystals may be the result of fracture strength optimisation (Gao et al. 2003).

These findings are in agreement with the work of Ya-Rong Zhang et al. (2014) who have examined both dentine and enamel and have come to the following conclusions:

1. The hardness and elastic modulus of enamel decrease gradually from the surface of the enamel to the dentine/enamel junction. A similar phenomenon is observed for dentine with the decrease of hardness and elastic moduli towards the pulp cavity. A hydrated environment influences the mechanical behaviour of dentine.
2. Crack resistance in enamel increases from outside to inside. For dentine, crack bridging, crack deflection and crack bifurcation in young, dehydrated dentine assist in energy dissipation and increase the crack growth resistance.

9.4.2 Piezoelectric Properties of Enamel and Dentine

The structure of dental hard tissues is similar to that of bone, with the difference that enamel does not contain collagen.

Measurements under compressive stress (Braden et al. 1966) have shown a piezoelectric response for dentine but not for enamel.

Reyes-Gasga et al. (2020) have studied the piezoelectric property of human dentine and enamel in both direct and converse mode. Their results indicate that the inorganic phase contributes to the piezoelectric property in dentine and to a lesser extent in enamel.

Marino and Gross (1989) have compared dentine and cementum from sperm whale teeth to human bones and obtained piezoelectric coefficients of 0.029pC/N and 0.028pC/N respectively and of 0.28pC/N for bone. They conclude that the strength of the piezoelectric effect was correlated to the capacities of the tissues to undergo adaptive remodelling which is consistent with the theory that piezoelectricity mediates orthodontically induced alveolar remodelling.

9.4.3 Mechanisms of the Carious Dissolution Process

Dental caries and periodontal diseases are the most common bacterial infections in humans (Loesche 1986).

9.4.4 The Carious Process at the Macroscale Level

Several models of dental caries have been proposed (Ten Cate and Featherstone 1991; Featherstone 1995; Ten Cate and Mundorff-Shrestha 1995; Zero 1995; Javed et al. 2020).

The eventual outcome of dental caries is determined by the dynamic balance between pathological factors that lead to demineralisation and protective factors that lead to remineralisation.

9.4.5 The Dental Caries at the Nanoscale Level

Human tooth enamel crystals are destroyed systematically: first, a central lesion elongated along the [11–20] direction appears on the basal faces (0001) planes of the crystals; secondly, this lesion develops anisotropically along the [0001] direction across the crystals and third the crystals break open (Fig. 9.3) (Voegel and Frank 1977).

No full explanation for the anisotropic crystal dissolution has yet been put forward. Neither the structural reason nor the implications for the mineralisation/demineralisation process have yet been fully understood. Nevertheless, several common-sense observations can be made:

1. There is a structural feature on the centre of the crystal basal faces different from the adjoining surfaces.
2. After dissolution, the lesion is hexagonal in shape and does not split the crystal along the [11–20] direction, as it would be observed from a grain boundary crossing the crystal parallel to [11–20].
3. The dissolution process propagates anisotropically along the [0001] direction, which indicates that the structural feature is three-dimensional.

Several approaches are followed on the ongoing research on that subject.

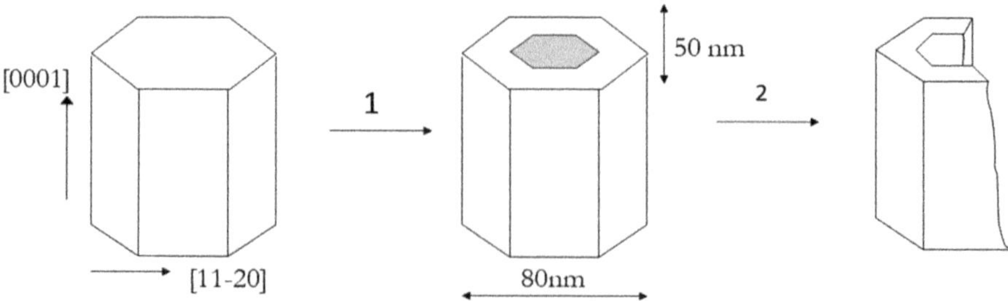

Fig. 9.3 Diagram of the tooth enamel crystals nanoscale carious dissolution mechanism

9.5 Defect Analysis

9.5.1 *Defects in Ionic Crystals*

9.5.1.1 Relationship to Crystal Growth

The importance of screw dislocations for crystal growth has been first shown by Frank (1949) who has shown that dislocations favour low saturation growth. The growth mechanism is favoured by the difficulty encountered by the crystal to balance electrical charges at the surface (Henricks 1955).

9.5.1.2 Surface Charge

In ionic crystals, jogs on surfaces and dislocations act as sources and sinks for point defects. The establishment of equilibria with the vacancies, solute ions and interstitials in the interior produces a net charge on the surface or dislocation, surrounded by a compensating space charge region. Within this space charge region, which typically has a thickness of the order of 10 nm, there are very large electric fields (up to about 105 V/cm) and very large perturbations of point defect concentrations (by factors of several hundred). Thus, ion diffusivities can be dramatically perturbed (Slifkin 1990; Amelinck 1979).

It is also likely that surface charges at the emergence of dislocations on ionic surfaces can favour the adhesion of proteins (Shtukenberg et al. 2017).

9.5.2 *Etch Pit Dislocation Analysis*

The analysis of emerging dislocations at crystal surfaces is a well-used method in materials science (Tolansky and Omar 1953; Sangwal 1987) and specifically for the analysis of ionic crystals where edge and screw dislocations etch differently (Gilman et al. 1958). In this technique, an etchant is applied on a crystal surface in order to reveal emerging dislocations or other surface defects. The subsequent analysis of the etch pit itself provides information on the nature of the dislocations themselves

9.5.3 *Etch Pit Dislocations in Apatite*

Lovell (1958) first observed etch pits on the natural apatite (0001) face and found that, within the limits of her model, each pit corresponded to one dislocation. This was confirmed by Patel et al. (1966), and Phakey and Leonard (1970), who, after acid etching and X-ray topography, showed that screw dislocations with Burgers vector b = c[0001] are present in the apatite lattice.

Arends (1973) has attributed the presence of etch pits on apatite (0001) surfaces to the existence of screw dislocations of Burgers vector parallel to the [0001] direction.

Using atomic force microscopy, Kwon et al. (2008) have shown asymmetrical hexagonal etch pits on the apatite surface.

More recently, Doroshkin (2012) has described eight dissolution models for the dissolution of apatite in acids amongst which the etch pit dissolution model.

9.5.3.1 Surface Energy Calculations

Based on surface energy calculations, anisotropy of the dissolution has been interpreted as being due to a screw dislocation of the Burgers vector parallel to the [0001] direction (Arends and Jongebloed 1977).

9.5.3.2 Crystal Defects in Enamel Crystals Analysis Using TEM

Defects in tooth enamel crystals are directly analysed using high-resolution electron microscopy (HREM). Brès et al. (1984) have identified three types of disorders that may act as dissolution sites: (a) regions of lattice buckling with departure from hexagonal symmetry, (b) dislocations, and (c) grain and twin boundaries. They have also suggested that the low angle grain boundary they have observed in crystals oriented a long [11–20] could be twisted a round [1–100] and give rise to a series of screw dislocations of Burgers vector parallel to [0001] and situated on the boundary (Brès et al. 1984).

What appears as a twist boundary with grains aligned along <-12-11> and <-24-23> rotated by an angle of 6.4° with respect to one another have been shown by Brès et al. (1990) (Fig. 9.4). Considerable strain is observed on both sides of the boundary, no amorphous layer is observed between the grains. As

Fig. 9.4 High-resolution electron micrograph of an enamel crystal showing a twist between grains oriented along the <-12-11> and <-24-23> directions

described by Amelinck (1979), after relaxation, twist boundaries give rise to a set of screw dislocations equally spaced at a right angle in the plane of the boundary, such dislocations are not observed, in the [11–20] direction.

Recent work by Gordon et al. (2015) and by Yun et al. (2020) have shown a high concentration of Na^+ and Mg^{2+} ions at the centre of the crystals. These findings have been precised by DeRocher et al. (2020) who have shown the existence of two nanometric layers enriched with magnesium flanking a core rich in sodium, fluoride and carbonate ions. This sandwich core being surrounded by a shell of mower concentration of substitutional defects. The authors propose that the residual stress created by the chemical gradients may favour the anisotropic carious dissolution. This is a "chicken and the egg" problem. Are the defects created by the chemical potential or is the chemical potential created by the crystal defect (Slifkin 1990; Amelinck 1979)? The question is open. However, it must be reminded that crystal defects, in particular in ionic crystals favour chemical segregation (Tochigi et al. 2017). Lattice strain has been observed by Brès et al. (1990).

All these works being still in progress, it must be noted that no explanation has yet been given of the elongated shape of the lesion along [11–20] and for the fact that it does not split open the crystals as would a grain boundary.

9.5.4 Central Dark Line

Another approach for understanding is the analysis of a contrast phenomenon observed at the of the initial dissolution sites of enamel crystals in the study of the Central Dark Line (CDL). Several authors have observed crystals aligned along the [11–20] or the [0001] on using transmission electron microscopy in phase-contrast mode this CDL which appears black or white depending on the microscope focus, the line disappearing at gaussian focus (Rönnholm 1962; Nylen et al. 1963; Frazier 1968; Marshall and Lawless 1981; Nakahara 1982; Nakahara and Kakei 1983, 1984; Brès et al. 1986; Reyes-Gasga et al. 2008, 2016; Reyes-Gasga and Brès 2015). The CDL is observed independently of specimen preparation methods and accelerating voltages. Because it is observed in both the [11–20] and [0001] directions, the CDL is actually a plane. The CDL has also been observed by TEM on grain boundaries between NiO ionic crystals and is attributed to a change of mean inner electric potential with respect to the bulk crystal (Rühle and Sass 1984; Cowley and Moodie 1957).

Reyes-Gasga et al. (2016) have shown using atomic resolution (better than 1Å resolution) TEM and STEM that the CDL is crystalline with a thickness of 2 Å and ran through the OH^- planes with the absence of any amorphous layer (Fig. 9.5).

The contrast line was also observed by atomic force microscopy (Robinson et al. 2004).

It is important to stress the fact that the CDL itself is not responsible for the anisotropic carious dissolution, but it is the same crystal defect that favours the

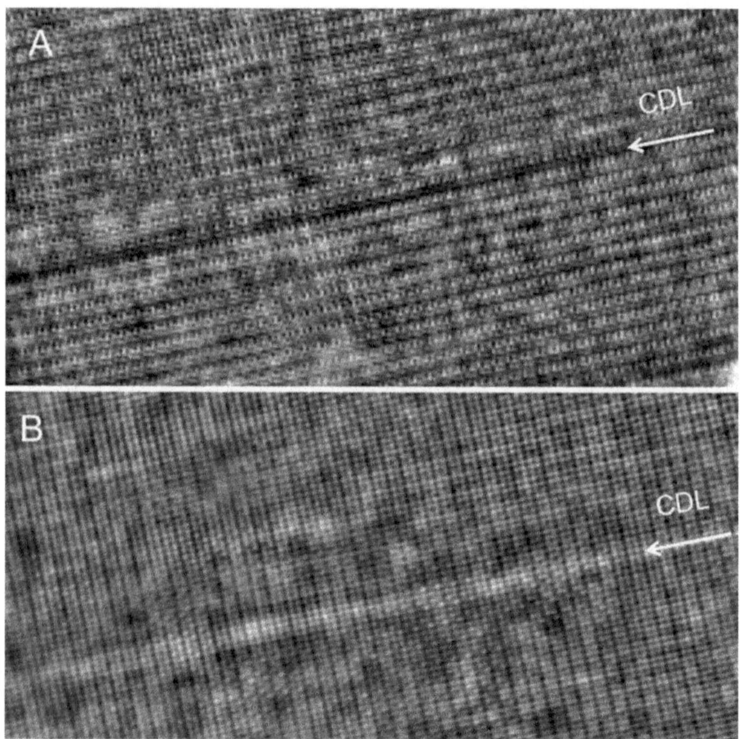

Fig. 9.5 Scanning transmission electron microscopy (STEM) images of the human tooth enamel crystal along the [11–20] zone axis. (**a**) Bright field-STEM image; (**b**) annular dark field-STEM image. The arrows indicate the central dark line (CDL) position. Note the dark (**a**) and white (**b**) contrast of the CDL, respectively. Electron beam damage is observed as background (Reyes-Gasga et al. 2016)

anisotropic dissolution that produces du CDL. Hence, any knowledge of the CDL can lead to a better understanding of the nanoscale dissolution process.

9.6 Conclusion

The piezoelectric control of its growth and remodelling and crystal defect control of the carious enamel dissolution process at the nanoscale make human teeth a sophisticated material indeed and a not ending subject of research.

Several fields of study are active notably the study of the relationship between the different architectural sub-elements and the fine nanoscale enamel carious dissolution mechanism; however, new information are required for a better understanding of these processes.

The development of these studies requires multidisciplinary approaches both in theoretical and experimental knowledge and the most sophisticated equipment.

Acknowledgements We are grateful to S. Tehuacanero-Cuapa, S. Tehuacanero-Nuñez, R. Hernández, P. López for their technical support. This work was partial financial support granted by DGAPA-UNAM through the project IN101319.

References

Ahn AC, Grodzinsky AJ (2009) Relevance of collagen piezoelectricity to "Wolff's Law": a critical review. Med Eng Phys 31:733–741. https://doi.org/10.1016/j.medengphy.2009.02.006

Amelinck S (1979) Dislocations in particular structures. In: Nabarro N (ed) Dislocations in solids, volume 2. Dislocations in crystals. North Holland, Amsterdam, pp 377–398

Arends J (1973) Dislocations and dissolution of enamel. Caries Res 7:261–268. https://doi.org/10.1159/000259848

Arends J, Jongebloed WL (1977) Dislocations and dissolution in apatites: theoretical considerations. Caries Res 11:186–188. https://doi.org/10.1159/000260266

Bertinetti L, Drouet C, Combes C, Rey C, Tampieri A, Coluccia S, Martra G (2009) Surface characteristics of nanocrystalline apatites: effect of Mg surface enrichment on morphology, surface hydration species, and cationic environments. Langmuir 25(10):5647–5654. https://doi.org/10.1021/la804230j

Bodier-Houllé P, Steuer P, Voegel JC, Cuisinier FJG (1998) First experimental evidence for human dentine crystal formation involving conversion of octacalcium phosphate to hydroxyapatite. Acta Cryst D54:1377–1381. https://doi.org/10.1107/S0907444998005769

Bouladjine A, Al-Kattan A, Dufour P, Drouet C (2009) New advances in nanocrystalline apatite colloids intended for cellular drug delivery. Langmuir 25(20):12256–12265. https://doi.org/10.1021/la901671j

Braden MA, Bairstow G, Beider I, Ritter BG (1966) Electrical and piezo-electrical properties of dental hard tissues. Nature 212:1565–1566. https://doi.org/10.1038/2121565a0

Brès EF, Hutchison JL (2002) Observation of the surface structure of biological calcium phosphate apatite crystals from human tooth enamel. J Biomed Mater Res Part B Appl Biomater 63:433–440. https://doi.org/10.1002/jbm.10254

Brès EF, Barry JC, Hutchison JL (1984) A structural basis for the carious dissolution of the apatite crystals of human tooth enamel. Ultramicroscopy 12:367–372. https://doi.org/10.1016/0304-3991(83)90250-4

Brès EF, Waddington WG, Voegel JC, Barry JC, Frank RM (1986) Theoretical simulation of a dark contrast line (DCL) in twinned apatite bicrystals and its possible correlation with the chemical properties of human dentine and enamel crystals. Biophys J 50:1185–1193

Brès EF, Hutchison JL, Voegel JC, Frank RM (1990) Observation of a low angle grain boundary in tooth enamel crystals using HREM. J Phys Colloq 51-Cl:97–102. https://doi.org/10.1002/jbm.10254

Brown WE, Schroeder LW, Ferris JS (1979) Interlayering of crystalline octacalcium phosphate and hydroxyapatite. J Phys Chem 83:1385–1388. https://doi.org/10.1021/j100474a006

Chappell H, Duer M, Groom N, Pickard C, Bristowe P (2008) Probing the surface structure of hydroxyapatite using NMR spectroscopy and first principles calculations. Phys Chem Chem Phys 10:600–606. https://doi.org/10.1039/b714512h

Cowley JM, Moodie AF (1957) The scattering of electrons by atoms and crystals. I. A new theoretical approach. Acta Cryst 10:609–619. https://doi.org/10.1107/S0365110X57002194

de Dios Teruel J, Alcolea A, Hernandez A, Ortiz Ruiz AJ (2015) Comparison of chemical composition of enamel and dentine in human, bovine, porcine and ovine teeth. Arch Oral Biol 30:768–775. https://doi.org/10.1016/j.archoralbio.2015.01.014

DeRocher KA, Smeets PJM, Goodge BH, Zachman MJ, Balachandran PV, Stegbauer L, Cohen MJ, Gordon LM, Rondinelli JM, Kourkoutis LF, Joester D (2020) Chemical gradients in human enamel crystallites. Nature 583:66–71. https://doi.org/10.1038/s41586-020-2433-3

Doroshkin SV (2012) Dissolution mechanism of calcium apatites in acids: a review of literature. World J Methodol 26:1–17. https://doi.org/10.5662/wjm.v2.i1.1

Featherstone JD (ed) (1995) Clinical aspects of de/remineralization of teeth. Adv Dent Res 9:1–340. https://doi.org/10.1177/154405910408301s08

Frank FC (1949) The influence of dislocations on crystal growth. Discuss Faraday Soc 5:48–66. https://doi.org/10.1039/DF9490500048

Frazier PD (1968) Adult enamel I: an electron microscopic study of crystallite size and morphology. J Ultrastruct Res 22:1–11. https://doi.org/10.1016/s0022-5320(68)90045-2

Fukuda E, Yasuda I (1957) On the piezoelectric effect of bone. Jpn J Appl Phys 12:1158–1162

Gao H, Ji B, Jäger IL, Arzt E, Fratzl P (2003) Materials become insensitive to flaws at nanoscale: lessons from nature. Proc Natl Acad Sci USA 100:5597–5600. https://doi.org/10.1073/pnas.0631609100

Gilman JJ, Johnston WG, Sears GW (1958) Dislocation etch pit formation in lithium fluoride. J Appl Phys 29:747–754. https://doi.org/10.1063/1.1723277

Giri B, Almer JD, Dong XN, Wanga X (2012) In situ mechanical behavior of mineral crystals in human cortical bone under compressive load using synchrotron X-ray scattering techniques. J Mech Behav Biomed Mater 14:101–112. https://doi.org/10.1016/j.jmbbm.2012.05.003

Gordon LM, Cohen MJ, MacRenaris KW, Pasteris JD, Seda T, Joester D (2015) Amorphous intergranular phases control the properties of rodent tooth enamel. Science 347:746–750. https://doi.org/10.1126/science.1258950

Henricks SP (1955) Screw dislocation and charge balance as a factor of crystal growth. J Am Mineral Soc 40:139–146

Javed S, Zakirulla M, Baig RU, Asif SM, Meere AB (2020) Development of artificial neural network model for prediction of post-Streptococcus mutans in dental caries. Comput Methods Programs Biomed 186:105198. https://doi.org/10.1016/j.cmpb.2019.105198

Kay MI, Young RA, Posner AS (1964) Crystal structure of hydroxyapatite. Nature 204:1050–1052. https://doi.org/10.1038/2041050a0

Kolmas J, Slosarczyk A, Wojtowicz A, Kolodziejski W (2007) Estimation of the specific surface area of apatites in human mineralized tissues using 31P MAS NMR. Solid State Nucl Magn Reson 32:53–58. https://doi.org/10.1016/j.ssnmr.2007.08.001

Kwon KY, Wang E, Chung A, Chang N, Saiz E, Choe UJ, Koobatian M, Lee SW (2008) Defect induced asymmetric pit formation on hydroxyapatite. Langmuir 24:11063–11066. https://doi.org/10.1021/la801735c

Leroy N, Brès E (2001) Structure and substitutions in fluorapatite. Eur Cells Mater 2:36–48. https://doi.org/10.22203/eCM.v002a05

Loesche WJ (1986) Role of streptococcus mutans in human dental decay. Microbiol Rev 50:353–380

Lovell LG (1958) Dislocation etch pits in apatite. Acta Metallurgica 6:775–777

Marino AA, Becker RO (1970) Piezoelectric effect and growth control in bone. Nature 228:473–474. https://doi.org/10.1038/228473a0

Marino AA, Gross BD (1989) Piezoelectricity in cementum, dentine, and bone. Arch Oral Biol 34:507–509. https://doi.org/10.1016/0003-9969(89)90087-3

Marshall AF, Lawless KR (1981) TEM study of the central dark line in enamel crystallites. J Dent Res 60:1779–1782. https://doi.org/10.1177/00220345810600100801

Matalová E, Lungová V, Sharpe P (2015) Development of tooth and associated structures. In: Vishwakarma A, Sharpe P, Shi S, Ramalingam M (eds) Stem cell biology and tissue engineering in dental sciences. Academic, London. https://doi.org/10.1016/B978-0-12-397157-9.00030-8

Mkhontoa D, Leeuw NH (2002) A computer modelling study of the effect of water on the surface structure and morphology of fluorapatite: introducing a $Ca_{10}(PO_4)_6F_2$ potential model. J Mater Chem 12:2633–2642. https://doi.org/10.1039/b204111a
Mohammadkhaha M, Marinkovica D, Zehna M, Checa S (2019) A review on computer modelling of bone piezoelectricity and its application to bone adaptation and regeneration. Bone 127:544–555. https://doi.org/10.1016/j.bone.2019.07.024
Mugnaioli E, Reyes-Gasga J, Garcia-García R, Kolb U, Hémmerlé J, Brès EF (2014) Evidence of non-centrosymmetry of human tooth hydroxyapatite crystals. Chem Eur J 20:6849–6852. https://doi.org/10.1002/chem.201402275
Nakahara H (1982) Electron microscope studies of the lattice image and the "central dark line" of crystallites in sound and carious human dentine. Bull Josai Dent Univ 11:209–215
Nakahara H, Kakei M (1983) The central dark line in developing enamel crystallites: an electron microscopic study. Bull Josai Dent Univ 12:1–7
Nakahara H, Kakei M (1984) Central dark line and carbonic anhydrase: problems to crystal nucleation in enamel. In: Fearnhead RW, Suga S (eds) Tooth enamel IV. Elsevier, Amsterdam, pp 42–46
Nelson DGA, Salimi H, Nancollas GH (1986) Octacalcium phosphate and apatite overgrowths: a crystallographic and kinetic study. J Colloid Interface Sci 110:32–39. https://doi.org/10.1016/0021-9797(86)90350-4
Nylen MU, Eanes ED, Omnell KA (1963) Crystal growth in rat enamel. J Cell Biol 18:109–123. https://doi.org/10.1083/jcb.18.1.109
Patel AR, Desai CC, Agarwal MK (1966) Cleavage and etching of prism faces in apatite. Acta Cryst 20:796–798. https://doi.org/10.1107/S0365110X66001877
Persson MA (2005) 100th anniversary: Sandstedt's experiments on tissue changes during tooth movement. J Orthod 32:27–28. https://doi.org/10.1179/146531205225020760
Peterson J, Wang Q, Dechow PC (2006) Material properties of the dentate maxilla. Anat Rep Part A 288A:962–972. https://doi.org/10.1002/ar.a.20358
Phakey PP, Leonard JR (1970) Dislocations and fault surfaces in natural apatites. J Appl Cryst 3:38–44. https://doi.org/10.1107/S0021889870005617
Rey C, Combes C, Drouet C, Cazalbou S, Grossin D, Brouillet F, Sarda S (2014) Surface properties of biomimetic nanocrystalline apatites; applications in biomaterials. Prog Cryst Growth Char Mater 60:63–73. https://doi.org/10.1016/J.PCRYSGROW.2014.09.005
Reyes-Gasga J, Brès ÉF (2015) Electron microscopy study of the Human tooth enamel. The central dark line. Encyclopaedia Anal Chem. https://doi.org/10.1002/9780470027318.a9495
Reyes-Gasga J, Carbajal de la Torre G, Bres E, Gil-Chavarria IM, Rodriguez-Hernandez AG, Garcia-Garcia R (2008) STEM-HAADF electron microscopy analysis of the central dark line defect of human tooth enamel crystallites. J Mater Sci Mater Med 19:877–882. https://doi.org/10.1007/s10856-007-3174-7
Reyes-Gasga J, Hemmerlé J, Brès EF (2016) Aberration corrected TEM study of the central dark line defect in human tooth enamel crystals. Microsc Microanal 22:1047–1055. https://doi.org/10.1017/S1431927616011648
Reyes-Gasga J, Galindo-Mentle M, Brès E, Vargas-Becerril N, Orozco E, Rodríguez-Gómez A, García-García R (2020) Detection of the piezoelectricity effect in nanocrystals from human teeth. J Phys Chem Solids 136:109140. https://doi.org/10.1016/j.jpcs.2019.109140
Robinson C, Connell SD (2017) Crystal initiation structures in developing enamel: possible implications for caries dissolution of enamel crystals. Front Physiol 8:405. https://doi.org/10.3389/fphys.2017.00405
Robinson C, Connell S, Kirkham J, Shore R, Smith A (2004) Dental enamel—a biological ceramic: regular substructures in enamel hydroxyapatite crystals revealed by atomic force microscopy. J Mater Chem 14:2242–2248. https://doi.org/10.1039/B401154F
Rönnholm E (1962) The amelogenesis of human teeth as revealed by electron microscopy II: the development of enamel crystallites. J Ultrastruct Res 6:249–303. https://doi.org/10.1016/s0022-5320(62)80036-7

Roy I, Mitra S, Maitra A, Mozumdar S (2003) Calcium phosphate nanoparticles as novel non-viral vectors for targeted gene delivery. Int J Pharm 250:25–33. https://doi.org/10.1016/s0378-5173(02)00452-0

Rühle M, Sass SL (1984) The detection of the change in mean inner potential at dislocations in grain boundaries in NiO. Philos Mag A 49:759–782. https://doi.org/10.1080/01418618408236562

Sangwal K (1987) Etching of the crystals. North-Holland, Amsterdam

Schwartz-Dabney CL, Dechow PC (2003) Variations in cortical material properties throughout the human dentate mandible. Am J Phys Anthropol 120:252–277. https://doi.org/10.1002/ajpa.10121

Shen WQ, Geng YM, Liu P, Huang XY, Li SY, Liu CD, Zhou Z, Xu PP (2017) Magnitude-dependent response of osteoblasts regulated by compressive stress. Sci Rep 7:44925. https://doi.org/10.1038/srep44925

Shtukenberg AG, Ward MD, Kahr B (2017) Crystal growth with macromolecular additives. Chem Rev 117:14042–14090. https://doi.org/10.1021/acs.chemrev.7b00285

Slifkin L (1990) Surface and dislocation effects on diffusion in ionic crystals. In: Laskar AL, Bocquet JL, Brebec G, Monty C (eds) Diffusion in materials. NATO ASI series (Series E: Applied sciences), vol 179. Springer, Dordrecht. https://doi.org/10.1007/978-94-009-1976-1_22

Tasker PW (1979) The stability of ionic crystal surfaces. J Phys C Solid State Phys 12:4977–4984. https://doi.org/10.1088/0022-3719/12/22/036

Ten Cate JM, Featherstone JD (1991) Mechanistic aspects of the interactions between fluoride and dental enamel. CRC Crit Rev Oral Biol Med 2:283–296. https://doi.org/10.1177/10454411910020030101

Ten Cate JM, Mundorff-Shrestha SA (1995) Working group report: laboratory models for caries (in vitro and animal models). Adv Dent Res 9:332–334. https://doi.org/10.1177/08959374950090032001

Tochigi E, Kezuka Y, Nakamura A, Nakamura A, Shibata N, Ikuhara Y (2017) Direct observation of impurity segregation at dislocation cores in an ionic crystal. Nano Lett 17:2908–2912. https://doi.org/10.1021/acs.nanolett.7b00115

Tolansky S, Omar M (1953) Etch-pits and dislocations. Nature 4343:171–172. https://doi.org/10.1038/171171a0

Verbruggen SW, Vaughan TJ, McNamara LM (2012) Strain amplification in bone mechanobiology: a computational investigation of the in vivo mechanics of osteocytes. J R Soc Interface 9:2735–2744. https://doi.org/10.1098/rsif.2012.0286

Voegel JC, Frank RM (1977) Stages in the dissolution of human enamel crystals in dental caries. Calcif Tissue Res 24:19–27. https://doi.org/10.1007/BF02223292

Voegel JC (1978) Le cristal d'apatite des tissus osseux et dentaires et sa destruction pathologique. PhD Dissertation, University Louis Pasteur, Strasbourg, France

Wang X, Zhang L, Zeng Q, Jiang G, Yang M (2018) First-principles study on the hydroxyl migration from inner to surface in hydroxyapatite. Appl Surf Sci 452:381–388. https://doi.org/10.1016/j.apsusc.2018.05.050

Weatherell JA (1975) Composition of dental enamel. Br Med Bull 31:115–119. https://doi.org/10.1093/oxfordjournals.bmb.a071263

Yilmaz ED, Schneider GA, Swain MV (2015) Influence of structural hierarchy on the fracture behaviour of tooth enamel. Philos Trans R Soc A 373:20140130. https://doi.org/10.1098/rsta.2014.0130

Yun F, Swain MV, Chen H, Cairney J, Qu J, Sha G, Liu H, Ringer SP, Han Y, Liu L, Zhang X, Zheng R (2020) Nanoscale pathways for human tooth decay – Central planar defect, organic rich precipitate and high-angle grain boundary. Biomaterials 235:119748. https://doi.org/10.1016/j.biomaterials.2019.119748

Zero DT (1995) In situ caries models. Adv Dent Res 9:214–230. https://doi.org/10.1177/08959374950090030501

Zhang YR, Du W, Zhou XD, Yu HY (2014) Review of research on the mechanical properties of the human tooth. Int J Oral Sci 6:61–69. https://doi.org/10.1038/ijos.2014.21

Chapter 10
Proteinases in Enamel Development

Shifa Shahid and John D. Bartlett

Abstract Proteinases are essential for proper enamel formation. During the secretory stage of enamel development, the ameloblasts responsible for enamel formation secrete enamel matrix proteins including amelogenin (AMELX), ameloblastin (AMBN), enamelin (ENAM), and matrix metalloproteinase-20 (MMP20, enamelysin). MMP20 cleaves the enamel matrix proteins soon after they are secreted and these cleavages are essential because the autosomal recessive inheritance of *MMP20* mutations can cause severe enamel defects. Later, during the transition to the early maturation stage of enamel development, kallikrein-related peptidase-4 (KLK4) is secreted by ameloblasts. KLK4 facilitates protein removal, which allows the enamel to mature into its final hardened form. Autosomal recessive inheritance of *KLK4* mutations can also cause severe enamel defects consisting of soft enamel that easily abrades from teeth. Recently, pre-secretory ameloblasts and secretory stage ameloblasts were demonstrated to express *A Disintegrin And Metallopeptidase* Domain-10 (ADAM10). Although ADAM10 may facilitate enamel development, mutations are not known to cause enamel defects in humans primarily because its inactivation is likely embryonic lethal as it is in mice. Nonetheless, as discussed below, the ability of ADAM10 to precisely locate on the cell membrane holds exciting possibilities for how cohorts of secretory stage ameloblasts move relative to one another as enamel development progresses.

10.1 Enamel Formation Overview

In unerupted teeth prior to mineralization, odontoblasts responsible for dentin formation secrete predentin, which is composed primarily of type I collagen. The basement membrane separating pre-ameloblasts from the predentin is then degraded. This occurs when ameloblasts extend finger-like cell protrusions through the

S. Shahid · J. D. Bartlett (✉)
Division of Biosciences, The Ohio State University, College of Dentistry, Columbus, OH, USA
e-mail: Shahid.30@osu.edu; Bartlett.196@osu.edu

M. Goldberg, P. Den Besten (eds.), *Extracellular Matrix Biomineralization of Dental Tissue Structures*, Biology of Extracellular Matrix 10,
https://doi.org/10.1007/978-3-030-76283-4_10

basement membrane and into the predentin (Smith et al. 2016). As the basement membrane is degraded, fragments are engulfed by the ameloblasts and are degraded in lysosomes (Sawada et al. 1990). However, once the basement membrane is completely removed, the ameloblast cell protrusions remain within the predentin. In teeth, dentin is the first mineral to form. It starts forming in the predentin near the dentin–enamel junction (DEJ). Soon after dentin at the DEJ is mineralized, the ameloblast protrusions initiate enamel formation by making mineralized ribbons that initiate on the mineralized collagen fibers within the dentin (Smith et al. 2016).

Ameloblast matrix secretion is necessary for proper ribbon formation. At this stage, these secreted proteins include amelogenin, ameloblastin, enamelin, and MMP20. As matrix proteins are secreted, the cell protrusions pull back to lengthen the mineral ribbons that will form the interrod enamel. As the tall columnar ameloblasts move back, they begin to form Tomes' processes, which are conical shaped structures present at the apical end of secretory stage ameloblasts (Fig. 10.1). The Tomes' process secretes enamel proteins primarily from one side, but not the other (Ronnholm 1962). This provides the Tomes' process with the ability to initiate enamel ribbons that will eventually coalesce into an enamel rod (Skobe 1977) with a direction that is angled by as much as ~40° from the path of the interrod enamel.

Once the forming enamel has reached its full thickness, the secretory stage is complete. This is when ameloblasts retract their Tomes' processes and transition (transition stage) into shorter cells that initiate the maturation stage of enamel development. This is also when ameloblasts stop secreting secretory stage enamel matrix proteins. Prior to the maturation stage, the enamel is protein rich and very soft. It is during the maturation stage that proteins are exported from the enamel as approximately 10,000 ribbons expand and coalesce into one enamel rod (Daculsi et al. 1984). Since this occurs for each enamel rod from both the rod and interrod enamel, the vast majority of mineral is formed during the maturation stage of enamel development. It is during the maturation stage that KLK4 is secreted to assist with the removal of enamel matrix proteins necessary for the enamel to reach its final hardened form (Simmer et al. 2009). Once enamel formation is complete, the ameloblasts become reduced in height and cover the enamel as the tooth erupts.

10.2 Matrix Metalloproteinase-20

The *Mmp20* mRNA was originally discovered by PCR-based homology cloning with a subsequent screening of a porcine enamel organ cDNA library for the full-length clone (Bartlett et al. 1996). Since that time, MMP20 has been well characterized for its role in enamel formation. MMP20 is a tooth-specific MMP because little expression is observed anywhere in normal tissues other than in the ameloblasts of the enamel organ and the odontoblasts of the pulp organ (Fukae et al. 1998; Turk et al. 2006). Odontoblasts display continuous low-level MMP20 expression. However, ameloblasts express MMP20 during the secretory stage through the transition

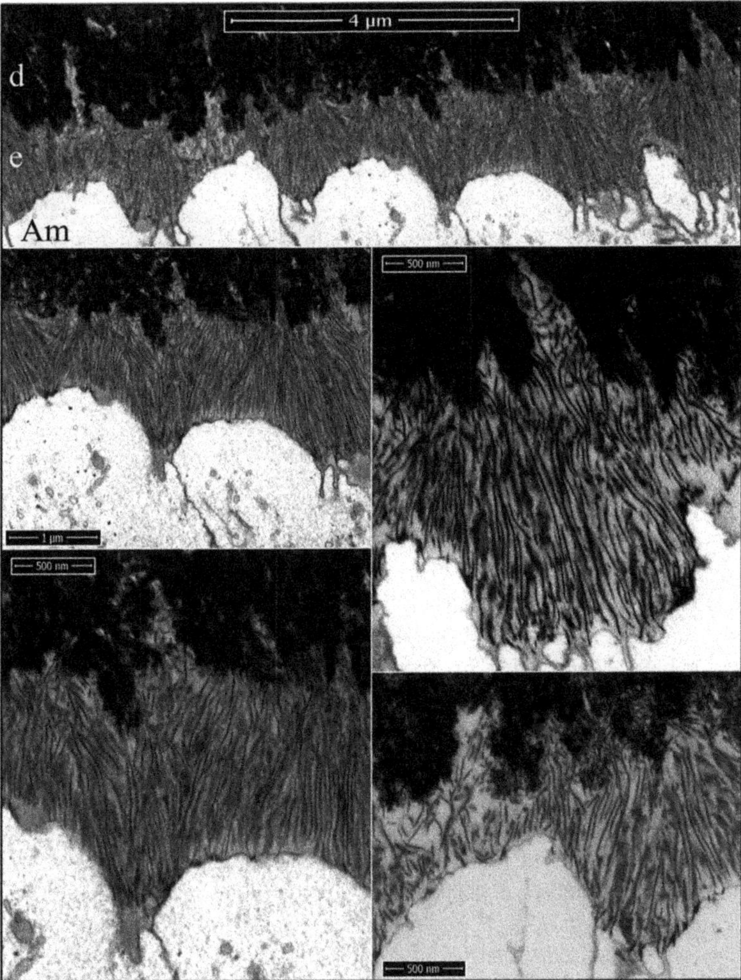

Fig. 10.1 Focused ion beam images of initial Tomes process formation in a wild-type mouse mandibular incisor. The initial enamel ribbons were continuous with dentin mineral and ran parallel to each other to the ameloblast membrane. Rod and interrod enamel form by the elongation of initial enamel ribbons. Key: Am, ameloblast; d, dentin; e, enamel. Reproduced from Smith et al. (2016)

stage with some MMP20 activity remaining in the early maturation stage (Bartlett 2013).

Two studies have demonstrated that amelogenin cleavage products purified from pig enamel were virtually identical to the cleavage products produced by the in vitro incubation of MMP20 with full-length recombinant porcine amelogenin. Since no other cleavage products were found in pig enamel that would indicate the presence of another proteinase, these results suggest that MMP20 is likely the only proteinase

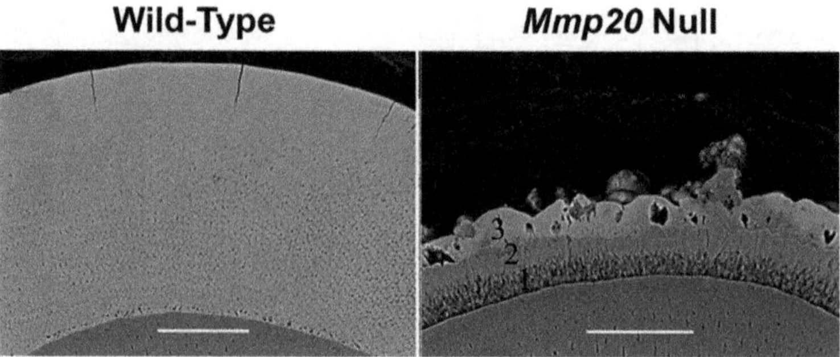

Fig. 10.2 Backscattered scanning electron microscopy analysis of mandibular incisor cross sections from 7-week-old wild-type and *Mmp20* null mice. The enamel layers of *Mmp20* null mice show a dark, poorly mineralized band at the dentin–enamel junction overlaid by three morphologically distinct mineralized layers. *Mmp20* null mice have rough enamel layers that are only a third as thick as the wild-type on average. Scale Bars, 50μm. Reproduced from Hu et al. (2016)

that cleaves enamel matrix proteins during the secretory stage of development (Nagano et al. 2009; Ryu et al. 1999).

Mice with homozygous inactivation of *Mmp20* have enamel that contains uncleaved amelogenins, have hypoplastic enamel with little to no rod pattern (Fig. 10.2), have enamel that breaks easily from the dentin surface, and have a deteriorating enamel and enamel organ morphology as development progresses (Caterina et al. 2002). Odontoblasts express MMP20, and dentin in *Mmp20* null mice have postponed mantle dentin mineralization compared to wild-type mice (Beniash et al. 2006). A recent study demonstrated that crown dentin density and root dentin thickness and density were significantly reduced in *Mmp20* null mice at 30 dpn, but at 60 dpn the dentin recovered and significant differences in these parameters were no longer observed. However, overall dentin thickness remained significantly reduced at both 30 and 60 dpn although the dentin thickness at 60 dpn had somewhat recovered (Wang et al. 2020).

Human *MMP20* mutations can cause autosomal recessive pigmented hypomaturation-type enamel defects termed amelogenesis imperfecta (AI). Currently, 18 different *MMP20* mutations are known to cause AI in humans. Interestingly, seven of these mutations are near or within the MMP20 hemopexin domain suggesting that it is required for proper enamel formation (Wang et al. 2020). All but three MMPs (MMP7, MMP23, MMP26) contain hemopexin domains (Fischer and Riedl 2020). The typical MMP domain structure consists of a signal peptide necessary for MMP secretion, a pro-peptide that is cleaved to catalytically activate the MMP, the catalytic domain, and a hinge region that connects the catalytic domain to the larger C-terminal hemopexin domain (Bartlett and Simmer 1999). For various MMPs the hemopexin domain has been shown to play a role in substrate specificity, in binding tissue inhibitors of MMPs, in binding with integrins, in proMMP activation, and pharmaceutical MMP modulators have been designed to target the hemopexin domain (Fischer and Riedl 2020; Wang et al. 2020). The function of the MMP20 hemopexin domain remains unknown, but because seven identified mutations near or in this domain cause AI, its function is likely essential.

Two extracted maxillary third molars from an individual with AI due to different *MMP20* mutations in each allele (p.Ala304Gly; p.Ala349Val) were assessed by nanohardness testing. The hardness of surface enamel was 55% normal. Enamel midway between the enamel surface and the DEJ was only 13% normal and the hardness of enamel nearest the DEJ was 61% normal. Strikingly, dentin close to the DEJ was also compromised. Its hardness was between 62% and 69% of normal (Wang et al. 2020). Therefore, MMP20 may play an essential role in dentin development that cannot be completely recovered in its absence by another proteinase(s) with overlapping functions.

MMP20 is the first identified enamel matrix proteinase and, although the function of its hemopexin domain remains unknown, MMP20 is otherwise well characterized for its function during enamel development.

10.3 Kallikrein-Related Peptidase-4

Like the *Mmp20* mRNA, the *Klk4* mRNA was originally discovered by PCR-based homology cloning with a subsequent screening of a porcine enamel organ cDNA library for the full-length clone (Simmer et al. 1998). The original name for KLK4 was Enamel Matrix Serine Proteinase-1 (EMSP1). The Human Gene Nomenclature Committee (London, UK), after several iterations, named it kallikrein-related peptidase-4 because it is present within the human kallikrein gene cluster on chromosome 19. Interestingly, *KLK4* occurred relatively recently during evolution and may be associated with the development of thicker enamel and earlier tooth eruption (Kawasaki et al. 2014).

KLK4 is expressed in maturation stage ameloblasts and is also expressed at very low levels in adult mice in the striated ducts of the submandibular salivary gland and in small patches of prostate epithelia. But unlike MMP20, KLK4 is not expressed in odontoblasts. Also for the *Klk4* null mice, only dental enamel was affected and no morphological deformities were observed in non-dental tissues (Simmer et al. 2009).

KLK4 cleaves recombinant amelogenin to generate 12 cleavage products (Ryu et al. 2002) and was demonstrated to further cleave the tyrosine-rich amelogenin peptide (TRAP) that is generated by MMP20 activity (Nagano et al. 2009). Since KLK4 is expressed in ameloblasts from the transition through the maturation stage of development (Hu et al. 2000), it is thought to cleave enamel matrix proteins to facilitate their removal from the hardening enamel. The soft, protein-rich enamel present in the *Klk4* null mouse strongly supports this conclusion.

Klk4 null mice were generated by the use of a LacZ reporter gene inserted at the natural *Klk4* translation initiation site. So, the LacZ reporter replaced the *Klk4* coding sequence but used the native *Klk4* promoter for its expression. Expression was assessed using β-galactosidase histochemistry (Simmer et al. 2009), which allowed assessment of the previously mentioned tissue-specific expression pattern. Enamel in

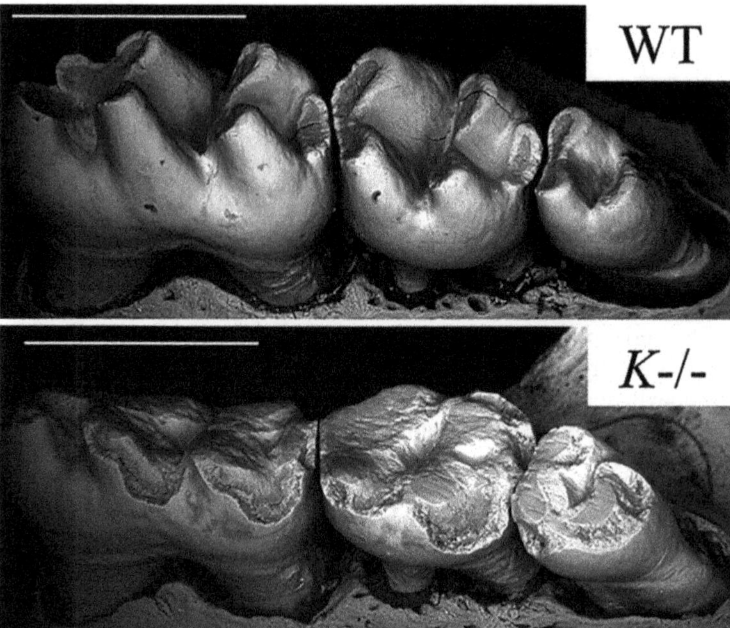

Fig. 10.3 Backscatter scanning electron microscopy of 9-week mandibular molar crowns. Shown are buccal-occlusal views of wild-type (WT) and *Klk4* null (K–/–) mice. These molars have been in occlusion for as much as 48 days (first molars) while the animals were maintained on a soft diet. The *Klk4* null molars show severe enamel abrasion that exposes the underlying dentin. Bars = 1 mm. Reproduced from Hu et al. (2016)

Klk4 null mice developed normally through the secretory stage of development. The null enamel had a normal prism/rod pattern and reached full thickness. However, as the ameloblasts progressed through the maturation stage, the enamel remained soft and upon tooth eruption, the compromised enamel abraded from their teeth (Fig. 10.3). Interestingly, the rod enamel sometimes pulled away from interrod enamel as if the rod enamel were pegs on a cribbage board and the interrod enamel was defining the peg holes. Another striking finding was that although the normal rod pattern was present in the *Klk4* ablated enamel, the 10,000 crystallites that the rod is composed of failed to interlock properly. This caused the crystallites to fall from the rods as would strands of angel hair spaghetti held in a circular arrangement if one's grip was slowly released (Simmer et al. 2009).

Four different *KLK4* mutations are known to cause human AI (Smith et al. 2017). The enamel on these teeth is of normal thickness but tends to be yellow-brown in color and chip from the teeth. No other phenotype resulted from these homozygous *KLK4* mutations. Teeth from the most recently identified human mutation (p. L211Rfs*37) were assessed for enamel properties. Outer enamel from the *KLK4* mutant teeth was significantly harder than the inner enamel layer whereas no difference was observed between these layers in control teeth (Smith et al. 2017).

This finding mirrors results observed in mouse *Klk4* null enamel (Smith et al. 2011). The reason for the harder outer layer and softer inner enamel layer in humans and mice with inactivated KLK4 is not known, but may be due to the shorter distance necessary for protein removal near the surface as compared to the longer distance necessary to remove proteins from the deeper enamel layers (Smith et al. 2011, 2017).

Since only enamel is affected and no other phenotype results from either MMP20 or KLK4 ablation, both MMP20 and KLK4 are tooth-specific proteinases. However, it is fascinating to note that while neither *Mmp20* heterozygous nor *Klk4* heterozygous mice have a distinct enamel phenotype, *Mmp20, Klk4* double heterozygous mice do have enamel malformation consisting of increased surface roughness with susceptibility to attrition and fractures. It appears that both proteinases at least partially function to help remove proteins from developing enamel and that a reduction in expression of both proteinases results in protein retention and enamel malformation in mice (Hu et al. 2016).

KLK4 is the second identified enamel matrix proteinase and, like MMP20, it is well characterized for its function during enamel development.

10.4 A Disintegrin and Metallopeptidase Domain-10

A Disintegrin And Metalloproteinase Domain 10 (ADAM10) belongs to a superfamily of zinc-dependent membrane-anchored metalloproteases and is involved in the cleavage of several transmembrane proteins from the surface of the plasma membrane, a process known as ectodomain shedding (Dempsey 2017; Parkin and Harris 2009). ADAM10 is one of the best characterized ADAMs; it is well studied in the context of Notch signaling and amyloid precursor protein cleavage. Crucial membrane-bound proteins such as adhesion molecules like E and N-cadherins, growth factors, and cytokines are also substrates of ADAM10 (Maretzky et al. 2005; Reiss and Saftig 2009; Schlöndorff and Blobel 1999).

ADAM10 activity is intrinsically regulated by cleavage of its prodomain revealing the active site as the enzyme is transported to the cell surface. Tetraspanins also mediate ADAM10 activity (Matthews et al. 2017). Tetraspanins (TSPANs) are a superfamily of transmembrane proteins that can partner with each other and other proteins to regulate their activity. ADAM10 partners with TSPAN5, TSPAN10, TSPAN14, TSPAN15, TSPAN17, and TSPAN33, collectively called the TspanC8 subfamily of tetraspanins because each has eight cysteines. These tetraspanins can regulate ADAM10 trafficking from the ER to the cell surface, select its localization on the plasma membrane as well as influence ADAM10 substrate specificity (Haining et al. 2012; Jorissen et al. 2010; Matthews et al. 2018; Weber et al. 2011). Direct association of ADAM10 with several tetraspanins indicates that this sheddase can exhibit differential activity depending on which tetraspanin it binds with (Jouannet et al. 2016).

ADAM10 was recently discovered in the ameloblasts of the enamel organ, specifically in the pre-secretory to secretory stage ameloblasts (Ikeda et al. 2019). During amelogenesis, cohorts of secretory ameloblasts slide by one another to form the decussating rod pattern present in rodent incisor enamel. ADAM10 is capable of cleaving cell–cell adhesions (Maretzky et al. 2005) and the intriguing possibility of TSPANs locating ADAM10 to ameloblasts at the ends of the moving cohorts provides a possible, though unproven, explanation for how these cells may move together as a group.

ADAM10 was also recently demonstrated to cleave the extracellular domain of recombinant RELT (Ikeda et al. 2019). RELT is a member of the tumor necrosis family receptor superfamily (TNFRSF), but its ligands remain unknown (Bossen et al. 2006). Mutations in RELT cause syndromic as well as non-syndromic amelogenesis imperfecta (Kim et al. 2019; Nikolopoulos et al. 2020). The involvement of RELT in amelogenesis imperfecta and its association with ADAM10 as a substrate suggest that ADAM10 may play an important role in enamel formation.

In Summary, both MMP20 and KLK4 are tooth-specific proteinases and are well characterized for their role in enamel development. In contrast, ADAM10 is expressed in an abundance of tissues and is one of the best-characterized ADAM family proteinases. MMP20 is expressed during the secretory to early maturation stages of enamel development. It cleaves enamel matrix proteins so that enamel will form properly and may also play a role in protein removal as evidenced by the *Mmp20, Klk4* double heterozygous mice that have susceptibility to enamel attrition and fractures. MMP20 is also expressed by odontoblasts and appears necessary for dentin to harden in a timely manner in mice and may be essential to fully mineralize dentin in humans. The predominant role of KLK4 in enamel formation is to further cleave the remaining enamel matrix proteins to facilitate their export out of the hardening enamel. This allows the mineralized enamel ribbons to grow in width and thickness and to interlock, which coalesces the enamel rods. KLK4 is not expressed in dentin so in its absence, dentin forms normally. Unlike MMP20 and KLK4, ADAM10 is a membrane-bound proteinase that is expressed in pre-ameloblasts and ameloblasts during the pre-secretory through the secretory stage of enamel development. It is also expressed sporadically in the pulp organ. ADAM10 has been postulated to play a role in the migration of cohorts of ameloblasts during the secretory stage and to potentially cleave the basement membrane that separates pre-ameloblasts from predentin. ADAM10 was also demonstrated to cleave the extracellular domain of the RELT TNF receptor. This links ADAM10 to enamel development because mutations in the *RELT* receptor can cause enamel malformation. Therefore, these three proteinases are essential components in enamel formation and we look forward to further defining their mechanistic function during amelogenesis.

References

Bartlett JD (2013) Dental enamel development: proteinases and their enamel matrix substrates. ISRN Dent 2013:684607

Bartlett JD, Simmer JP (1999) Proteinases in developing dental enamel. Crit Rev Oral Biol Med 10 (4):425–441

Bartlett JD, Simmer JP, Xue J, Margolis HC, Moreno EC (1996) Molecular cloning and mrna tissue distribution of a novel matrix metalloproteinase isolated from porcine enamel organ. Gene 183 (1–2):123–128

Beniash E, Skobe Z, Bartlett JD (2006) Formation of the dentino-enamel interface in enamelysin (mmp-20)-deficient mouse incisors. Eur J Oral Sci 114(Suppl 1):24–29

Bossen C, Ingold K, Tardivel A, Bodmer JL, Gaide O, Hertig S, Ambrose C, Tschopp J, Schneider P (2006) Interactions of tumor necrosis factor (tnf) and tnf receptor family members in the mouse and human. J Biol Chem. 281(20):13964–13971

Caterina JJ, Skobe Z, Shi J, Ding Y, Simmer JP, Birkedal-Hansen H, Bartlett JD (2002) Enamelysin (matrix metalloproteinase 20)-deficient mice display an amelogenesis imperfecta phenotype. J Biol Chem 277(51):49598–49604

Daculsi G, Menanteau J, Kerebel LM, Mitre D (1984) Length and shape of enamel crystals. Calcif Tissue Int 36(5):550–555

Dempsey PJ (2017) Role of adam10 in intestinal crypt homeostasis and tumorigenesis. Biochim Biophys Acta Mol Cell Res 1864(11 Pt B):2228–2239

Fischer T, Riedl R (2020) Challenges with matrix metalloproteinase inhibition and future drug discovery avenues. Expert Opin Drug Discov 1–14

Fukae M, Tanabe T, Uchida T, Lee SK, Ryu OH, Murakami C, Wakida K, Simmer JP, Yamada Y, Bartlett JD (1998) Enamelysin (matrix metalloproteinase-20): localization in the developing tooth and effects of ph and calcium on amelogenin hydrolysis. J Dent Res 77(8):1580–1588

Haining EJ, Yang J, Bailey RL, Khan K, Collier R, Tsai S, Watson SP, Frampton J, Garcia P, Tomlinson MG (2012) The tspanc8 subgroup of tetraspanins interacts with a disintegrin and metalloprotease 10 (adam10) and regulates its maturation and cell surface expression. J Biol Chem 287(47):39753–39765

Hu JC, Ryu OH, Chen JJ, Uchida T, Wakida K, Murakami C, Jiang H, Qian Q, Zhang C, Ottmers V et al (2000) Localization of emsp1 expression during tooth formation and cloning of mouse cdna. J Dent Res 79(1):70–76

Hu Y, Smith CE, Richardson AS, Bartlett JD, Hu JC, Simmer JP (2016) Mmp20, klk4, and mmp20/klk4 double null mice define roles for matrix proteases during dental enamel formation. Mol Genet Genomic Med 4(2):178–196

Ikeda A, Shahid S, Blumberg BR, Suzuki M, Bartlett JD (2019) Adam10 is expressed by ameloblasts, cleaves the relt tnf receptor extracellular domain and facilitates enamel development. Sci Rep 9(1):14086

Jorissen E, Prox J, Bernreuther C, Weber S, Schwanbeck R, Serneels L, Snellinx A, Craessaerts K, Thathiah A, Tesseur I et al (2010) The disintegrin/metalloproteinase adam10 is essential for the establishment of the brain cortex. J Neurosci 30(14):4833–4844

Jouannet S, Saint-Pol J, Fernandez L, Nguyen V, Charrin S, Boucheix C, Brou C, Milhiet PE, Rubinstein E (2016) Tspanc8 tetraspanins differentially regulate the cleavage of adam10 substrates, notch activation and adam10 membrane compartmentalization. Cell Mol Life Sci 73 (9):1895–1915

Kawasaki K, Hu JC, Simmer JP (2014) Evolution of klk4 and enamel maturation in eutherians. Biol Chem 395(9):1003–1013

Kim JW, Zhang H, Seymen F, Koruyucu M, Hu Y, Kang J, Kim YJ, Ikeda A, Kasimoglu Y, Bayram M et al (2019) Mutations in relt cause autosomal recessive amelogenesis imperfecta. Clin Genet 95(3):375–383

Maretzky T, Reiss K, Ludwig A, Buchholz J, Scholz F, Proksch E, de Strooper B, Hartmann D, Saftig P (2005) Adam10 mediates e-cadherin shedding and regulates epithelial cell-cell adhesion, migration, and beta-catenin translocation. Proc Natl Acad Sci USA 102(26):9182–9187

Matthews AL, Szyroka J, Collier R, Noy PJ, Tomlinson MG (2017) Scissor sisters: regulation of adam10 by the tspanc8 tetraspanins. Biochem Soc Trans 45(3):719–730

Matthews AL, Koo CZ, Szyroka J, Harrison N, Kanhere A, Tomlinson MG (2018) Regulation of leukocytes by tspanc8 tetraspanins and the "molecular scissor" adam10. Front Immunol 9:1451

Nagano T, Kakegawa A, Yamakoshi Y, Tsuchiya S, Hu JC, Gomi K, Arai T, Bartlett JD, Simmer JP (2009) Mmp-20 and klk4 cleavage site preferences for amelogenin sequences. J Dent Res 88 (9):823–828

Nikolopoulos G, Smith CEL, Brookes SJ, El-Asrag ME, Brown CJ, Patel A, Murillo G, O'Connell MJ, Inglehearn CF, Mighell AJ (2020) New missense variants in relt causing hypomineralised amelogenesis imperfecta. Clin Genet 97(5):688–695

Parkin E, Harris B (2009) A disintegrin and metalloproteinase (adam)-mediated ectodomain shedding of adam10. J Neurochem 108(6):1464–1479

Reiss K, Saftig P (2009) The "a disintegrin and metalloprotease" (adam) family of sheddases: physiological and cellular functions. Semin Cell Dev Biol 20(2):126–137

Ronnholm E (1962) An electron microscopic study of the amelogenesis in human teeth. I. The fine structure of the ameloblasts. J Ultrastruct Res 6:229–248

Ryu OH, Fincham AG, Hu CC, Zhang C, Qian Q, Bartlett JD, Simmer JP (1999) Characterization of recombinant pig enamelysin activity and cleavage of recombinant pig and mouse amelogenins. J Dent Res 78(3):743–750

Ryu O, Hu JC, Yamakoshi Y, Villemain JL, Cao X, Zhang C, Bartlett JD, Simmer JP (2002) Porcine kallikrein-4 activation, glycosylation, activity, and expression in prokaryotic and eukaryotic hosts. Eur J Oral Sci 110(5):358–365

Sawada T, Yamamoto T, Yanagisawa T, Takuma S, Hasegawa H, Watanabe K (1990) Evidence for uptake of basement membrane by differentiating ameloblasts in the rat incisor enamel organ. J Dent Res 69(8):1508–1511

Schlöndorff J, Blobel CP (1999) Metalloprotease-disintegrins: modular proteins capable of promoting cell-cell interactions and triggering signals by protein-ectodomain shedding. J Cell Sci 112(Pt 21):3603–3617

Simmer JP, Fukae M, Tanabe T, Yamakoshi Y, Uchida T, Xue J, Margolis HC, Shimizu M, DeHart BC, Hu CC et al (1998) Purification, characterization, and cloning of enamel matrix serine proteinase 1. J Dent Res 77(2):377–386

Simmer JP, Hu Y, Lertlam R, Yamakoshi Y, Hu JC (2009) Hypomaturation enamel defects in klk4 knockout/lacz knockin mice. J Biol Chem 284(28):19110–19121

Skobe Z (1977) Enamel rod formation in the monkey observed by scanning electron microscopy. Anat Rec 187(3):329–334

Smith CE, Richardson AS, Hu Y, Bartlett JD, Hu JC, Simmer JP (2011) Effect of kallikrein 4 loss on enamel mineralization: comparison with mice lacking matrix metalloproteinase 20. J Biol Chemi 286(20):18149–18160

Smith CE, Hu Y, Hu JC, Simmer JP (2016) Ultrastructure of early amelogenesis in wild-type, amelx (-/-), and enam(-/-) mice: enamel ribbon initiation on dentin mineral and ribbon orientation by ameloblasts. Mol Genet Genomic Med 4(6):662–683

Smith CEL, Kirkham J, Day PF, Soldani F, McDerra EJ, Poulter JA, Inglehearn CF, Mighell AJ, Brookes SJ (2017) A fourth klk4 mutation is associated with enamel hypomineralisation and structural abnormalities. Front Physiol 8:333

Turk BE, Lee DH, Yamakoshi Y, Klingenhoff A, Reichenberger E, Wright JT, Simmer JP, Komisarof JA, Cantley LC, Bartlett JD (2006) Mmp-20 is predominately a tooth-specific enzyme with a deep catalytic pocket that hydrolyzes type v collagen. Biochemistry (Mosc) 45 (12):3863–3874

Wang SK, Zhang H, Chavez MB, Hu Y, Seymen F, Koruyucu M, Kasimoglu Y, Colvin CD, Kolli TN, Tan MH et al (2020) Dental malformations associated with biallelic mmp20 mutations. Mol Genet Genomic Med 8(8):e1307

Weber S, Niessen MT, Prox J, Lüllmann-Rauch R, Schmitz A, Schwanbeck R, Blobel CP, Jorissen E, de Strooper B, Niessen CM et al (2011) The disintegrin/metalloproteinase adam10 is essential for epidermal integrity and notch-mediated signaling. Development 138(3):495–505

Chapter 11
Enamel Matrix Biomineralization: The Role of pH Cycling

Wu Li, Yan Zhang, Sylvie Babajko, and Pamela Den Besten

Abstract Dental enamel, which is the most highly mineralized tissue in the human body, is critically important for the protection of the underlying tooth structures throughout a lifetime. When enamel is absent or poorly formed due to genetic defects, called amelogenesis imperfecta (AI), or through environmental or epigenetic effects such as fluorosis, or lost due to dental caries, the entire tooth structure may be compromised. Ameloblasts differentiate through secretory, transition, and maturation stages to generate the fully mineralized enamel matrix. Enamel matrix proteins produced during the secretory stage undergo limited hydrolyzed by matrix metalloproteinase 20 (MMP20) to initiate crystal growth. Then during maturation, the remaining matrix proteins are completely hydrolyzed by kallikrein 4 (KLK4). The protein fragments are endocytosed and this allows complete mineralization of the enamel space. Enamel mineralization is unique in that it requires pH cycling, in which the ameloblasts modulate the matrix pH between an acidic and neutral pH. In this chapter, Part 1 is an overview of enamel formation, and how pH cycling influences amelogenin hydrolysis by KLK4 and enamel mineralization. Part 2 describes how ion-pumps and transporters regulate the enamel matrix pH as mineralization occurs.

W. Li (✉) · Y. Zhang · P. D. Besten
Department of Orofacial Sciences, School of Dentistry, University of California at San Francisco, San Francisco, CA, USA
e-mail: Wu.Li@ucsf.edu; Yan.zhang2@ucsf.edu; pamela.denbesten@ucsf.edu

S. Babajko
Centre de Recherche des Cordeliers INSERM UMRS 1138 Université de Paris, Sorbonne Université Molecular Oral Pathophysiology Laboratory, Paris, France
e-mail: sylvie.babajko@inserm.fr

M. Goldberg, P. Den Besten (eds.), *Extracellular Matrix Biomineralization of Dental Tissue Structures*, Biology of Extracellular Matrix 10,
https://doi.org/10.1007/978-3-030-76283-4_11

11.1 Part 1: Secretion of Matrix Proteins, Enamel Mineralization, and pH Regulation

11.1.1 Introduction

In the secretion stage of enamel formation, epithelial-derived ameloblasts secrete matrix proteins, to form a matrix with a cheese-like texture (Aoba and Moreno 1987; Smith and Nanci 1989). Enamel crystal formation is initiated in this matrix, and the crystals grow primarily in length and reach their full length at the end of the secretory stage. The secretory stage is followed by the transition stage, where the protein components in the enamel matrix dramatically decrease while the mineral components rapidly increase, resulting in a chalky appearing enamel matrix (Daculsi et al. 1984; Simmer et al. 2009). This transition leads to the maturation stage where the crystals continue to thicken and widen as proteins are simultaneously removed after degradation and replaced by hydroxyapatite (HAP) mineral. Following maturation, tooth enamel is a highly mineralized extracellular biomaterial, consisting of approximately 96–98% minerals and only 2–4% organic materials and water (He et al. 2011; Robinson 2014).

11.1.2 Secretory Stage Enamel

Enamel proteins are synthesized by the cells of the enamel organ, including ameloblasts. Ameloblasts elongate and form Tomes' processes, through which enamel proteins are secreted in a prism-like structure, on top of the dentin matrix (Kallenbach 1973). The full thickness of the enamel matrix is established by matrix proteins that include amelogenins, ameloblastin, enamelin, tuftelin, and keratins (Robinson et al. 1998).

Amelogenins, which are transcribed from multiple alternatively spliced variants of the amelogenin gene, are the primary structural proteins, constituting 90–95% of total proteins in the secretory enamel protein matrix (Gibson 2011). Amelogenins play a central role in crystal growth and enamel thickness (Gibson et al. 2001; Robinson et al. 1998). The amelogenin knockout mouse has an enamel layer that is much thinner than that in the wild-type mouse, and the enamel crystals are disorganized, deformed, and disoriented (Gibson et al. 2001). The absence of ameloblastin (*Ambn*) results in a detachment of ameloblasts from the underlying matrix at the start of the secretory stage and generation of a thin enamel with an irregular prism pattern (Wazen et al. 2009). When enamelin is absent, ameloblasts are unable to adhere to the underlining enamel surface, and they prematurely undergo apoptosis (Hu et al. 2011). Mutations in keratin 75 are associated with an increased incidence of dental caries (Duverger et al. 2014).

The secretory matrix also contains proteinases, including matrix metalloproteinases (MMPs) (Bartlett 2013; Wöltgens et al. 1991), MMP2, MMP3,

MMP9, MMP12, MMP13 and MMP20 (Bartlett 2013; DenBesten and Heffernan 1989; Goldberg et al. 2003; Llano et al. 1997; Wöltgens et al. 1991). MMP20 is the predominant enamel matrix proteinase in the secretory stage and cleaves enamel matrix proteins immediately after their secretion (Bartlett et al. 1998; Simmer and Hu 2002).

Mineral formed in the enamel matrix is initially amorphous calcium, which subsequently transforms to hydroxyapatite (HAP) (Beniash et al. 2009). The phosphorylated N terminal serine 16 on alternatively spliced amelogenins enhances the stabilization of amorphous calcium phosphate (ACP) mineral precursors and modulates the timing of conversion of ACP to apatite mineral (Shin et al. 2020). At the end of the secretory stage, long thin crystals of HAP extend the full length of the enamel matrix. The secretory stage ends as capillaries invaginate into the stellate reticulum layer of the enamel organ, overlying ameloblasts. Up to 50% of the ameloblasts undergo apoptosis during this transition so that there is no longer one ameloblast overlaying each enamel prism. The ameloblasts then further differentiate into maturation stage ameloblasts (Smith 1998).

11.1.3 Maturation Stage Enamel

Maturation ameloblasts cyclically modulate between ruffle-ended (RE) cells and smooth-ended (SE) cells (Fig. 11.1). Matrix proteins secreted at the maturation stage include amelotin (AMTN), odontogenic ameloblast-associated protein (ODAM) (Moffatt et al. 2008), and secretory calcium-binding phosphoprotein proline-glutamine rich 1 (SCPPPQ1). These proteins are thought to participate in structuring an extracellular matrix with the distinctive capacity of attaching epithelial cells to mineralized surfaces (Fouillen et al. 2017), and the formation of the final layer of aprismatic enamel (Abbarin et al. 2015).

Kallikrein 4 (KLK4), a serine proteinase previously named enamel matrix serine proteinase (EMSP1), first identified and purified from porcine enamel matrix by Simmer and colleagues (Simmer et al. 1998), is the predominant enamel matrix proteinase in the maturation stage (Simmer and Hu 2002). KLK4 hydrolyzes matrix proteins to create space for the mineralizing enamel crystals to expand in width.

11.1.4 Discovery of pH Cycling

Takano and coworkers first reported the visualization of ameloblast modulation on the enamel surface, by staining with the calcium-binding dye glyoxal bi (2-hydroxyanil) (GBHA). Enamel under SE stains red by GBHA, whereas enamel under RE stains a yellow/green color, or is unstained (Takano et al. 1982a) (see Fig.11.1). However, an unsolved question at that time was that Ca^{45} uptake is not correlated with RE and SE morphologies (Takano et al. 1982b).

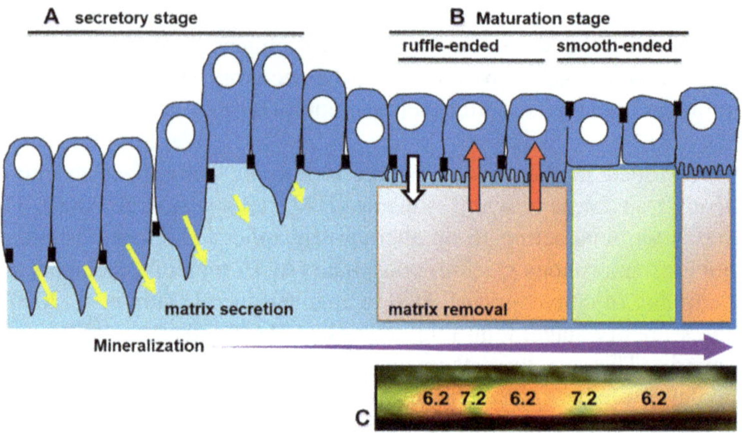

Fig. 11.1 Enamel formation and ameloblast modulation in the rodent incisor. In the secretory stage (**a**), ameloblasts secrete a protein-rich enamel matrix (yellow arrows) into which long thin crystals are formed. At the maturation stage (**b**) KLK4 is secreted (white arrow) and hydrolyzes matrix proteins. The hydrolyzed peptide fragments are endocytosed (red arrows) and mineral content (purple arrow) increases. In the maturation stage, ameloblasts become ruffle ended, facing an acidic enamel matrix (orange), and then cycle into smooth-ended ameloblasts, facing neutral enamel (light green). (**c**) A mouse incisor stained with pH indicator dye shows cycles of matrix acidification and neutralization in the maturation stage

Sasaki stained the enamel surface with a pH indicator dye and found that the enamel had alternating acidic and neutral pH. Furthermore, these pH cycles of acidic (5.8–6.0) and neutral (7.0–7.2) (see Fig. 11.1) were located at the same area stained by GBHA, indicating that enamel under RE ameloblasts was more acidic, and was more neutral under SE ameloblasts (Sasaki et al. 1991b) (see Figs. 11.1 and 11.2).

The correlation of enamel staining by GBHA and pH indicator dye later became clearer, as it was noted that GBHA stains poorly at acid pH. Therefore, ionic calcium staining by GBHA in the developing enamel matrix is also pH dependent. Calcein, another dye that stains ionic calcium, is also pH sensitive, and so has a similar property to that of GBHA, of only staining the neutral enamel matrix (Josephsen 1983; Picard et al. 2000).

Takagi et al. further characterized the profiles of the enamel matrix proteins collected from enamel either with acidic or neutral pH (Takagi et al. 1998). They discovered that the neutral enamel matrix contains relatively intact amelogenin and enamelin proteins, while the acid zones contain mostly lower molecular weight amelogenins and enamelin fragments. This suggests that amelogenin hydrolysis is more effective in the acidic conditions underlying RE ameloblasts.

We do not yet understand what drives the cyclic modulation of maturation ameloblasts. However, we do know that ion transporters localized to the ameloblast apical plasma membranes transport calcium into the enamel matrix to form HAP, while modulating matrix pH to direct this process. Bicarbonate ion transporters, including Ae2 and NBCe1, neutralize the acidified matrix. Matrix acidification is

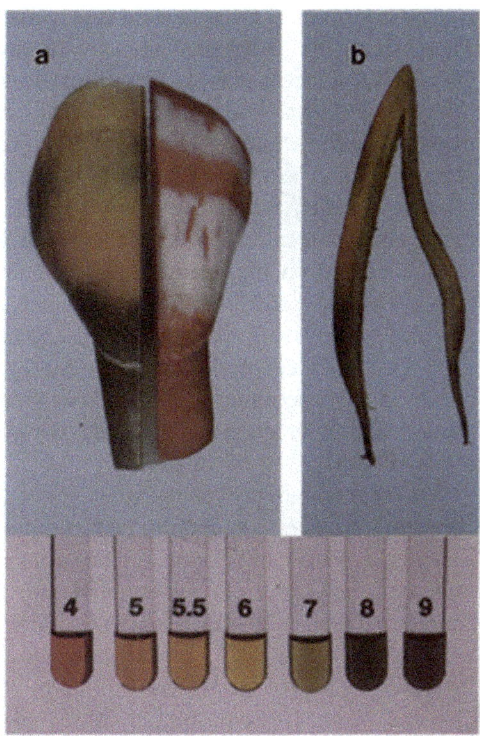

Fig. 11.2 Staining of bovine tooth enamel surfaces. (**a**) A bovine developing tooth sagittally cut in half. The left half was stained with a pH Universal indicator mixture for 1–2 min. Alternate stripes of orange correspond to pH 5.5–6.0 and green to pH 7.0. The right half was stained with GBHA solution. Red stripes of staining correspond to the neutral bands of green staining with the pH indicator and unstained white zones to acidic orange zones. (**b**) Staining with the Universal indicator of a bovine enamel slice. Orange or green coloration occurs not only on the forming surface but also in depth. Color standards at different pH values are shown at the bottom (Sasaki et al. 1991b)

driven by the formation of HAP crystals which release protons into the matrix (Bronckers 2017; Smith 1998). Matrix acidification may also be controlled by proton secretion by V-ATPase translocated to the RE apical plasma membrane (Damkier et al. 2014; Toei et al. 2010). Modulations of ion transporter activities by either genetic defects, or environmental factors, such as excess fluoride, affect pH cycling and enamel matrix formation (Bronckers et al. 2015; DenBesten and Li 2011). Therefore, pH cycling, which is related to both matrix protein endocytosis and matrix mineralization is a unique mechanism that drives the formation of biomineralized enamel.

11.1.5 Amelogenin-Mediated Enamel Matrix Mineralization

The amelogenin protein structure is highly pH dependent. At pH 3.0 amelogenin is globally unstructured, while at pH 5.6, it forms elaborated structures with increased oligomerization and nanosphere formation. At pH 7.2 amelogenin assembles into highly organized structures, binding together to form branched chains (Beniash et al. 2012). In addition, at neutral pH, changes in electrolyte concentration dramatically affect protein adsorption onto the HAP surface (Shimabayashi et al. 1997).

The C terminus of amelogenin is a primary site for amelogenin binding to HAP. Mice lacking the 13 terminal amino acids have poorly formed enamel (Pugach et al. 2010). Glutamate-containing proteins interact strongly with Ca^{2+} ions on the surface of the HAP crystal, and acidic amino acids have been considered as major determinants of protein binding (Jaeger et al. 2005). It is therefore likely that the acid glutamates and aspartate (2 each) in the amelogenin C terminus may facilitate binding to the HAP surface.

When amelogenins assemble in a neutral aqueous environment, the hydrophilic C termini extend on the outside of the amelogenin nanostructures (Margolis et al. 2006). The exposed C terminal amino acids can bind to the apatite crystal surfaces, and also to positively charged protons and calcium in solution (Le et al. 2006; Ryu et al. 1998). Moreover, the positively charged groups of amelogenin (such as amine groups at the N-terminus and side chains) may also contribute to the apatite binding through their interactions with the negatively charged groups, such as phosphates, on the HAP surface.

The concentration of amelogenin influences HAP crystal growth. HAP crystal formation requires a supersaturated calcium phosphate environment (Eanes 1976; Eanes et al. 1965). In solution, amelogenins inhibit crystal growth in the presence of high concentrations of calcium and phosphate (Aoba et al. 1989; Iijima and Moradian-Oldak 2004). However, Beniash and co-workers observed that pre-assembled full-length amelogenin (amelogenin lacking the exon 4 transcript) increases the size of formed crystals at low concentrations of calcium and phosphate (1.5 and 2.5 mM) (Beniash et al. 2005). In the presence of calcium and phosphate concentrations similar to those measured in enamel fluid (0.5 and 2.5 mM, respectively); crystals do not grow well on an apatite surface (Aoba and Moreno 1987; Habelitz et al. 2005). However, with the addition of amelogenin at low concentration (0.4 mg/ml), a few nanospheres assemble with minimal crystals formation, and at higher concentrations of amelogenin (1.6 mg/ml) more nanospheres assemble and longer fibrous crystals are formed (Habelitz et al. 2004).

At these higher amelogenin concentrations, amelogenin nanospheres assemble as a monolayer on the crystal surface. It appears that this monolayer of amelogenin nanospheres functions differently from amelogenin monomers in solution (Aoba et al. 1989) or in an amelogenin gel (Iijima et al. 2002). The amelogenin monolayer nanostructure may create a highly charged local environment to attract calcium and phosphate ions to form a saturated niche on the crystal surface for crystal growth.

Additional multilayers of amelogenin nanospheres physically protect the fragile crystals, but they also inhibit further crystal growth. Therefore, the layer of amelogenin adsorbed on the crystal surface needs to be preferentially removed to release the space for crystal growth.

11.1.5.1 Disassembly of Amelogenin Bound on HAP Crystals

When bound to solid surfaces, protein conformations usually change (Hlady and Buijs 1996). Proteins, such as amelogenin, which can easily adsorb onto solid surfaces, have large changes in conformation upon binding, and less internal stability (Gray 2004). Tarasevich et al. found that at neutral pH, self-assembled amelogenin that was bound onto a single fluoroapatite (FAP) crystal, formed much smaller structures on the FAP surfaces, than the original nanospheres present in solution. Their studies provide strong evidence that amelogenin nanospheres undergo possible quaternary structural changes upon interacting with apatite surfaces (Tarasevich et al. 2009a, b). These adsorption-induced conformational changes of the amelogenin protein may further alter the subsequent interactions between amelogenin and proteinases.

Tao et al. further quantitatively analyzed the dynamics of this adsorption-induced amelogenin nanosphere disassembly. They reported that the amelogenin nanospheres disassembled onto the HAP surface to form oligomeric adsorbates (25-mer), the subunits of the larger nanosphere. They concluded that a surface-triggered disassembly mechanism actually reversed the process of oligomer nanosphere self-assembly (Tao et al. 2015). This supports the possibility that amelogenin–crystal interactions are involved in determining the extent to which oligomers adsorb, assemble, and disassemble.

Chen et al. used in situ atomic force microscopy (AFM) to show that on a positively charged surface, amelogenin first assembles as a relatively uniform population of decameric oligomers, and then becomes two main populations: higher-order assemblies of oligomers and amelogenin monomers. On negatively charged surfaces, amelogenin nanostructure disassembles into a film of monomers (Chen et al. 2011) (see Fig. 11.3). An acidic pH of the enamel matrix, resulted from HAP formation, may further disassemble amelogenin nanospheres (Beniash et al. 2012).

Therefore, a possible sequence for amelogenin and pH-mediated enamel matrix mineralization is as follows. Amelogenins bind to HAP crystals and then disassemble to form monolayers that attract calcium and phosphate ions to form a saturated niche on the crystal surface for crystal growth. As crystal growth is initiated, the matrix begins to acidify and the N-terminus of amelogenins, which contain more positively charged histidines, reduces its binding affinity to HAP crystals, while the hydrophilic amelogenin C termini containing more acidic amino acids, still hangs on the surface of crystals. This allows more access sites for proteinases, facilitating the digestion of bound amelogenin. As the matrix neutralizes, HAP synthesis begins, and as the crystals grow, the surface of newly formed crystals will contact another layer of proteins in the enamel matrix. This then initiates amelogenin disassembly again, and the cycle repeats itself during enamel maturation.

11.1.6 Enamel Matrix Proteinases, MMP-20, and KLK4

The transition of enamel matrix from protein-dominated content to more than 95% mineralized tissue requires hydrolysis of matrix proteins to allow continued crystal growth as enamel matures. MMP20 and KLK4 are two major proteinases involved

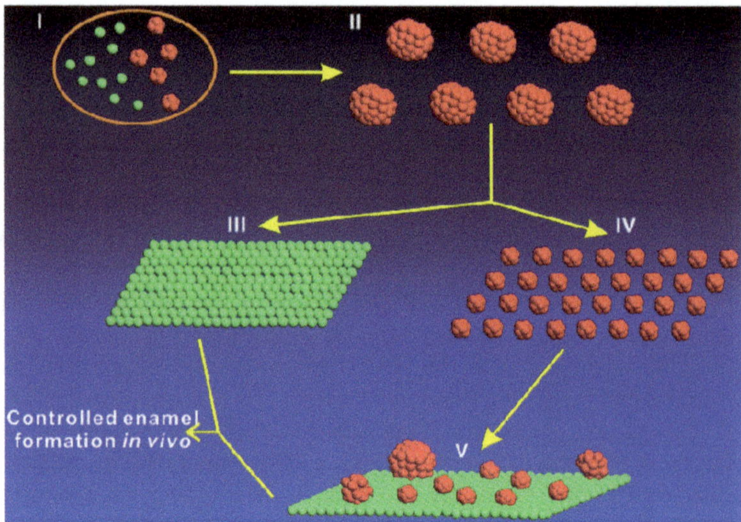

Fig. 11.3 A proposed pathway of amelogenin self-assembly and structural dynamics in vivo. Intracellular amelogenin monomers (green) and hexamers (red) in ameloblast cells (I). After amelogenin proteins are secreted into the matrix, they assemble into nanospheres (II). A monolayer of amelogenin monomers forms after the nanospheres interact with a negatively charged hydrophilic surface (III). Decameric amelogenin oligomers form following the disassembly of nanospheres through interaction with positively charged surfaces (IV). Decameric oligomers exhibit unexpected structural dynamics on positively charged surfaces in situ and form a mixture of higher-order assemblies of oligomers and monomers (V) (Chen et al. 2011)

in the hydrolysis of amelogenins and other enamel matrix proteins, to provide space for enamel crystals growth (Bartlett et al. 1998; Bartlett and Simmer 1999; Caterina et al. 1999; Fukae et al. 1998; Hu et al. 2002; Ryu et al. 2002).

MMP20, first reported by Bartlett and co-workers (Bartlett et al. 1998; Llano et al. 1997), belongs to the MMP superfamily and shares a similar structure with most other family members (Massova et al. 1998). Recombinant porcine, mouse, bovine, and human MMP20s have been expressed, purified, and activated in vitro and can hydrolyze both recombinant and native amelogenins (Li et al. 1999; Llano et al. 1997).

Amelogenin hydrolysis is initiated in secretory stage enamel at its C-terminal telopeptide. The most prevalent splice variants code the full length 25 kDa amelogenin that lacks exon 4 and the leucine-rich amelogenin polypeptide (LRAP) precursor. The 25 kDa amelogenin is cleaved into products of 20 kDa amelogenin and then a 5.0 kDa tyrosine-rich amelogenin polypeptide (TRAP). Similarly, the LRAP precursor is cleaved at the C terminus to form a 6.5 kDa LRAP (Fincham et al. 1981, 1989). Twenty kDa amelogenin, LRAP, and TRAP are the three major hydrolytic products in the secretory enamel protein matrix (Brookes et al. 1995).

KLK4 is upregulated during the transition stage. KLK4 rapidly cleaves amelogenins at multiple sites, accelerating amelogenin degradation during enamel

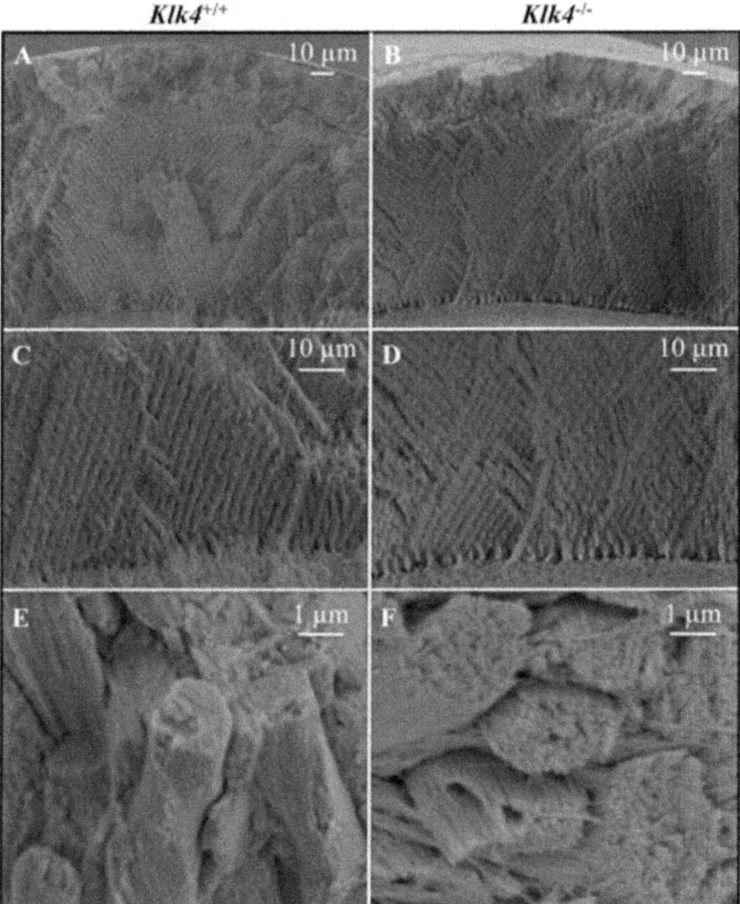

Fig. 11.4 Comparison of enamel from the wild-type and *Klk4* null mice. (**a**, **b**) SEM at the same scale showing mandibular incisor enamel from the DEJ (bottom) to the surface (top) that has been fractured in the erupted portion by pressing on it with a knife. There is no observable difference in the overall thickness of the enamel layer between the wild-type and *Klk4* null mice. (**c**, **d**) Higher magnification showing the decussating patterns of enamel rods just above the dentin enamel junction. (**e**, **f**) enamel rods in the wild-type mice have tightly packed crystallites that lose some aspect of their individuality. Enamel rods in the *Klk4* null mice are composed of distinctly individual crystallites resembling angel hair spaghetti. Holes or vacancies in some rods give the impression that smaller bundles of crystallites broke at a slightly deeper level and slid out of the rod (Simmer et al. 2009)

maturation, to allow the final thickening of enamel crystals. The importance of KLK4 in the final stage of enamel mineralization is shown in the KLK4 knockout mouse, where the enamel layer is normal in thickness with an intact decussating structure but the crystals remain separated, resembling a pattern of uncooked angel hair spaghettis (Fig.11.4) (Bartlett and Simmer 1999; Ryu et al. 2002; Simmer et al. 2009).

11.1.6.1 Optimal Conditions for Proteinase Activity in the Enamel Matrix

In the secretory matrix, MMP-20 cleaves amelogenin at neutral pH (Fukae et al. 1998), whereas KLK4 has an optimal pH close to 6.1 (Lu et al. 2008). In vitro, optimal KLK4 activity against a peptide substrate occurs even lower, around pH 5, and at neutral pH, KLK4 activity was reduced by almost half (Fig. 11.5, unpublished data).

This acidic optimal pH for KLK4 suggests the importance of acidic conditions in the degradation of amelogenin and other enamel proteins during tooth enamel maturation (Sasaki et al. 1991a). In addition, MMP20 is reported to be able to activate KLK4 (Yamakoshi et al. 2013). Interestingly, some MMPs are reported to be activated by low pH, which plays an important role in caries formation (Amaral et al. 2018). Therefore, the sequential activation of KLK4 by MMP20 remaining in the maturation stage matrix may also contribute to the tooth enamel maturation.

The absorption of amelogenin onto hydroxyapatite also affects its hydrolysis. When amelogenin is absorbed onto hydroxyapatite it is readily hydrolyzed and removed by enamel matrix proteinases (88% for MMP20 and 98% for KLK4), at significantly higher rates as compared to hydrolysis of amelogenin in solution. There are also a greater number of cleavage sites as identified by LC-MALDI MS/MS, when amelogenin is hydrolyzed after absorption onto HAP as compared to those in solution (Zhu et al. 2014). These results suggest that the adsorption of amelogenin to HAP results in their preferential and selective degradation and removal from HAP by both MMP20 and KLK4. It may be that the disassembly of nanospheres into monomers at low pH and through amelogenin–crystal interactions results in more cleavage sites accessible to enzymes and lead to a higher rate of hydrolysis.

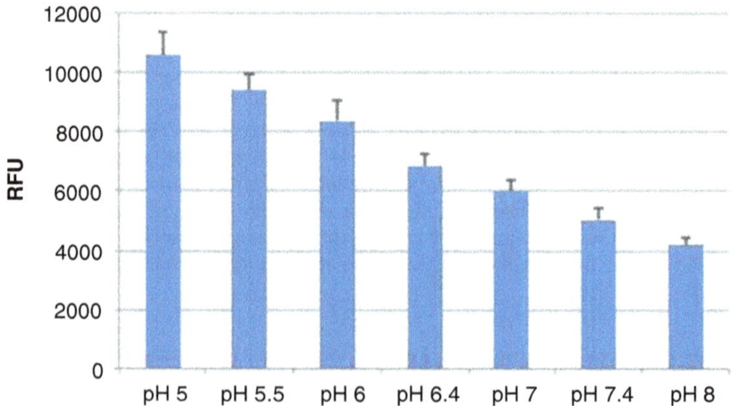

Fig. 11.5 Increased KLK4 activities with reduced pH. KLK4 activities reduced with increased pH, analyzed using a quenched fluorescent substrate for KLK4

11.1.7 A Model for pH Cycling and KLK4-Regulated Enamel Maturation

How does pH cycling affect enamel matrix mineralization? A possible cyclic sequence of mineralization and protein removal is in Fig. 11.6.

Adsorption of amelogenins onto the crystal surface results in conformational changes of the bound proteins, which then disassemble into oligomers or monomers on the surface of HAP crystals. These conformational changes may both increase local supersaturation of calcium and phosphate at the HAP surface, and expose more amelogenin cleavage sites to proteinases.

The hydrolysis and removal of amelogenins from the crystal surface open up the space surrounding the crystals, possibly with rapid precipitation of amorphous calcium phosphate (ACP), which then subsequently transforms to HAP. HAP formation releases protons contributing to matrix acidification, which then enhances amelogenin disassembly and KLK4 activity resulting in more protein hydrolysis. When the pH is regulated by ameloblasts to neutral, nascent crystals growing into this new space come into contact with the next layer of amelogenin nanospheres, and yet another cycle of the interaction-mediated preferential removal of bound amelogenin and crystal growth is initiated. The cycles of binding-growth-hydrolysis repeat until most of the matrix proteins are removed and the HAP crystals grow to fill up the entire enamel space to form the hardest tissue in our body.

11.2 Part 2: Regulations of Ion-Pumps, Transporters and Carbonic Anhydrases (CAs) Involved in Enamel Mineralization and pH Modulation

Maturation stage ameloblasts are responsible for the transport of 86% of minerals that are required for enamel matrix biomineralization (Smith 1998). As maturation stage ameloblasts cycle between smooth- and ruffle-ended ameloblasts, the

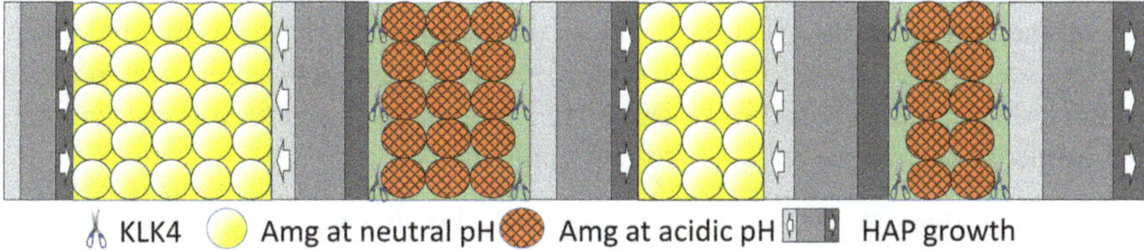

Fig. 11.6 Diagram showing a hypothetical sequence of the effect of pH-cycling on HAP crystal growth in maturing tooth enamel. When amelogenins (depicted as spheres) move from neutral (yellow) to acidic (orange) KLK4 activity is increased and amelogenins are hydrolyzed. When the matrix returns to neutral pH, HAP crystal growth (arrows) increases in width, and then this process repeats as the matrix acidifies, to inhibit HAP growth and increase amelogenin hydrolysis.

expression and spatial cellular distribution of ion-transporters and carbonic anhydrases modulate to direct extracellular matrix pH cycling (Smith et al. 1996) while also transporting calcium and phosphate into the enamel matrix for crystal growth. Genes coding ion pumps (V-type-H^+ATPase), transporters (CFTR, SLC26A, AE2, NHE1, NBCe1, NKCC1), and CAs (CA2 and CA6) are upregulated in ruffle-ended ameloblasts (Lacruz et al. 2011; Kim and Hong 2018). Reciprocal interactions between maturation ameloblasts and underlying enamel matrix contribute to the changes in gene expression between ruffle-ended and smooth-ended ameloblast in this unique process of enamel biomineralization.

11.2.1 Regulation of Matrix pH

11.2.1.1 Matrix Acidification

Cyclic acidification of the enamel matrix is associated with matrix mineralization (see Part 1). Matrix acidification occurs with the formation of hydroxyapatite crystals, which generate protons as the crystals form, and may also possibly be enhanced by extrusion of protons into the extracellular enamel matrix by V-H^+ATPase and NHE1 (SLC(A1).

V-Type H^+-ATPase subunits (Atp6v0d2, Atp6v1b2, Atp6v1c1 and Atp6v1e1) are highly upregulated in maturation stage ameloblasts. These subunits are involved in vesical acidification required for endocytosis of matrix proteins. The presence of subunits ATP6v1c1, ATP6v1e1, and Atp6v0a1 at the apical membrane of ruffle-ended maturation ameloblasts (Damkier et al. 2014; Josephsen et al. 2010; Sarkar et al. 2016), suggests an additional possibility, that these subunits of V-type H^+-ATPase may export protons to assist the enamel matrix acidification. However, only a relatively small amount of bicarbonate is available for buffering the protons that are rapidly released as hydroxyapatite crystals form (Smith 1998). Thus, protons adding to the enamel matrix through V-type H^+-ATPase may have only a limited role in further acidifying the enamel matrix.

NHE1, which is encoded by the SLC9A1 gene extrudes protons into the matrix as it mediates Na^+ influx and H^+ efflux. NHE1 as an acid extruder was identified in the lateral membrane during both the secretory and the maturation stages of ameloblasts (Josephsen et al. 2010). SLC9A1 is widely expressed with a preferential expression in the kidney, intestine, and liver, and NHE1 protein has a central role in regulating pH homeostasis, cell migration, and cell volume.

Calcium Transport Acidification of the enamel matrix occurs when calcium and phosphate are transported into the matrix to form hydroxyapatite crystals. Maturation stage ameloblasts are responsible for the deposition of more than 70% of the calcium in enamel tissues (Smith 1998). This transport of calcium by the ameloblasts requires tight control of intracellular calcium concentrations ($[Ca^{2+}]$) to avoid Ca^{2+} toxicity (Hubbard 2000).

Calcium transport into ameloblasts is mediated by store-operated calcium entry **(SOCE)** (Eckstein et al. 2017; Nurbaeva et al. 2017), ORAI, as well as by the cation channel **TRPM7** (transient receptor potential cation channel subfamily M) (Faouzi et al. 2017; Nakano et al. 2016) and **connexin 43** (Al-Ansari et al. 2018; Toth et al. 2010). The majority of Ca^{2+} is stored in the endoplasmic reticulum lumen, and the rest of excess Ca^{2+} binds to cytoplasmic calcium-binding proteins, including **calbindin-D9k, -D28k, calmodulin** (Berdal et al. 1993; Berdal et al. 1996; Hubbard 1995).

Calcium extrudes from ameloblasts into the enamel matrix through the basolateral and apical transmembrane Ca^{2+} pump, cotransporters/ion-exchangers, including **Ca^{2+}-ATPase (PMCA), potassium-independent Na^+/Ca^{2+} (NCX) and potassium-dependent Na^+/Ca^{2+} exchangers (NCKX)** (Nurbaeva et al. 2017). Among the four types of PMCA subunits, PMCA-1 (ATP2B1) and PMCA-4 (ATP2B4) are the predominant isoforms of PMCA in human and mouse ameloblasts. Though the expression levels of these two isoforms remain comparable between secretory and maturation stage ameloblasts (Borke et al. 1993; Borke et al. 1995; Zaki et al. 1996), in maturation stage ameloblasts PMCA has been immunolocalized at the apical surface (Salama et al. 1987; Zaki et al. 1996).

As compared to PMCA, ion exchangers **NCX** and **NCKX** have a higher turnover rate of delivering calcium (Eisenmann et al. 1982), which is critical for providing a large amount of calcium for the rapid growth of hydroxyapatite crystals. The NCX isoform exchanges one Ca^{2+} outward for three Na^+ inward (Yu and Choi 1997). *Ncx1* and *Ncx3* have been found to be equally expressed by secretory and maturation stage ameloblasts (Lacruz et al. 2012), suggesting that NCXs may act as housekeeping calcium transport in both secretory and maturation stage ameloblasts.

Whole transcriptome analysis shows that NCKX4 is the dominant isoform of the NCKX family expressed by ameloblasts, and is significantly upregulated in maturation ameloblasts as compared to secretory ameloblasts (Hu et al. 2012; Lacruz et al. 2012). NCKX4 can extrude one Ca^{2+} and one K^+ in exchange for four Na^+ inward (Lytton 2007). In humans, the *NCKX4* gene has been identified as a causal gene of amelogenesis imperfecta (Jalloul et al. 2016; Parry et al. 2013; Smith et al. 2017).

Phosphate Transport

Hydroxyapatite has a calcium/phosphate ratio of 1.67, and therefore, also requires phosphate to be transported to the enamel matrix. A Na^+-dependent Pi transporter such as **NaPi-2b (SLC34A2),** which is located in ruffle-ended ameloblasts, may operate in a coordinated way with NCKX4 to direct phosphate into the cells (Bronckers 2017; Bronckers et al. 2015).

11.2.1.2 Buffering to Neutralize Matrix pH

Secretory Stage In the secretory stage enamel, where hydroxyapatite crystal formation is initiated to form long thin crystals, matrix acidification as a result of crystal formation, is likely buffered by the amelogenins in the secretory enamel matrix (Guo et al. 2015). Amelogenins are rich in histidine, which can absorb protons (Bansal et al. 2012; Simmer and Fincham 1995).

Maturation Stage In the maturation stage, there are open junctions between smooth-ended ameloblasts, whereas ruffled-ended ameloblasts have tight apical junctions and open basal junctions (Skobe et al. 1985). The intercellular space between smooth-ended ameloblasts allows the free paracellular movement of fluid, ions, and small molecules, which might be able to correct the imbalance of pH immediately in the confined enamel space (Hanawa et al. 1990; Kawamoto and Shimizu 1997; McKee et al. 1986; Smith et al. 1987; Takano 1995; Takano et al. 1982b). This has been described as a "fluid flush" between smooth-ended ameloblasts, which brings in "base" and takes away "acid" to aid in the pH regulation activities.

In addition to this possibility, it is widely accepted that extrusion of bicarbonate by ameloblasts into the extracellular spaces buffers the intense acid loading during the rapid crystal growth in mineralizing enamel (Lacruz et al. 2010; Smith 1998; Varga et al. 2018; Yin and Paine 2017). Whole transcriptome analyses have shown that as compared to secretory ameloblasts, maturation ameloblasts have significantly upregulated genes involved in bicarbonate transport. These genes include: the chloride/bicarbonate channel **CFTR** (cystic fibrosis transmembrane conductance regulator); **NBCe1** (sodium bicarbonate exchanger, solute carrier family 4 member 4); **CA2** (carbonic anhydrase CAII) and **CA6** (CAVI); **AE2** (anion exchange protein 2) and the bicarbonate chloride exchangers **SLC26A3/A4/A6/A7** (solute carrier 26, members A3, A4, A, and A7) (Bronckers 2017; Kim and Hong 2018; Lacruz et al. 2012; Simmer et al. 2014).

Ion Transporters Ion transporters localized on *apical ameloblast membranes* include CFTR (Bronckers et al. 2010), SLC26 members (Yin et al. 2015), and CA (Bronckers 2017). CFTR likely contributes to bicarbonate transport in epithelial cells both directly by permeation through the channel, and indirectly by facilitating the function of Slc26 (Fong 2012). Using large-scale transcriptomic analysis, SLC26A4 has been shown as a target gene modulated by NaF and bisphenol in dental epithelia of rats (Jedeon et al. 2016b). To maintain bicarbonate supplies for extracellular transport, maturation-stage ameloblasts use **carbonic anhydrases,** CA2 and CA6 to catalyze the hydration of CO_2. CA2 generates bicarbonate in the cytosol of ameloblasts (Josephsen et al. 2010; Lin et al. 1994; Toyosawa et al. 1996), and CA6, the secreted form of carbonic anhydrase, may generate bicarbonate in the enamel space (Smith et al. 2006).

Ion transporters localized on the *basal ameloblast membranes* include AE2, NBCe1, NHE1, and NKCC1. AE2 encoded by the SLC4A2 gene is a $Cl^-/HCO3^-$ exchanger that exports bicarbonate into the extracellular space, in exchange for Cl^- (Lyaruu et al. 2008; Paine et al. 2008). Bicarbonate can also be transported into ameloblasts by NBCe1, encoded by the SLC9A1 gene, which also mediates inward Na^+ influx and H^+ efflux (Josephsen et al. 2010; Lacruz et al. 2010; Paine et al.

2008). NKCC1, encoded by the SLC12A2 gene, mediates sodium and chloride transport and reabsorption.

pH Sensing Proteins pH sensing proteins that may direct ameloblast regulation of matrix pH include the G-protein-coupled receptor 68 (GPR68). Defects in GPR68 are associated with amelogenesis imperfecta, suggesting a functional role of this pH sensor in ameloblast modulation of matrix pH cycling (Varga et al. 2018). GPR68 can also regulate steroid receptors, also present in maturation-stage ameloblasts (Houari et al. 2016). Steroid hormones and their receptors were reported to regulate CFTR, SLC26A6, NHE, and CA in several tissues (Gholami et al. 2013).

External Factors Regulating Ion Transport by Ameloblasts Ameloblasts targeted by environmental toxicants and chemicals with endocrine-disrupting activities result in alterations in enamel formation (Babajko et al. 2017; Jedeon et al. 2013, 2016a, b) (see also Chap. 12). All of the transporters described in this chapter, with the possible exception of AE2, can be modulated by hormones or molecules with endocrine activity.

Hormonal changes that may affect enamel formation also include possible effects on the function of the HPA (hypothalamic–pituitary–adrenal) axis during enamel formation. Evidence for this possibility is the finding by Boyce et al., that children entering kindergarten who had increased salivary cortisol reactivity, had primary (deciduous) mandibular incisors with thinner and more hypomineralized enamel (Boyce et al. 2010). Cortisol reactivity in young children is associated with maternal prenatal stress (Luecken et al. 2013), and as the primary mandibular incisors are formed during the prenatal period of life, this suggests a possible link between alternations in HPA axis regulations and enamel formation.

11.2.2 Conclusions

Ameloblast Regulation of pH Cycling Our understanding of how ameloblasts modulate and regulate pH cycling, as described in this chapter, has been advanced through immunolocalization of channels, transporters, and exchangers on ameloblast membranes, as well as in the use of cell culture models to investigate calcium transport and intracellular pH regulation (Bori et al. 2016; Bronckers 2017; Bronckers et al. 2015; Nurbaeva et al. 2015) (see Fig. 11.7).

However, much remains to be understood about the regulation of pH cycling by ameloblasts. As we further understand the mechanisms by which pH is regulated in enamel we will be better able to ensure optimal amelogenesis, enamel formation and mineralization.

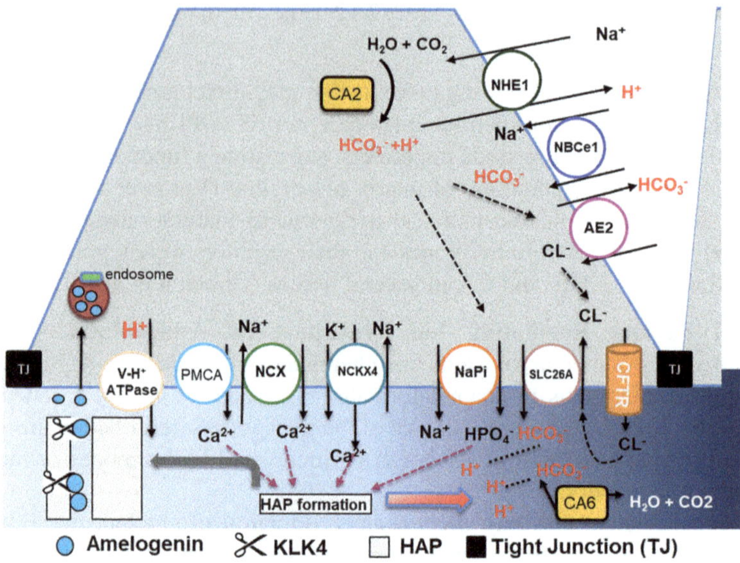

Fig. 11.7 Model of ion transporters in maturation stage ameloblasts to regulate matrix pH from acidic (light blue) to basic (dark blue), as HAP is formed

Acknowledgements The work was supported by NIH/NIDCR fundings R01DE027971, R01DE025709 and R01DE027076.

References

Abbarin N, San Miguel S, Holcroft J, Iwasaki K, Ganss B (2015) The enamel protein amelotin is a promoter of hydroxyapatite mineralization. J Bone Miner Res 30:775–785. https://doi.org/10.1002/jbmr.2411

Al-Ansari S et al (2018) The importance of Connexin 43 in enamel development and mineralization. Front Physiol 9:750. https://doi.org/10.3389/fphys.2018.00750

Amaral SF, Scaffa PMC, Rodrigues RDS, Nesadal D, Marques MM, Nogueira FN, Sobral MAP (2018) Dynamic influence of pH on metalloproteinase activity in human coronal and radicular dentin. Caries Res 52:113–118. https://doi.org/10.1159/000479825

Aoba T, Moreno EC (1987) The enamel fluid in the early secretory stage of porcine amelogenesis: chemical composition and saturation with respect to enamel mineral. Calcif Tissue Int 41:86–94. https://doi.org/10.1007/BF02555250

Aoba T, Moreno EC, Kresak M, Tanabe T (1989) Possible roles of partial sequences at N- and C-termini of amelogenin in protein-enamel mineral interactions. J Dent Res 68:1331–1336. https://doi.org/10.1177/00220345890680090901

Babajko S, Jedeon K, Houari S, Loiodice S, Berdal A (2017) Disruption of steroid axis, a new paradigm for molar incisor hypomineralization (MIH). Front Physiol 8:343. https://doi.org/10.3389/fphys.2017.00343

Bansal AK, Shetty DC, Bindal R, Pathak A (2012) Amelogenin: a novel protein with diverse applications in genetic and molecular profiling. J Oral Maxillofac Pathol 16:395–399. https://doi.org/10.4103/0973-029X.102495

Bartlett JD (2013) Dental enamel development: proteinases and their enamel matrix substrates. ISRN dentistry 2013:684607. https://doi.org/10.1155/2013/684607

Bartlett JD, Simmer JP (1999) Proteinases in developing dental enamel. Crit Rev Oral Biol Med 10:425–441

Bartlett JD, Ryu OH, Xue J, Simmer JP, Margolis HC (1998) Enamelysin mRNA displays a developmental defined pattern of expression and encodes a protein which degrades amelogenin. Connect Tissue Res 39:101–109. https://doi.org/10.3109/03008209809023916

Beniash E, Simmer JP, Margolis HC (2005) The effect of recombinant mouse amelogenins on the formation and organization of hydroxyapatite crystals in vitro. J Struct Biol 149:182–190. https://doi.org/10.1016/j.jsb.2004.11.001

Beniash E, Metzler RA, Lam RS, Gilbert PU (2009) Transient amorphous calcium phosphate in forming enamel. J Struct Biol 166:133–143. https://doi.org/10.1016/j.jsb.2009.02.001

Beniash E, Simmer JP, Margolis HC (2012) Structural changes in amelogenin upon self-assembly and mineral interactions. J Dent Res 91:967–972. https://doi.org/10.1177/0022034512457371

Berdal A, Hotton D, Pike JW, Mathieu H, Dupret JM (1993) Cell- and stage-specific expression of vitamin D receptor and calbindin genes in rat incisor: regulation by 1,25-dihydroxyvitamin D3. Dev Biol 155:172–179. https://doi.org/10.1006/dbio.1993.1016

Berdal A, Hotton D, Saffar JL, Thomasset M, Nanci A (1996) Calbindin-D9k and calbindin-D28k expression in rat mineralized tissues in vivo. J Bone Miner Res 11:768–779. https://doi.org/10.1002/jbmr.5650110608

Bori E et al (2016) Evidence for bicarbonate secretion by ameloblasts in a novel cellular model. J Dent Res 95:588–596. https://doi.org/10.1177/0022034515625939

Borke JL, Zaki AE, Eisenmann DR, Ashrafi SH, Ashrafi SS, Penniston JT (1993) Expression of plasma membrane Ca^{++} pump epitopes parallels the progression of enamel and dentin mineralization in rat incisor. J Histochem Cytochem 41:175–181

Borke JL, Zaki A-M, Eisenmann DR, Mednieks MI (1995) Localization of plasma membrane Ca^{2+} pump mRNA and protein in human ameloblasts by in situ hybridization and immunohistochemistry. Connect Tissue Res 33:139–144

Boyce WT, Den Besten PK, Stamperdahl J, Zhan L, Jiang Y, Adler NE, Featherstone JD (2010) Social inequalities in childhood dental caries: the convergent roles of stress, bacteria and disadvantage. Soc Sci Med 71:1644–1652. https://doi.org/10.1016/j.socscimed.2010.07.045

Bronckers AL (2017) Ion transport by ameloblasts during amelogenesis. J Dent Res 96:243–253. https://doi.org/10.1177/0022034516681768

Bronckers A et al (2010) The cystic fibrosis transmembrane conductance regulator (CFTR) is expressed in maturation stage ameloblasts, odontoblasts and bone cells. Bone 46:1188–1196. https://doi.org/10.1016/j.bone.2009.12.002

Bronckers AL, Lyaruu D, Jalali R, Medina JF, Zandieh-Doulabi B, DenBesten PK (2015) Ameloblast modulation and transport of Cl(−), Na(+), and K(+) during amelogenesis. J Dent Res 94:1740–1747. https://doi.org/10.1177/0022034515606900

Brookes SJ, Robinson C, Kirkham J, Bonass WA (1995) Biochemistry and molecular biology of amelogenin proteins of developing dental enamel. Arch Oral Biol 40:1–14. https://doi.org/10.1016/0003-9969(94)00135-X

Caterina J, Shi J, Krakora S, Bartlett JD, Engler JA, Kozak CA, Birkedal-Hansen H (1999) Isolation, characterization, and chromosomal location of the mouse enamelysin gene. Genomics 62:308–311. https://doi.org/10.1006/geno.1999.5990

Chen C-L, Bromley KM, Moradian-Oldak J, JJ DY (2011) In situ AFM study of amelogenin assembly and disassembly dynamics on charged surfaces provides insights on matrix protein self-assembly. J Am Chem Soc 133:17406–17413. https://doi.org/10.1021/ja206849c

Daculsi G, Kerebel B, Kerebel LM, Mitre D (1984) High-resolution study by transmission electron microscopy of a microhypoplasia of the human enamel surface. Arch Oral Biol 29:201–203. https://doi.org/10.1016/0003-9969(84)90055-4

Damkier HH, Josephsen K, Takano Y, Zahn D, Fejerskov O, Frische S (2014) Fluctuations in surface pH of maturing rat incisor enamel are a result of cycles of H(+)-secretion by ameloblasts and variations in enamel buffer characteristics. Bone 60:227–234. https://doi.org/10.1016/j.bone.2013.12.018

DenBesten PK, Heffernan LM (1989) Separation by polyacrylamide gel electrophoresis of multiple proteases in rat and bovine enamel. Arch Oral Biol 34:399–404. https://doi.org/10.1016/0003-9969(89)90117-9

DenBesten P, Li W (2011) Chronic fluoride toxicity: dental fluorosis. Monogr Oral Sci 22:81–96. https://doi.org/10.1159/000327028

Duverger O et al (2014) Hair keratin mutations in tooth enamel increase dental decay risk. J Clin Investig 124:5219–5224. https://doi.org/10.1172/JCI78272

Eanes ED (1976) The interaction of supersaturated calcium phosphate solutions with apatitic substrates. Calcif Tissue Res 20:75–89. https://doi.org/10.1007/BF02546399

Eanes ED, Gillessen IH, Posner AS (1965) Intermediate states in the precipitation of hydroxyapatite. Nature 208:365–367. https://doi.org/10.1038/208365a0

Eckstein M et al (2017) Store-operated Ca(2+) entry controls ameloblast cell function and enamel development. JCI Insight 2:e91166. https://doi.org/10.1172/jci.insight.91166

Eisenmann DR, Ashrafi S, Zaki AE (1982) Multi-method analysis of calcium localization in the secretory ameloblast. J Dent Res Spec No:1555-1562

Faouzi M, Kilch T, Horgen FD, Fleig A, Penner R (2017) The TRPM7 channel kinase regulates store-operated calcium entry. J Physiol 595:3165–3180. https://doi.org/10.1113/JP274006

Fincham AG, Belcourt AB, Termine JD, Butler WT, Cothran WC (1981) Dental enamel matrix: sequences of two amelogenin polypeptides. Biosci Rep 1:771–778. https://doi.org/10.1007/BF01114799

Fincham AG, Hu YY, Pavlova Z, Slavkin HC, Snead ML (1989) Human amelogenins: sequences of "TRAP" molecules. Calcif Tissue Int 45:243–250. https://doi.org/10.1007/bf02556044

Fong P (2012) CFTR-SLC26 transporter interactions in epithelia. Biophys Rev 4:107–116. https://doi.org/10.1007/s12551-012-0068-9

Fouillen A, Dos Santos NJ, Mary C, Castonguay JD, Moffatt P, Baron C, Nanci A (2017) Interactions of AMTN, ODAM and SCPPPQ1 proteins of a specialized basal lamina that attaches epithelial cells to tooth mineral. Sci Rep 7:46683. https://doi.org/10.1038/srep46683

Fukae M et al (1998) Enamelysin (matrix metalloproteinase-20): localization in the developing tooth and effects of pH and calcium on amelogenin hydrolysis. J Dent Res 77:1580–1588. https://doi.org/10.1177/00220345980770080501

Gholami K, Muniandy S, Salleh N (2013) In-vivo functional study on the involvement of CFTR, SLC26A6, NHE-1 and CA isoenzymes II and XII in uterine fluid pH, volume and electrolyte regulation in rats under different sex-steroid influence. Int J Med Sci 10:1121–1134. https://doi.org/10.7150/ijms.5918

Gibson CW (2011) The amelogenin proteins and enamel development in humans and mice. J Oral Biosci 53:248–256. https://doi.org/10.2330/joralbiosci.53.248

Gibson CW et al (2001) Amelogenin-deficient mice display an amelogenesis imperfecta phenotype. J Biol Chem 276:31871–31875. https://doi.org/10.1074/jbc.M104624200

Goldberg M, Septier D, Bourd K, Hall R, George A, Goldberg H, Menashi S (2003) Immunohistochemical localization of MMP-2, MMP-9, TIMP-1, and TIMP-2 in the forming rat incisor. Connect Tissue Res 44:143–153. https://doi.org/10.1080/03008200390223927

Gray JJ (2004) The interaction of proteins with solid surfaces. Curr Opin Struct Biol 14:110–115. https://doi.org/10.1016/j.sbi.2003.12.001

Guo J, Lyaruu DM, Takano Y, Gibson CW, DenBesten PK, Bronckers AL (2015) Amelogenins as potential buffers during secretory-stage amelogenesis. J Dent Res 94:412–420. https://doi.org/10.1177/0022034514564186

Habelitz S, Kullar A, Marshall SJ, DenBesten PK, Balooch M, Marshall GW, Li W (2004) Amelogenin-guided crystal growth on fluoroapatite glass-ceramics. J Dent Res 83:698–702. https://doi.org/10.1177/154405910408300908

Habelitz S, Denbesten PK, Marshall SJ, Marshall GW, Li W (2005) Amelogenin control over apatite crystal growth is affected by the pH and degree of ionic saturation. Orthod Craniofac Res 8:232–238. https://doi.org/10.1111/j.1601-6343.2005.00343.x

Hanawa M, Takano Y, Wakita M (1990) An autoradiographic study of calcium movement in the enamel organ of rat molar tooth germs. Arch Oral Biol 35:899–906

He B, Huang S, Zhang C, Jing J, Hao Y, Xiao L, Zhou X (2011) Mineral densities and elemental content in different layers of healthy human enamel with varying teeth age. Arch Oral Biol 56:997–1004. https://doi.org/10.1016/j.archoralbio.2011.02.015

Hlady VV, Buijs J (1996) Protein adsorption on solid surfaces. Curr Opin Biotechnol 7:72–77. https://doi.org/10.1016/s0958-1669(96)80098-x

Houari S, Loiodice S, Jedeon K, Berdal A, Babajko S (2016) Expression of steroid receptors in ameloblasts during amelogenesis in rat incisors. Front Physiol 7:503. https://doi.org/10.3389/fphys.2016.00503

Hu JC, Sun X, Zhang C, Liu S, Bartlett JD, Simmer JP (2002) Enamelysin and kallikrein-4 mRNA expression in developing mouse molars. Eur J Oral Sci 110:307–315. https://doi.org/10.1034/j.1600-0722.2002.21301.x

Hu JC, Lertlam R, Richardson AS, Smith CE, McKee MD, Simmer JP (2011) Cell proliferation and apoptosis in enamelin null mice. Eur J Oral Sci 119(Suppl 1):329–337. https://doi.org/10.1111/j.1600-0722.2011.00860.x

Hu P, Lacruz RS, Smith CE, Smith SM, Kurtz I, Paine ML (2012) Expression of the sodium/calcium/potassium exchanger, NCKX4, in ameloblasts. Cells Tissues Organs 196:501–509. https://doi.org/10.1159/000337493

Hubbard MJ (1995) Calbindin28kDa and calmodulin are hyperabundant in rat dental enamel cells. Identification of the protein phosphatase calcineurin as a principal calmodulin target and of a secretion-related role for calbindin28kDa. Eur J Biochem 230:68–79. https://doi.org/10.1111/j.1432-1033.1995.tb20535.x

Hubbard MJ (2000) Calcium transport across the dental enamel epithelium. Crit Rev Oral Biol Med 11:437–466. https://doi.org/10.1177/10454411000110040401

Iijima M, Moradian-Oldak J (2004) Interactions of amelogenins with octacalcium phosphate crystal faces are dose dependent. Calcif Tissue Int 74:522–531. https://doi.org/10.1007/s00223-002-0011-3

Iijima M, Moriwaki Y, Wen HB, Fincham AG, Moradian-Oldak J (2002) Elongated growth of octacalcium phosphate crystals in recombinant amelogenin gels under controlled ionic flow. J Dent Res 81:69–73. https://doi.org/10.1177/002203450208100115

Jaeger C, Groom NS, Bowe EA, Horner A, Davies ME, Murray RC, Duer MJ (2005) Investigation of the nature of the protein–mineral interface in bone by solid-state NMR. Chem Mater 17:3059–3061. https://doi.org/10.1021/cm050492k

Jalloul AH, Rogasevskaia TP, Szerencsei RT, Schnetkamp PP (2016) A functional study of mutations in K^+-dependent Na^+-Ca^{2+} exchangers associated with amelogenesis imperfecta and non-syndromic oculocutaneous albinism. J Biol Chem 291:13113–13123. https://doi.org/10.1074/jbc.M116.728824

Jedeon K et al (2013) Enamel defects reflect perinatal exposure to bisphenol A. Am J Pathol 183:108–118. https://doi.org/10.1016/j.ajpath.2013.04.004

Jedeon K, Berdal A, Babajko A (2016a) Impact of three endocrine disruptors, Bisphenol A, Genistein and Vinclozolin on female rat enamel. Bull Group Int Rech Sci Stomatol Odontol 53:e28

Jedeon K, Houari S, Loiodice S, Thuy TT, Le Normand M, Berdal A, Babajko S (2016b) Chronic exposure to bisphenol A exacerbates dental fluorosis in growing rats. J Bone Miner Res 31:1955–1966. https://doi.org/10.1002/jbmr.2879

Josephsen K (1983) Indirect visualization of ameloblast modulation in the rat incisor using calcium-binding compounds. Scand J Dent Res 91:76–78. https://doi.org/10.1111/j.1600-0722.1983.tb00780.x

Josephsen K, Takano Y, Frische S, Praetorius J, Nielsen S, Aoba T, Fejerskov O (2010) Ion transporters in secretory and cyclically modulating ameloblasts: a new hypothesis for cellular control of preeruptive enamel maturation. Am J Physiol Cell Physiol 299:C1299–C1307. https://doi.org/10.1152/ajpcell.00218.2010

Kallenbach E (1973) The fine structure of tomes' process of rat incisor ameloblasts and its relationship to the elaboration of enamel. Tissue Cell 5:501–524. https://doi.org/10.1016/S0040-8166(73)80041-2

Kawamoto T, Shimizu M (1997) Pathway and speed of calcium movement from blood to mineralizing enamel. J Histochem Cytochem 45:213–230. https://doi.org/10.1177/002215549704500207

Kim HE, Hong JH (2018) The overview of channels, transporters, and calcium signaling molecules during amelogenesis. Arch Oral Biol 93:47–55. https://doi.org/10.1016/j.archoralbio.2018.05.014

Lacruz RS, Nanci A, Kurtz I, Wright JT, Paine ML (2010) Regulation of pH during amelogenesis. Calcif Tissue Int 86:91–103. https://doi.org/10.1007/s00223-009-9326-7

Lacruz RS, Smith CE, Chen Y-B, Hubbard MJ, Hacia JG, Paine ML (2011) Gene-expression analysis of early- and late-maturation-stage rat enamel organ. Eur J Oral Sci 119:149–157

Lacruz RS et al (2012) Identification of novel candidate genes involved in mineralization of dental enamel by genome-wide transcript profiling. J Cell Physiol 227:2264–2275. https://doi.org/10.1002/jcp.22965

Le TQ, Gochin M, Featherstone JD, Li W, DenBesten PK (2006) Comparative calcium binding of leucine-rich amelogenin peptide and full-length amelogenin. Eur J Oral Sci 114(Suppl 1):320–326. https://doi.org/10.1111/j.1600-0722.2006.00313.x

Li W, Machule D, Gao C, DenBesten PK (1999) Activation of recombinant bovine matrix metalloproteinase-20 and its hydrolysis of two amelogenin oligopeptides. Eur J Oral Sci 107:352–359

Lin HM, Nakamura H, Noda T, Ozawa H (1994) Localization of H(+)-ATPase and carbonic anhydrase II in ameloblasts at maturation. Calcif Tissue Int 55:38–45

Llano E et al (1997) Identification and structural and functional characterization of human enamelysin (MMP-20). Biochemistry 36:15101–15108. https://doi.org/10.1021/bi972120y

Lu Y, Papagerakis P, Yamakoshi Y, Hu JCC, Bartlett JD, Simmer JP (2008) Functions of KLK4 and MMP-20 in dental enamel formation. Biol Chem 389:695–700. https://doi.org/10.1515/BC.2008.080

Luecken LJ, Lin B, Coburn SS, MacKinnon DP, Gonzales NA, Crnic KA (2013) Prenatal stress, partner support, and infant cortisol reactivity in low-income Mexican American families. Psychoneuroendocrinology 38:3092–3101. https://doi.org/10.1016/j.psyneuen.2013.09.006

Lyaruu DM et al (2008) The anion exchanger Ae2 is required for enamel maturation in mouse teeth. Matrix Biol 27:119–127. https://doi.org/10.1016/j.matbio.2007.09.006

Lytton J (2007) Na^+/Ca^{2+} exchangers: three mammalian gene families control Ca^{2+} transport. Biochem J 406:365–382. https://doi.org/10.1042/BJ20070619

Margolis HC, Beniash E, Fowler CE (2006) Role of macromolecular assembly of enamel matrix proteins in enamel formation. J Dent Res 85:775–793. https://doi.org/10.1177/154405910608500902

Massova I, Kotra LP, Fridman R, Mobashery S (1998) Matrix metalloproteinases: structures, evolution, and diversification. FASEB J 12:1075–1095

McKee MD, Martineau-Doize B, Warshawsky H (1986) Penetration of various molecular-weight proteins into the enamel organ and enamel of the rat incisor. Arch Oral Biol 31:287–296. https://doi.org/10.1016/0003-9969(86)90042-7

Moffatt P, Smith CE, St-Arnaud R, Nanci A (2008) Characterization of Apin, a secreted protein highly expressed in tooth-associated epithelia. J Cell Biochem 103:941–956. https://doi.org/10.1002/jcb.21465
Nakano Y et al (2016) A critical role of TRPM7 as an ion channel protein in mediating the mineralization of the craniofacial hard tissues. Front Physiol 7:258. https://doi.org/10.3389/fphys.2016.00258
Nurbaeva MK et al (2015) Dental enamel cells express functional SOCE channels. Sci Rep 5:15803. https://doi.org/10.1038/srep15803
Nurbaeva MK, Eckstein M, Feske S, Lacruz RS (2017) Ca(2+) transport and signalling in enamel cells. J Physiol 595:3015–3039. https://doi.org/10.1113/JP272775
Paine ML et al (2008) Role of NBCe1 and AE2 in secretory ameloblasts. J Dent Res 87:391–395
Parry DA et al (2013) Identification of mutations in SLC24A4, encoding a potassium-dependent sodium/calcium exchanger, as a cause of amelogenesis imperfecta. Am J Hum Genet 92:307–312. https://doi.org/10.1016/j.ajhg.2013.01.003
Picard V, Govoni G, Jabado N, Gros P (2000) Nramp 2 (DCT1/DMT1) expressed at the plasma membrane transports iron and other divalent cations into a calcein-accessible cytoplasmic pool. J Biol Chem 275:35738–35745. https://doi.org/10.1074/jbc.M005387200
Pugach MK et al (2010) The amelogenin C-terminus is required for enamel development. J Dent Res 89:165–169. https://doi.org/10.1177/0022034509358392
Robinson C (2014) Enamel maturation: a brief background with implications for some enamel dysplasias. Front Physiol 5:388. https://doi.org/10.3389/fphys.2014.00388
Robinson C, Brookes SJ, Shore RC, Kirkham J (1998) The developing enamel matrix: nature and function. Eur J Oral Sci 106(Suppl 1):282–291. https://doi.org/10.1111/j.1600-0722.1998.tb02188.x
Ryu OH, Hu CC, Simmer JP (1998) Biochemical characterization of recombinant mouse amelogenins: protein quantitation, proton absorption, and relative affinity for enamel crystals. Connect Tissue Res 38:207–214
Ryu O et al (2002) Porcine kallikrein-4 activation, glycosylation, activity, and expression in prokaryotic and eukaryotic hosts. Eur J Oral Sci 110:358–365
Salama AH, Zaki AE, Eisenmann DR (1987) Cytochemical localization of Ca^{2+}-Mg^{2+} adenosine triphosphatase in rat incisor ameloblasts during enamel secretion and maturation. J Histochem Cytochem 35:471–482
Sarkar J, Wen X, Simanian EJ, Paine ML (2016) V-type ATPase proton pump expression during enamel formation. Matrix Biol 52-54:234–245. https://doi.org/10.1016/j.matbio.2015.11.004
Sasaki S, Takagi T, Suzuki M (1991a) Amelogenin degradation by an enzyme having acidic pH optimum and the presence of acidic zones in developing bovine enamel. In: Suga S, Nakahara H (eds) Mechanisms and phylogeny of mineralization in biological systems. Springer Japan, Tokyo, pp 79–81
Sasaki S, Takagi T, Suzuki M (1991b) Cyclical changes in pH in bovine developing enamel as sequential bands. Arch Oral Biol 36(3):227–231
Shimabayashi S, Uno T, Nakagaki M (1997) Formation of a surface complex between polymer and surfactant and its effect on the dispersion of solid particles. Colloids Surf A Physicochem Eng Aspects 123–124:283–295. https://doi.org/10.1016/S0927-7757(96)03820-4
Shin NY et al (2020) Amelogenin phosphorylation regulates tooth enamel formation by stabilizing a transient amorphous mineral precursor. J Biol Chem 295:1943–1959. https://doi.org/10.1074/jbc.RA119.010506
Simmer JP, Fincham AG (1995) Molecular mechanisms of dental enamel formation. Crit Rev Oral Biol Med 6:84–108
Simmer JP, Hu JCC (2002) Expression, structure, and function of enamel proteinases. Connect Tissue Res 43:441–449. https://doi.org/10.1080/03008200290001159
Simmer JP et al (1998) Purification, characterization, and cloning of enamel matrix serine proteinase 1. J Dent Res 77:377–386. https://doi.org/10.1177/00220345980770020601

Simmer JP, Hu Y, Lertlam R, Yamakoshi Y, Hu JC (2009) Hypomaturation enamel defects in Klk4 knockout/LacZ knockin mice. J Biol Chem 284:19110–19121. https://doi.org/10.1074/jbc.M109.013623

Simmer JP, Richardson AS, Wang SK, Reid BM, Bai Y, Hu Y, Hu JC (2014) Ameloblast transcriptome changes from secretory to maturation stages. Connect Tissue Res 55(Suppl 1):29–32. https://doi.org/10.3109/03008207.2014.923862

Skobe Z, LaFrazia F, Prostak K (1985) Correlation of apical and lateral membrane modulations of maturation ameloblasts. J Dent Res 64:1055–1061. https://doi.org/10.1177/00220345850640080601

Smith CE (1998) Cellular and chemical events during enamel maturation. Crit Rev Oral Biol Med 9:128–161. https://doi.org/10.1177/10454411980090020101

Smith CE, Nanci A (1989) Secretory activity as a function of the development and maturation of ameloblasts. Connect Tissue Res 22:147–156

Smith CE, McKee MD, Nanci A (1987) Cyclic induction and rapid movement of sequential waves of new smooth-ended ameloblast modulation bands in rat incisors as visualized by polychrome fluorescent labeling and GBHA-staining of maturing enamel. Adv Dent Res 1:162–175

Smith CE, Issid M, Margolis HC, Moreno EC (1996) Developmental changes in the pH of enamel fluid and its effects on matrix-resident proteinases. Adv Dent Res 10:159–169

Smith CE, Nanci A, Moffatt P (2006) Evidence by signal peptide trap technology for the expression of carbonic anhydrase 6 in rat incisor enamel organs. Eur J Oral Sci 114:147–153. https://doi.org/10.1111/j.1600-0722.2006.00273.x

Smith CEL, Poulter JA, Antanaviciute A, Kirkham J, Brookes SJ, Inglehearn CF, Mighell AJ (2017) Amelogenesis imperfecta; genes, proteins, and pathways. Front Physiol 8:435. https://doi.org/10.3389/fphys.2017.00435

Takagi T, Ogasawara T, Tagami J, Akao M, Kuboki Y, Nagai N, LeGeros RZ (1998) pH and carbonate levels in developing enamel. Connect Tissue Res 38:181–187. discussion 201-185. https://doi.org/10.3109/03008209809017035

Takano Y (1995) Enamel mineralization and the role of ameloblasts in calcium transport. Connect Tissue Res 33:127–137

Takano Y, Crenshaw MA, Bawden JW, Hammarström L, Lindskog S (1982a) The visualization of the patterns of ameloblast modulation by the glyoxal bis(2-hydroxyanil) staining method. J Dent Res Spec No:1580-1587

Takano Y, Crenshaw MA, Reith EJ (1982b) Correlation of 45Ca incorporation with maturation ameloblast morphology in the rat incisor. Calcif Tissue Int 34:211–213. https://doi.org/10.1007/BF02411236

Tao J, Buchko GW, Shaw WJ, De Yoreo JJ, Tarasevich BJ (2015) Sequence-defined energetic shifts control the disassembly kinetics and microstructure of amelogenin adsorbed onto hydroxyapatite (100). Langmuir 31:10451–10460. https://doi.org/10.1021/acs.langmuir.5b02549

Tarasevich BJ, Lea S, Bernt W, Engelhard M, Shaw WJ (2009a) Adsorption of amelogenin onto self-assembled and fluoroapatite surfaces. J Phys Chem B 113:1833–1842. https://doi.org/10.1021/jp804548x

Tarasevich BJ, Lea S, Bernt W, Engelhard MH, Shaw WJ (2009b) Changes in the quaternary structure of amelogenin when adsorbed onto surfaces. Biopolymers 91:103–107. https://doi.org/10.1002/bip.21095

Toei M, Saum R, Forgac M (2010) Regulation and isoform function of the V-ATPases. Biochemistry 49:4715–4723. https://doi.org/10.1021/bi100397s

Toth K, Shao Q, Lorentz R, Laird DW (2010) Decreased levels of Cx43 gap junctions result in ameloblast dysregulation and enamel hypoplasia in Gja1Jrt/+ mice. J Cell Physiol 223:601–609. https://doi.org/10.1002/jcp.22046

Toyosawa S, Ogawa Y, Inagaki T, Ijuhin N (1996) Immunohistochemical localization of carbonic anhydrase isozyme II in rat incisor epithelial cells at various stages of amelogenesis. Cell Tissue Res 285:217–225

Varga G, DenBesten P, Racz R, Zsembery A (2018) Importance of bicarbonate transport in pH control during amelogenesis - need for functional studies. Oral Dis 24:879–890. https://doi.org/10.1111/odi.12738

Wazen RM, Moffatt P, Zalzal SF, Yamada Y, Nanci A (2009) A mouse model expressing a truncated form of ameloblastin exhibits dental and junctional epithelium defects. Matrix Biol 28:292–303. https://doi.org/10.1016/j.matbio.2009.04.004

Wöltgens JH, Lyaruu DM, Bervoets TJ (1991) Possible functions of alkaline phosphatase in dental mineralization: cadmium effects. J Biol Buccale 19:125–128

Yamakoshi Y, Simmer JP, Bartlett JD, Karakida T, Oida S (2013) MMP20 and KLK4 activation and inactivation interactions in vitro. Arch Oral Biol 58:1569–1577. https://doi.org/10.1016/j.archoralbio.2013.08.005

Yin K, Paine ML (2017) Bicarbonate transport during enamel maturation. Calcif Tissue Int 101:457–464. https://doi.org/10.1007/s00223-017-0311-2

Yin K et al (2015) SLC26A gene family participate in pH regulation during enamel maturation. PLoS One 10:e0144703. https://doi.org/10.1371/journal.pone.0144703

Yu SP, Choi DW (1997) Na(+)-Ca2+ exchange currents in cortical neurons: concomitant forward and reverse operation and effect of glutamate. Eur J Neurosci 9:1273–1281. https://doi.org/10.1111/j.1460-9568.1997.tb01482.x

Zaki AE, Hand AR, Mednieks MI, Eisenmann DR, Borke JL (1996) Quantitative immunocytochemistry of Ca(2+)-Mg2+ ATPase in ameloblasts associated with enamel secretion and maturation in the rat incisor. Adv Dent Res 10:245–251

Zhu L, Liu H, Witkowska HE, Huang Y, Tanimoto K, Li W (2014) Preferential and selective degradation and removal of amelogenin adsorbed on hydroxyapatites by MMP20 and KLK4 in vitro. Front Physiol 5:268. https://doi.org/10.3389/fphys.2014.00268

Chapter 12
Environmental Factors and Enamel/Dentin Defects

Sylvie Babajko and Pamela Den Besten

Abstract The exposome that can also affect tooth formation is constituted by (1) general external factors (such as climate or pollution), (2) internal factors (including hormones, oxidative stress, metabolic factors), and (3) specific external environmental factors including contaminants, drugs, and lifestyles. These environmental factors include large numbers of molecules that are able to interact with various tissues. This review will concentrate on agents that have been identified as acting directly on dental tissues resulting in a disruption of tooth development and acquired pathologies diagnosed by clinical observation. Among these factors, **fluoride** is the most studied as its excessive absorption may lead to dental fluorosis. Another important group of molecules that can affect mineralized tissues are **endocrine disruptors,** which may contribute to the more recently described Molar Incisor Hypomineralization (MIH). These molecules alone or also in combination with other factors such as drugs including corticoids, tobacco-related contaminants, heavy metals, and probably many others, alter tooth formation and mineralization of tooth tissues. Environmental factors that affect the oral cavity include "contaminants" which are not known to the individual exposed to them (most notably in water or food). Others are "toxics" when they are consumed with knowledge of their side effects or the risks associated with their consumption, such as alcohol, tobacco, some medications, or even excessive sugar. Questions about the capacity of environmental factors for disruption and their mechanisms of action within the oral cavity (mineralized matrices, soft tissues, microbiota) are of critical importance for optimal oral health and should be the subject of increased research in the coming years.

S. Babajko (✉)
Molecular Oral Pathophysiology Laboratory, Centre de Recherche des Cordeliers, INSERM, Université de Paris, Sorbonne Université, Paris, France
e-mail: sylvie.babajko@inserm.fr

P. D. Besten
Department of Orofacial Sciences, University of California, San Francisco, San Francisco, CA, USA

M. Goldberg, P. Den Besten (eds.), *Extracellular Matrix Biomineralization of Dental Tissue Structures*, Biology of Extracellular Matrix 10,
https://doi.org/10.1007/978-3-030-76283-4_12

12.1 Fluoride and Fluorosis

Among the molecules with chronic exposure in the oral cavity, fluoride is one of the most widespread, as it is naturally present in drinking water and food, and is also commonly used to prevent caries. Due to its high reactivity, fluorine is never found isolated in its natural state but is always bound to various electropositive elements such as lithium, sodium, potassium, calcium, and magnesium. When combined with mono-charged cations, it is highly soluble in water. The high reactivity of the fluoride ion results in the enhanced mineral formation, and because of this, fluoride exposure to tooth enamel prevents caries by enhancing remineralization of early caries lesions. At higher concentrations, fluoride can also inhibit bacterial enolase activity (Hamilton 1990; Robinson et al. 2004).

Fluoride exposure to tissues and organs of the body can occur by ingestion, such as from food or water, or inhalation such as from associated with coal burning or inhalation anesthetics. Exposure to an excess of fluoride leads to the development of dental and bone fluorosis, which has been described since the first half of the twentieth century. Recent estimates suggest that around 200 million people from among 25 nations are subjected to the risk of fluorosis (Rasool et al. 2018).

Fluorosis is characterized by increasing white opacity, corresponding to increased hypomineralization and porosity. One of the first signs of fluorosis is the change in enamel color caused by lighter horizontal white lines (striae of Retzius) following the perikymata at the enamel surface. When fluorosis is more severe, the enamel is more extensively hypomineralized, and teeth may erupt with defects which posteruptively result in a loss of enamel from the tooth surface. More extensive defects can result in the loss of enamel from the dentin surface, and a secondary brown coloration as the dentin surface is stained posteruptively. The severity of the pathology depends on the dose of fluoride ingested, the time window of exposure relative to tooth formation, and the genetic background that may predispose to fluorosis. Recent hypotheses include the possibility that combinations of fluoride with other environmental factors present in drinking waters or food may aggravate the pathology or lower the sensitivity to fluoride (Jedeon et al. 2016a).

The mechanisms by which fluoride alters mineralization are likely caused by both extracellular and intracellular effects (Jalali et al. 2017). Maturation stage ameloblasts and enamel matrix are most affected by fluoride exposure. Rats exposed to increasing doses of fluoride (from 25 to 100 ppm) show no differences as compared to controls in the enamel secretion phase; but in the maturation stage the number of ameloblast modulations is reduced, and mineralization is affected (DenBesten 1986). The consequence is that fluoride results in an increase in the amount of matrix proteins retained in the maturation stage, related to a delay in the degradation of amelogenins during enamel maturation resulting in inhibition of crystal growth (DenBesten and Li 2011).

The mechanism by which fluoride reduces ameloblast modulations is not yet understood, However, what is known is that altered ameloblast modulation is associated with changes in the enamel matrix pH. As hydroxyapatite is formed, protons are released with acidification of matrix pH. A fluoride-mediated acceleration of the rate of hydroxyapatite formation results in hypomineralized bands of enamel, and then, as calcium is depleted, this is followed by a subsequent band of hypomineralized matrix. This pattern of fluoride-mediated hydroxyapatite crystal formation disrupts matrix pH modulation (DenBesten 1986; Lyaruu et al. 2014; Guo et al. 2015; Bronckers et al. 2009).

In addition, to these effects, fluoride may directly affect ameloblast function and modulation. In ameloblasts, fluoride can alter TGF-β signaling to affect the expression of the matrix proteinase, KLK4 (Suzuki et al. 2014; Le et al. 2017). Fluoride also induces the downregulation of the collagen related small leucine-rich peptide, asporin (Houari et al. 2014) in ameloblasts, and protein retention may be mediated by oxidative stress and ER stress in fluoride-treated ameloblasts (Sharma et al. 2008; Suzuki and Bartlett 2014; Suzuki et al. 2017).

Other cellular effects of fluoride have been identified in various model systems, including processes related to inflammation, metabolism, epithelium integrity, cell migration (see Barbier et al. 2010; Ji et al. 2018 for reviews), and neurotoxicity (Bashash et al. 2018; Green et al. 2019). Despite intensive research on fluoride effects in mineralized tissues and others, fluoride mechanisms of action are still unclear. Fluoride has been listed as an endocrine disruptor chemical, altering several endocrine axes including the androgen and thyroid axis (Duan et al. 2016; Yang et al. 2018; Chaitanya et al. 2018). Fluoride action in ameloblasts may involve steroid receptors (Le et al. 2017).

Fluoride can also alter bone and dentin mineralization and hardness (Vieira et al. 2005). Fluorotic dentin presents as hypermineralized sclerotic peritubular dentin, with hypomineralized areas. The tubules have an irregular distribution with a narrow, discontinuous lumen as compared to physiological dentin (Rojas-Sánchez et al. 2007). Severe fluorotic dentin shows an increased susceptibility to decay due to the enlargement of the size of its crystallites. These crystallites are not homogeneously arranged and the collagen fibrils are distributed anarchically (Waidyasekera et al. 2010). Rats chronically exposed to high-dose fluoride present with hypomineralized enamel and dentin as measured by EDS, with a decreased Ca/P ratio (Houari et al. 2014).

In addition to extracellular effects, fluoride is also able to modulate gene expression in odontoblasts, with effects on genes involved in mineralization including asporin (Houari et al. 2014). Interestingly, fluoride modulates asporin expression inversely in dentin as compared to enamel with both resulting in hypomineralization. Asporin promotes mineralization by interacting with collagen 1, and in dentin, which has high concentrations of type 1 collagen, fluoride-induced decreased asporin synthesis results in hypomineralization. Conversely, fluoride increases asporin synthesis in the enamel which contains minimal amounts of collagen, also resulting in hypomineralization (Houari et al. 2014). It has also recently been shown that excess fluoride alters the initiation of dentin mineralization in mice by reducing Wnt/β-catenin signaling to downregulate dentin-sialo-phosphoprotein production (Kao et al. 2020). As in ameloblasts, high doses of fluoride induce rapidly cell apoptosis in vitro (Karube et al. 2009).

12.2 Endocrine Disruptors and Enamel Hypomineralization

In the last several decades, environmental conditions are dramatically changing, with the accumulation of multiple, still unidentified, pollutants directly or indirectly contributing to many pathologies in humans and animals. Among them, molecules presenting endocrine-disrupting activities are suspected to increase the prevalence and the severity of many pathologies such as hormone-dependant cancers, fertility, behavior, diabetes, and obesity (Gore et al. 2015). They have also been reported to disrupt dental development (Jedeon et al. 2015).

Endocrine disruptors (ED) are widely present in cosmetics, food, medical devices, and equipment, though contamination is mostly oral, through food. EDs are a large heterogeneous group of organic and inorganic compounds (Karthikeyan et al. 2019). Among these, one of the most prevalent is bisphenols, which are found in the majority of the population (Philips et al. 2018). Most EDs have structural properties that allow them to passively cross the cell membranes and the placental barrier, making the fetal and early-post-natal time periods as particularly critical for ED actions on organ development. They are thus able to impact oral tissues and microbiota directly, or indirectly by when circulating after transmucosal absorption (Gayrard et al. 2013), dermal and oral ingestion.

Among the hundreds of environmental factors with endocrine-disrupting activity, **bisphenol A** (BPA) is one of the most studied with more than 10,000 publications referenced on PubMed during the last 20 years. More importantly, BPA is one of the 8 molecules identified by the European Union as an endocrine disruptor for health. Its use is highly regulated but populations are still contaminated. Several substitutes with similar structures are now proposed to replace BPA, notably BPS, though these may also present endocrine-disrupting activity despite different toxicokinetic properties (Gayrard et al. 2019).

In 2013, Jedeon and co-workers reported BPA impact on amelogenesis of rats daily exposed to low doses of BPA (5μg/kg/day) from conception until 30 or 100 days after birth. This dose is in the same range than environmental exposure with the tolerated daily intake now fixed at 4μg/kg/day. At 30 days, 75% of male rats and 30% of females presented opacities on their incisors. Conversely, 100 days old rats did not present any defects indicating a window of susceptibility to low-dose BPA. Enamel defects in rats exposed to BPA were only present on the teeth developing during the perinatal period and appeared as similar to those seen in humans with Molar Incisor Hypomineralization (MIH) (Jedeon et al. 2013). When enamel defects in rats exposed to BPA and in humans with MIH were compared by scanning electron microscopy and EDX, they were found to have similar features, including increased carbon content, consistent with a relative increase in organic material. In both rats and humans, albumin was found to be present in enamel, suggesting the possibility of ameloblast damage that would allow serum proteins to move past the otherwise closed apical gap junctions to the enamel surface (Farah et al. 2010).

MIH is an enamel pathology described in 2001 and is diagnosed in children around 6–8 years of age, presenting random white opacities on the enamel of first permanent molars and sometimes also on permanent incisors (Jälevik 2001; Weerheijm et al. 2001). MIH prevalence seems to increase in parallel to other pathologies associated with environmental factors cited above. MIH affects now more than 14% of children aged 6 to 10 years old all over the world (Zhao et al. 2018) whereas it was almost absent until the decade of the 1980s. Its etiology is likely multifactorial with a possible genetic or epigenetic susceptibility. Several causal factors have been proposed such as prematurity, extended breastfeeding, viral or bacterial infections, respiratory diseases, asthma, smoking, amoxicillin, vitamin deficiency (Alaluusua 2010; Serna et al. 2016; Silva et al. 2016). None of these factors is satisfactory to fully explain the occurrence of MIH, and changing environmental conditions and lifestyle may also be considered as possible causal factors. MIH also appears as a heterogeneous pathology with possibly multiple causal factors some of them being found in the environment.

BPA is able to bind to estrogen receptors (ERα and ERβ) (Delfosse et al. 2012), which are present in ameloblasts (Ferrer et al. 2005; Jedeon et al. 2014a), as well as to GPRE and ERRγ, also present in ameloblasts (personal communication). BPA can also modulate the action of many steroid receptors especially the androgen receptor (AR) and the glucocorticoid receptor (GR) both of which are highly expressed in dental epithelium and during amelogenesis (Houari et al. 2016). All these reasons, including common features between human enamel defects and those of rats exposed to BPA, led us to propose endocrine disruptors such as BPA as possible candidates contributing to MIH.

BPA is not the sole ED reported to disrupt amelogenesis. Similar data were reported for rats exposed to **vinclozolin**, a common dicarboximide fungicide (Ronilan) with a powerful anti-androgenic activity. Vinclozolin prevents AR nuclear translocation in ameloblasts and thus modulates the expression of enamel key genes involved in the mineralization process such as SLC24A6 (Jedeon et al. 2014b, 2016b).

Exposure to **2,3,7,8-tetrachlorodibenzo-p-dioxin (TCDD)** has also been associated with dental defects in humans (Alaluusua et al. 2004), and in experimental rodent models (Gao et al. 2004). TCDD belongs to the dioxin family constituted by dibenzo-p-dioxins (PCDD/TCDD), Polychlorodibenzofurans or furans (PCDF), and Polychlorated and polybromated Biphenyls (PCB/PBB). Contrary to bisphenols which have short half-lives, are eliminated in few hours and thus do not accumulate in contaminated organisms, these organic pollutants are persistent in bodies and the environment. Their mechanisms of action are still unclear but most of their genomic and non-genomic effects are transmitted by the ubiquitous Aryl Hydrocarbon receptor (AhR) involved in a broad spectrum of pathologies (Bock 2019 for review). AhR is able to interact with ER thus presenting endocrine-disrupting activity also.

The best-known dioxin accident took place in Seveso, Italy, in 1976. More than 30 kg of TCDD were spread over an area of approximately 18 km^2 with over a thousand people exposed to TCDD. Blood samples were collected for clinical chemistry tests and to quantify individual TCDD exposures, and these reports included an association between levels of TCDD contamination and occurrence of dental defects. In vitro experiments supported these human observations, with teeth having delayed enamel and dentin mineralization (Gao et al. 2004) and dysregulated expression of dentin sialophosphoprotein (Kiukkonen et al. 2006).

Several epidemiological studies reported dental defects in children exposed to **PCBs** in Japan as well as in Slovakia (Jan et al. 2007). Careful analysis of teeth shows that accumulation of PCBs may chronically affect dental tissues (Jan et al. 2013).

Interestingly, many EDs have been reported for their ability to disrupt steroidogenesis and corticoid axis (Gore et al. 2015). In fact, the question of corticoids' ability to alter enamel matrix is often raised and debated, notably for children with prescriptions for inhaled asthma drugs before the age of 3 years who seem to present more opacities due to enamel hypomineralization (Wogelius et al. 2010). Corticoids and EDs acting through their receptors (GR) may disrupt amelogenesis either by directly modulating enamel key genes such as amelogenin which harbors a GR responsive element in its promoter sequence (Gibson et al. 1997) or indirectly by their well-known action on circulating calcium and phosphate status and bone demineralization. More studies are required to completely understand the action of drugs and hormones acting through the corticoid axis including cortisol on enamel synthesis, including their ability to accumulate in enamel and interact with ameloblasts that express these receptors (Houari et al. 2016).

12.3 Combinations of Environmental Factors

Studies on the effects of EDs in dental development require further research to understand their mechanisms of action. While it is well known that steroid hormones (estrogens, androgens, retinoids, corticoids) and peptide hormones (GH for example) can impact dental development and affect enamel and dentin mineralization, EDs, may interact with and disrupt the activity of hormone receptors involved in dental development in a more powerful fashion than endogenous hormones.

Another important topic concerns combinatory effects between already known environmental factors impacting dental development like fluoride and EDs, and also other toxicants that can accumulate in enamel, such as heavy metals (Cd, Hg, or Pb). These toxicants may directly affect mineralized tissue formation, or alternatively, indirectly affect formation through the gut–brain axis (Li et al. 2021 for review). The gut–brain axis modulates neurotransmitters such as serotonin, which can affect dental development. In fact, dental cells express many receptors that transmit their effects with differential expression during cell differentiation and tooth development.

The combinatory effects of environmental factors may explain their effects at low doses and the apparent increase in sensitivity in the presence of multiple EDs. Thus, the sensitivity to fluoride may be increased if the drinking waters contain other pollutants with hypomineralizing effects such as BPA (Jedeon et al. 2016a) or lead

(Leite et al. 2011). It has also been reported that a combination of fluoride with amoxicillin (Sahlberg et al. 2013) or with TCDD (Salmela et al. 2011) has additional effects on rodent dental development.

12.4 Conclusions

Enamel synthesis, which occurs during the perinatal period of time, from the third trimester of fetal life to 6 years after birth, follows a well-known spatial-temporal sequence of ameloblast proliferation, differentiation, maturation, and death, characterized by specific gene expression pattern (Lacruz et al. 2017). Finally, ameloblasts are lost during tooth eruption making enamel defects (if any) irreparable and thus irreversible, but also leaving a mineralized record of environmental influences during the time of enamel formation. The combined approach of identifying associations between EDs and enamel defects in humans, and then using animal model systems such as the continuously growing rat incisor to investigate molecular mechanisms will allow us to understand how changes in the environment irreversibly affect tooth organ development.

Conflict of Interest No conflict of interest to declare.

References

Alaluusua S (2010) Aetiology of molar-incisor hypomineralisation: a systematic review. Eur Arch Paediatr Dent 11(2):53–58. https://doi.org/10.1007/BF03262713

Alaluusua S, Calderara P, Gerthoux PM, Lukinmaa PL, Kovero O, Needham L, Patterson DG Jr, Tuomisto J, Mocarelli P (2004) Developmental dental aberrations after the dioxin accident in Seveso. Environ Health Perspect 112(13):1313–1318. https://doi.org/10.1289/ehp.6920

Barbier O, Arreola-Mendoza L, Del Razo LM (2010) Molecular mechanisms of fluoride toxicity. Chem Biol Interact 188(2):319–333. https://doi.org/10.1016/j.cbi.2010.07.011

Bashash M, Marchand M, Hu H, Till C, Martinez-Mier EA, Sanchez BN, Basu N, Peterson KE, Green R, Schnaas L, Mercado-García A, Hernández-Avila M, Téllez-Rojo MM (2018) Prenatal fluoride exposure and attention deficit hyperactivity disorder (ADHD) symptoms in children at 6-12 years of age in Mexico City. Environ Int 121(Pt 1):658–666. https://doi.org/10.1016/j.envint.2018.09.017

Bock KW (2019) Aryl hydrocarbon receptor (AHR): from selected human target genes and crosstalk with transcription factors to multiple AHR functions. Biochem Pharmacol 168:65–70. https://doi.org/10.1016/j.bcp.2019.06.015

Bronckers AL, Lyaruu DM, DenBesten PK (2009) The impact of fluoride on ameloblasts and the mechanisms of enamel fluorosis. J Dent Res 88(10):877–893. https://doi.org/10.1177/0022034509343280

Chaitanya NCSK, Karunakar P, Allam NSJ, Priya MH, Alekhya B, Nauseen S (2018) A systematic analysis on possibility of water fluoridation causing hypothyroidism. Indian J Dent Res 29 (3):358–363. https://doi.org/10.4103/ijdr.IJDR_505_16

Delfosse V, Grimaldi M, Pons JL, Boulahtouf A, le Maire A, Cavailles V, Labesse G, Bourguet W, Balaguer P (2012) Structural and mechanistic insights into bisphenols action provide guidelines

for risk assessment and discovery of bisphenol A substitutes. Proc Natl Acad Sci USA 109 (37):14930–14935. https://doi.org/10.1073/pnas.1203574109
DenBesten PK (1986) Effects of fluoride on protein secretion and removal during enamel development in the rat. J Dent Res 65(10):1272–1277
DenBesten P, Li W (2011) Chronic fluoride toxicity: dental fluorosis. Monogr Oral Sci 22:81–96. https://doi.org/10.1159/000327028
Duan L, Zhu J, Wang K, Zhou G, Yang Y, Cui L, Huang H, Cheng X, Ba Y (2016) Does fluoride affect serum testosterone and androgen binding protein with age-specificity? A population-based cross-sectional study in Chinese male farmers. Biol Trace Elem Res 174(2):294–299. https://doi.org/10.1007/s12011-016-0726-z
Farah RA, Monk BC, Swain MV, Drummond BK (2010) Protein content of molar-incisor hypomineralisation enamel. J Dent 38(7):591–596. https://doi.org/10.1016/j.jdent.2010.04.012
Ferrer VL, Maeda T, Kawano Y (2005) Characteristic distribution of immunoreaction for estrogen receptor alpha in rat ameloblasts. Anat Rec A Discov Mol Cell Evol Biol 284(2):529–536. https://doi.org/10.1002/ar.a.20190
Gao Y, Sahlberg C, Kiukkonen A, Alaluusua S, Pohjanvirta R, Tuomisto J, Lukinmaa PL (2004) Lactational exposure of Han/Wistar rats to 2,3,7,8-tetrachlorodibenzo-p-dioxin interferes with enamel maturation and retards dentin mineralization. J Dent Res 83(2):139–144. https://doi.org/10.1177/154405910408300211
Gayrard V, Lacroix MZ, Collet SH, Viguié C, Bousquet-Melou A, Toutain PL, Picard-Hagen N (2013) High bioavailability of bisphenol A from sublingual exposure. Environ Health Perspect 121(8):951–956. https://doi.org/10.1289/ehp.1206339
Gayrard V, Lacroix MZ, Grandin FC, Collet SH, Mila H, Viguié C, Gély CA, Rabozzi B, Bouchard M, Léandri R, Toutain PL, Picard-Hagen N (2019) Oral systemic bioavailability of bisphenol A and bisphenol S in pigs. Environ Health Perspect 127(7):77005. https://doi.org/10.1289/EHP4599
Gibson CW, Collier PM, Yuan ZA, Chen E, Adeleke-Stainback P, Lim J, Rosenbloom J (1997) Regulation of amelogenin gene expression. Ciba Found Symp 205:187–197.; discussion 197-9. https://doi.org/10.1002/9780470515303.ch13
Gore AC, Chappell VA, Fenton SE, Flaws JA, Nadal A, Prins GS, Toppari J, Zoeller RT (2015) EDC-2: the endocrine society's second scientific statement on endocrine-disrupting chemicals. Endocr Rev 36(6):E1–E150. https://doi.org/10.1210/er.2015-1010
Green R, Lanphear B, Hornung R, Flora D, Martinez-Mier EA, Neufeld R, Ayotte P, Muckle G, Till C (2019) Association between maternal fluoride exposure during pregnancy and IQ scores in offspring in Canada. JAMA Pediatr 173(10):940–948. https://doi.org/10.1001/jamapediatrics.2019.1729
Guo J, Lyaruu DM, Takano Y, Gibson CW, DenBesten PK, Bronckers AL (2015) Amelogenins as potential buffers during secretory-stage amelogenesis. J Dent Res 94(3):412–420. https://doi.org/10.1177/0022034514564186
Hamilton IR (1990) Biochemical effects of fluoride on oral bacteria. J Dent Res. https://doi.org/10.1177/00220345900690S128
Houari S, Wurtz T, Ferbus D, Chateau D, Dessombz A, Berdal A, Babajko S (2014) Asporin and the mineralization process in fluoride-treated rats. J Bone Miner Res 29(6):1446–1455. https://doi.org/10.1002/jbmr.2153
Houari S, Loiodice S, Jedeon K, Berdal A, Babajko S (2016) Expression of steroid receptors in ameloblasts during amelogenesis in rat incisors. Front Physiol 7:503. https://doi.org/10.3389/fphys.2016.00503
Jalali R, Guy F, Ghazanfari S, Lyaruu D, van Ruijven L, DenBesten P, Martignon S, Castiblanco G, Bronckers ALJJ (2017) Mineralization-defects are comparable in fluorotic impacted human teeth and fluorotic mouse incisors. Arch Oral Biol 83:214–221. https://doi.org/10.1016/j.archoralbio.2017.07.018
Jälevik B (2001) Enamel hypomineralization in permanent first molars. A clinical, histomorphological and biochemical study. Swed Dent J Suppl 149:1–86

Jan J, Sovcikova E, Kocan A, Wsolova L, Trnovec T (2007) Developmental dental defects in children exposed to PCBs in eastern Slovakia. Chemosphere 67(9):S350–S354. https://doi.org/10.1016/j.chemosphere.2006.05.148

Jan J, Uršič M, Vrecl M (2013) Levels and distribution of organochlorine pollutants in primary dental tissues and bone of lamb. Environ Toxicol Pharmacol 36(3):1040–1045. https://doi.org/10.1016/j.etap.2013.09.005

Jedeon K, De la Dure-Molla M, Brookes SJ, Loiodice S, Marciano C, Kirkham J, Canivenc-Lavier MC, Boudalia S, Bergès R, Harada H, Berdal A, Babajko S (2013) Enamel defects reflect perinatal exposure to bisphenol A. Am J Pathol 183(1):108–118. https://doi.org/10.1016/j.ajpath.2013.04.004

Jedeon K, Loiodice S, Marciano C, Vinel A, Canivenc Lavier MC, Berdal A, Babajko S (2014a) Estrogen and bisphenol A affect male rat enamel formation and promote ameloblast proliferation. Endocrinology 155(9):3365–3375. https://doi.org/10.1210/en.2013-2161

Jedeon K, Marciano C, Loiodice S, Boudalia S, Canivenc Lavier MC, Berdal A, Babajko S (2014b) Enamel hypomineralization due to endocrine disruptors. Connect Tissue Res 55(Suppl 1):43–47. https://doi.org/10.3109/03008207.2014.923857

Jedeon K, Berdal A, Babajko S (2015) The tooth, target organ of bisphenol A, could be used as a biomarker of exposure to this agent. In: Editor YG (ed) Bisphenol A: sources, risks of environmental exposure and human health effects. Nova Science Publishers, New York, pp 205–225

Jedeon K, Houari S, Loiodice S, Thuy TT, Le Normand M, Berdal A, Babajko S (2016a) Chronic exposure to bisphenol A exacerbates dental fluorosis in growing rats. J Bone Miner Res 31(11):1955–1966. https://doi.org/10.1002/jbmr.2879

Jedeon K, Loiodice S, Salhi K, Le Normand M, Houari S, Chaloyard J, Berdal A, Babajko S (2016b) Androgen receptor involvement in rat amelogenesis: an additional way for endocrine-disrupting chemicals to affect enamel synthesis. Endocrinology 157(11):4287–4296. https://doi.org/10.1210/en.2016-1342

Ji M, Xiao L, Xu L, Huang S, Zhang D (2018) How pH is regulated during amelogenesis in dental fluorosis. Exp Ther Med 16(5):3759–3765. https://doi.org/10.3892/etm.2018.6728

Kao YH, Igarashi N, Abduweli Uyghurturk D, Li Z, Zhang Y, Ohshima H, MacDougall M, Takano Y, Den Besten P, Nakano Y (2020 Oct 28) Fluoride alters signaling pathways associated with the initiation of dentin mineralization in enamel fluorosis susceptible mice. Biol Trace Elem Res. https://doi.org/10.1007/s12011-020-02434-y

Karthikeyan BS, Ravichandran J, Mohanraj K, Vivek-Ananth RP, Samal A (2019) A curated knowledgebase on endocrine disrupting chemicals and their biological systems-level perturbations. Sci Total Environ 692:281–296. https://doi.org/10.1016/j.scitotenv.2019.07.225

Karube H, Nishitai G, Inageda K, Kurosu H, Matsuoka M (2009) NaF activates MAPKs and induces apoptosis in odontoblast-like cells. J Dent Res 88(5):461–465. https://doi.org/10.1177/0022034509334771

Kiukkonen A, Sahlberg C, Lukinmaa PL, Alaluusua S, Peltonen E, Partanen AM (2006) 2,3,7,8-tetrachlorodibenzo-p-dioxin specifically reduces mRNA for the mineralization-related dentin sialophosphoprotein in cultured mouse embryonic molar teeth. Toxicol Appl Pharmacol 216(3):399–406. https://doi.org/10.1016/j.taap.2006.06.015

Lacruz RS, Habelitz S, Wright JT, Paine ML (2017) Dental enamel formation and implications for oral health and disease. Physiol Rev 97(3):939–993. https://doi.org/10.1152/physrev.00030.2016

Le MH, Nakano Y, Abduweli Ugyghurturk D, Zhu L, Den Besten PK (2017) Fluoride alters Klk4 expression in maturation ameloblasts through androgen and progesterone receptor signaling. Front Physiol 8:925

Leite GA, Sawan RM, Teófilo JM, Porto IM, Sousa FB, Gerlach RF (2011) Exposure to lead exacerbates dental fluorosis. Arch Oral Biol 56(7):695–702. https://doi.org/10.1016/j.archoralbio.2010.12.011

Li N, Li J, Zhang Q, Gao S, Quan X, Liu P, Xu C (2021) Effects of endocrine disrupting chemicals in host health: three-way interactions between environmental exposure, host phenotypic responses, and gut microbiota. Environ Pollut 271:116387. https://doi.org/10.1016/j.envpol.2020.116387

Lyaruu DM, Medina JF, Sarvide S, Bervoets TJ, Everts V, Denbesten P, Smith CE, Bronckers AL (2014) Barrier formation: potential molecular mechanism of enamel fluorosis. J Dent Res 93(1):96–102

Philips EM, Jaddoe VWV, Asimakopoulos AG, Kannan K, Steegers EAP, Santos S, Trasande L (2018) Bisphenol and phthalate concentrations and its determinants among pregnant women in a population-based cohort in the Netherlands, 2004-5. Environ Res 161:562–572. https://doi.org/10.1016/j.envres.2017.11.051

Rasool A, Farooqi A, Xiao T, Ali W, Noor S, Abiola O, Ali S, Nasim W (2018) A review of global outlook on fluoride contamination in groundwater with prominence on the Pakistan current situation. Environ Geochem Health 40(4):1265–1281. https://doi.org/10.1007/s10653-017-0054-z

Robinson C, Connell S, Kirkham J, Brookes SJ, Shore RC, Smith AM (2004) The effect of fluoride on the developing tooth. Caries Res 38(3):268–276. https://doi.org/10.1159/000077766

Rojas-Sánchez F, Alaminos M, Campos A, Rivera H, Sánchez-Quevedo MC (2007) Dentin in severe fluorosis: a quantitative histochemical study. J Dent Res 86(9):857–861. https://doi.org/10.1177/154405910708600910

Sahlberg C, Pavlic A, Ess A, Lukinmaa PL, Salmela E, Alaluusua S (2013) Combined effect of amoxicillin and sodium fluoride on the structure of developing mouse enamel in vitro. Arch Oral Biol 58(9):1155–1164. https://doi.org/10.1016/j.archoralbio.2013.03.007

Salmela E, Lukinmaa PL, Partanen AM, Sahlberg C, Alaluusua S (2011) Combined effect of fluoride and 2,3,7,8-tetrachlorodibenzo-p-dioxin on mouse dental hard tissue formation in vitro. Arch Toxicol 85(8):953–963. https://doi.org/10.1007/s00204-010-0619-4

Serna C, Vicente A, Finke C, Ortiz AJ (2016) Drugs related to the etiology of molar incisor hypomineralization: a systematic review. J Am Dent Assoc 147(2):120–130. https://doi.org/10.1016/j.adaj.2015.08.011

Sharma R, Tsuchiya M, Bartlett JD (2008) Fluoride induces endoplasmic reticulum stress and inhibits protein synthesis and secretion. Environ Health Perspect 116(9):1142–1146. https://doi.org/10.1289/ehp.11375

Silva MJ, Scurrah KJ, Craig JM, Manton DJ, Kilpatrick N (2016) Etiology of molar incisor hypomineralization - a systematic review. Community Dent Oral Epidemiol 44(4):342–353. https://doi.org/10.1111/cdoe.12229

Suzuki M, Bartlett JD (2014) Sirtuin1 and autophagy protect cells from fluoride-induced cell stress. Biochim Biophys Acta 1842(2):245–255. https://doi.org/10.1016/j.bbadis.2013.11.023

Suzuki M, Shin M, Simmer JP, Bartlett JD (2014) Fluoride affects enamel protein content via TGF-beta1-mediated KLK4 inhibition. J Dent Res 93(10):1022–1027. https://doi.org/10.1177/0022034514545629

Suzuki M, Everett ET, Whitford GM, Bartlett JD (2017) 4-phenylbutyrate mitigates fluoride-induced cytotoxicity in ALC cells. Front Physiol 8:302. https://doi.org/10.3389/fphys.2017.00302

Vieira A, Hancock R, Dumitriu M, Schwartz M, Limeback H, Grynpas M (2005) How does fluoride affect dentin microhardness and mineralization? J Dent Res 84(10):951–957. https://doi.org/10.1177/154405910508401015

Waidyasekera K, Nikaido T, Weerasinghe D, Watanabe A, Ichinose S, Tay F, Tagami J (2010) Why does fluorosed dentine show a higher susceptibility for caries: an ultra-morphological explanation. J Med Dent Sci 57(1):17–23

Weerheijm KL, Jälevik B, Alaluusua S (2001) Molar-incisor hypomineralisation. Caries Res 35 (5):390–391. https://doi.org/10.1159/000047479

Wogelius P, Haubek D, Nechifor A, Nørgaard M, Tvedebrink T, Poulsen S (2010) Association between use of asthma drugs and prevalence of demarcated opacities in permanent first molars in 6-to-8-year-old Danish children. Community Dent Oral Epidemiol 38(2):145–151. https://doi.org/10.1111/j.1600-0528.2009.00510.x

Yang L, Jin P, Wang X, et al (2018) Fluoride activates microglia, secretes inflammatory factors and influences synaptic neuron plasticity in the hippocampus of rats. Neurotoxicology 69:108–120

Zhao D, Dong B, Yu D, Ren Q, Sun Y (2018) The prevalence of molar incisor hypomineralization: evidence from 70 studies. Int J Paediatr Dent 28(2):170–179. https://doi.org/10.1111/ipd.12323

www.ingramcontent.com/pod-product-compliance
Ingram Content Group UK Ltd.
Pitfield, Milton Keynes, MK11 3LW, UK
UKHW021833270726
14058UKWH00001B/126

* 9 7 8 3 0 3 0 7 6 2 8 5 8 *